Computational Microelectronics

Edited by S. Selberherr

The Monte Carlo Method for Semiconductor Device Simulation

C. Jacoboni and P. Lugli

Springer-Verlag Wien New York

Prof. Dr. Carlo Jacoboni
Dipartimento di Fisica
Università di Modena, Italy

Prof. Paolo Lugli
Dipartimento di Ingegneria Meccanica
II Università di Roma "Tor Vergata", Italy

Typeset by Asco Trade Typesetting Ltd., Hong Kong
Printed in Austria by Novographic, Ing. W. Schmid, A-1238 Wien

With 228 Figures

Library of Congress Cataloging-in-Publication Data. Jacoboni, Carlo. The Monte Carlo method for semiconductor device simulation / C. Jacoboni, P. Lugli. p. cm.—(Computational microelectronics, ISSN 0179-0307) ISBN 0-387-82110-4 (U.S.) 1. Semiconductors—Mathematical models. 2. Monte Carlo method. I. Lugli, P. (Paolo), 1956– . II. Title. III. Series. TK7871.85.J23.1989. 621.381'52'015118—dc20. 89-21588

ISSN 0179-0307
ISBN 3-211-82110-4 Springer-Verlag Wien-New York
ISBN 0-387-82110-4 Springer-Verlag New York-Wien

Preface

The field of microelectronics has witnessed astonishing advances in the last decade, mainly pushed by the tremendous progress in semiconductor technology. Better materials have become available and novel device structures have been proposed and realized. With respect to semiconductor devices and circuits, which are the building blocks of today's microelectronics, the main trend is towards miniaturization. Submicron dimensions are a common feature of prototype as well as production devices. Such a reduction of dimensions allows the fabrication of chips with more than one million components. In addition, it opens the way to new physical phenomea, which were not present in traditional structures. In this context, the Monte Carlo simulation has particular significance.

Traditionally, the design and manufacture of electronic devices have been quite independent of attempts to simulate their physical behavior. Engineers in industrial laboratories had their own "recipes" and considered such attempts as academic exercises, too idealized with respect to the "dirty" experimental conditions of real systems. In the last decade the situation has greatly improved due to several factors: i) Physicists have gained a deeper knowledge of the microscopic transport processes occurring within the device; ii) Engineers have attained better control of their working conditions, mainly from the point of view of material science; iii) Simulation programs have become more efficient and accurate, mainly due to the availability of very powerful computational tools; iv) The high degree of sophistication of the last generation of submicron devices does not allow a reasonable design without a deep theoretical knowledge of their behavior. These occurrences are not independent of each other. On the contrary, they represent a good example of the positive feedback between science and technology.

Several families of device simulators are today available. Some of them are "mature", others are highly advanced, but not yet sufficiently reliable or efficient. With regard to simulators based on Monte Carlo technique, they are usually considered to be the most accurate, since they contain the "true" charge transport, including its non-local features and its natural fluctuations, but they are also thought to be too expensive in terms of computer time. In our opinion, the trends in microelectronics pointed out above and the availability of very fast computers (with parallel processors and advanced graphic capabilities) at reasonable prices, are transforming the situation very rapidly. This book reflects our belief that Monte Carlo will be the theoretical tool for tomorrow's microelectronics, and hopes to offer a contribution in just this direction.

The book contains all the information we consider necessary to set up a Monte Carlo simulator, starting from the knowledge of basic transport physics and continuing up to the general principles of the Monte Carlo technique and its applications to the simulation of semiconductor devices. It is therefore aimed at those who work in this field or wish to enter it, ranging from graduate students in physics or electrical engineering, to expert researchers who will find a unified presentation of subjects otherwise spread over a vast range of literature. It is a book that does not require a deep knowledge of advanced physics or electronics. Any graduate student in physics or electrical engineering should be able to understand it without particular difficulty. Some material is occasionally presented at a more advanced level but skipping these sections will not prevent an overall understanding of the rest of the book.

A brief introduction is contained in chapter 1. Chapters 2 and 3 present the basic principles of charge transport in semiconductors (Chapter 2) and of the Monte Carlo technique (Chapter 3), respectively. Chapter 4 contains a review of semiconductor devices and forms a bridge to the last part of the book, devoted to the simulation of electronics devices. Chapter 5 presents the general techniques, while the final Chapter 6 contains several applications to specific devices.

This book is the result of many years of research performed by the group working in Modena on Monte Carlo simulation of charge transport in semiconductors. Therefore we gratefully acknowledge the contributions of all our collegues and students.

One of us (P.L.) would also like to acknowledge the collaboration with D. K. Ferry and the other coworkers during his stays at Colorado State University and Arizona State University.

Modena, July 1989 C. Jacoboni and P. Lugli

Contents

Introduction 1

The name of the Monte Carlo method is inspired by the gambling casinos at the city of Monte Carlo in Monaco. The mathematical techniques used by this method are in fact based on the selection of random numbers [1–4]. In its present form, the method is attributed to Fermi, Von Neumann, and Ulam, who developed it for the solution of problems related to neutron transport during the secret research at Los Alamos for the construction of the atomic bomb during world war II. There are, however, indications of previous uses of methods based on selections of random numbers. In particular the name of Lord Kelvin is mentioned for a paper of 1901 [5], and Gosset (better known with the pseudonym Student) used experimental sampling to support his well known theoretical studies of statistical distributions. Fermi himself used already Monte Carlo techniques in the 30's in connection with neutron transport [6].

In principle, the Monte Carlo (MC) method can be considered as a very general mathematical tool for the solution of a great variety of problems. To someone who is not at all familiar with the method it may seem strange that the generation of random numbers can be used to solve a well defined mathematical problem, but a simple example will convince us that it is actually so.

Let us consider the evaluation of a definite integral

$$\int_a^b f(x)\, dx, \tag{1.1}$$

where f is a positive function, as shown in Fig. 1.1. Let M be a number greater than the maximum of $f(x)$ in the interval (a, b). Now we generate pairs of numbers x_r and y_r, evenly distributed in the intervals (a, b) and $(0, M)$, respectively. For each pair we consider the point of coordinates x_r and y_r. The fraction of points that fall below the curve $f(x)$, i.e. such that $y_r \leqslant f(x_r)$, gives the ratio between the area of interest and the area of the box $ABCD$ and therefore it yields the value of the integral in Eq. (1.1).

The method sketched above is obviously very rough, and sophisticated MC techniques for the evaluation of integrals are much more efficient. The example, however, points out an important feature of all MC techniques:

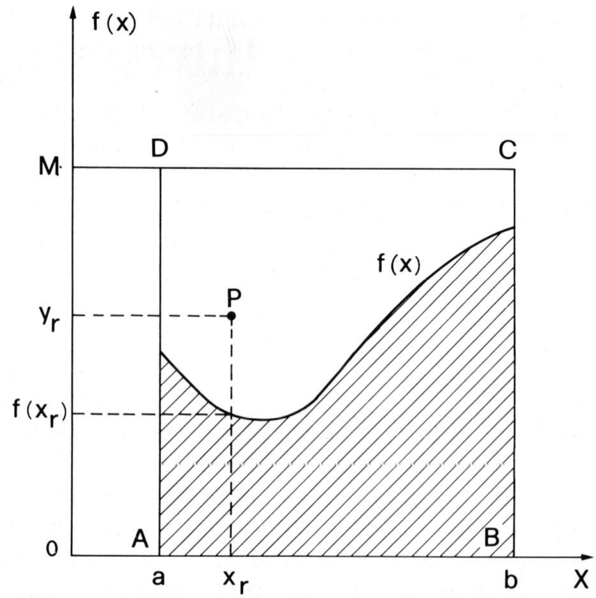

Fig. 1.1. Example of Monte Carlo evaluation of an integral (see text)

more precise results can be obtained by generating a larger number of points. More generally, being based on random numbers, the results obtained with a MC procedure are never exact, but rigorous in a statistical sense: the exact result lies in given intervals with given probabilities. The uncertainty on the results is strictly related to the variance of the possible outcomes and it is smaller if the size of the sample (i.e. the amount of computations devoted to the solution of the problem) is larger. It is then clear that (i) a great effort is devoted by MC experts to devise variance-reducing techniques and (ii) the enormous development of MC applications that occurred in the last decades is due to, and has pushed for, the even more extraordinary development of electronic computers.

The applications of MC methods can be divided into two major groups. One consists of direct simulations of systems that are already statistical in their nature; in such cases it is not even necessary to have well defined mathematical equations that describe the behaviour of the system. The second group consists of MC methods devised for the solution of well defined mathematical equations. In such cases the methods are used to solve the equations that describe the problem of interest.

The majority of real cases are, as it is often the case, a mixture of the two extreme limits indicated above. The application to the simulation of semiconductor devices, considered in this book, is a good example. In fact, transport problems are statistical in nature, but are also described, to some extent of accuracy, by well defined transport equations. A MC method applied to the solution of these equations may or may not correspond to a direct simulation of the physical system under examination. In the former case, the simulation may yield more information than that obtainable from the solution of the equations. For instance, the MC solution of the Boltzmann transpor equation (BE) not only gives the distribution function that verifies the equation, but also yields information on its fluctuations that is lost in BE itself.

On the other hand, the direct simulation of a statistical problem is at times very inefficient. Examples of this type will be encountered in this book. One important case is the analysis of situations that are rare in the actual physical system. If a direct simultation is performed, a large amount of computation is devoted to the "uninteresting" normal situations and very little to the rare occurrances of interest. In such cases it is necessary to distort the simulation by applying some more sophisticated MC techniques that reduce the variance of the quantity of interest, giving up the advantages offered by the direct simulation.

The application of MC techniques to high-field transport in semiconductors was introduced by Kurosawa [7] at the Semiconductor Conference held in Kyoto in 1966.[1] Since then the method has been greatly improved and widely used to obtain results for various situations in practically all materials of interest [9–12]. Among the most significant developments of the MC technique we mention the work of Price [13], of the Malvern group [14–19] with the introduction of the self-scattering scheme and the extension of the method to many-particle simulations [20], to harmonic fields [21] and to degenerate statistics [22, 23].

Before leaving this brief introduction to the MC method as a general technique for solving problems of great mathematical complexity, we would like to stress a point that is often underestimated. MC calculations can be considered as simulated experiments. As in real experiments, the final results of the simulation are not the end of the story: they must be interpreted in physical terms in order to obtain a better understanding of the problem at hand. In this respect MC is again a useful tool. In fact, since the system in which the "measurement" is performed is idealized in the model, it can be changed at will. In particular, physical parameters can be modified in such a way as to give greater emphasis to the special effects under examination.

The application of MC techniques to the simulation of semiconductor devices started quite soon after the introduction of the method [24, 25] but has received great attention only recently [26] since it requires very large amount of computations, made possible only by the most recent computers. The history of device simulation is interesting in itself: In the past decades people working in actual design and manufacturing of new devices and those who were simulating them with numerical techniques could hardly com-

---

[1] In fact, a paper by Lüthi and Wyder [8] did appear in 1960 with an application of the MC method to the problem of electrical conductivity in thin metallic wires. The method introduced there is very similar to the one proposed later and independently by Kurosawa.

municate with each other. The former accused the theoreticians of studying systems that were too much idealized, too far away from real devices; the latter complained that the device makers were not even able to say exactly how their systems were made. In spite of the fact that they were both right, the simulation activity was not at all useless. The output of this work was more a "taste of what is going on" in devices, rather than an actual prediction of the characteristics of a new device.

The technological improvements on the one hand [27–30], and the availability of fast computers on the other, have narrowed the gap. With the recent advances in material growth, contact deposition and impurity control, devices have become more transparent from the physical point of view. Incidentally, this has provided the theoreticians with physical systems of extreme interest. At the same time, MC algorithms have gained in sophistication and are now able to handle phenomena and systems of great complexity. These are two fundamental steps since, as it will be stressed throughout the book, the necessary input for a MC simulation of semiconductor materials and devices is the physical system under investigation. Many semiconductor devices can nowadays be simulated with the MC method, which is increasingly becoming a very useful modelling tool.

References

[1] Hammersley, J. M., Handscomb, D. C.: Monte Carlo Methods. London: Methuen. 1964.

[2] Shreider, Yu. A. (ed.): The Monte Carlo Method. Oxford: Pergamon. 1966.

[3] Rubinstein, R. Y.: Simulation and the Monte Carlo Method. New York: Wiley. 1981.

[4] Spanier, J., Gelbard, M.: Monte Carlo Principles and Neutron Transport Problems. Reading, Mass.: Addison-Wesley. 1969.

[5] Lord Kelvin: Phil. Mag. (6) *2*, 1 (1901).

[6] Anderson, H. L.: J. Stat. Phys. *43*, 731 (1986).

[7] Kurosawa, T.: Proc., 8th Int. Conf. Phys. Semic., Kyoto. J. Phys. Soc. Japan, Suppl. *24*, 424 (1966).

[8] Lüthi, B., Wyder, P.: Helv. Phys. Acta *33*, 667 (1960).

[9] Alberigi-Quaranta, A., Jacoboni, C., Ottaviani, G.: La Rivista del Nuovo Cimento *1*, 445 (1971).

[10] Jacoboni, C., Reggiani, L.: Advances in Physics *28*, 493 (1979).

[11] Price, P. J.: Semiconductors and Semimetals *14*, 249 (1979).

[12] Jacoboni, C., Reggiani, L.: Rev. Mod. Phys. *55*, 645 (1983).

[13] Price, P. J.: In: Proc., 9th Int. Conf. Phys. Semic. (Ryukin, S. M., ed.), p. 753. Leningrad: Nauka. 1968.

[14] Rees, H. D.: Phys. Lett. *a26*, 416, (1968).

[15] Rees, H. D.: J. Phys. Chem. Solids *30*, 643 (1969).

[16] Fawcett, W., Hilsum, C., Rees, H. D.: Solid State Commun. *7*, 1257 (1969).

[17] Fawcett, W., Boardman, D. A., Swain, S.: J. Phys. Chem. Solids *31*, 1963 (1970).

[18] Fawcett, W., Rees, H. D.: Phys. Lett. *11*, 731 (1969).

[19] Fawcett, W.: In: Electrons in Crystalline Solids (Salam, A., ed.), p. 531. Vienna: IAEA. 1973.

[20] Lebwohl, P. A., Price, P. J.: Solid State Commun. *9*, 1221 (1971) and Appl. Phys. Lett. *19*, 530 (1971).

[21] Price, P. J.: IBM J. Res. Dev. *17*, 39 (1973).
[22] Bosi, S., Jacoboni, C.: J. Phys. *C9*, 315 (1976).
[23] Lugli, P., Ferry, D. K.: IEEE Trans. Electron Dev. *ED-32*, 2431 (1985).
[24] Hockney, R. W., Warriner, R. A., Reiser, M.: Electron. Lett. *10*, 484 (1974).
[25] Baccarani, G., Jacoboni, C., Mazzone, A. M.: Solid State Electr. *20*, 5 (1977).
[26] Lugli, P., Jacoboni, C.: In: ESSDERC 87, Proc., 17th European Solid State Device Research Conference (Calzolari, P. U., Soncini, G., eds.), p. 97 and references therein. Bologna: Tecnoprint. 1987.
[27] Sze, S. M.: Physics of Semiconductor Devices, 2nd edn. New York: Wiley. 1982.
[28] Solomon, P.: Proc. IEEE *70*, 489 (1982).
[29] IEEE J. Quantum Electronics *QE-22*, n. 9 (1986), special issue on semiconductor quantum wells and superlattices: physics and applications.
[30] Selberherr, S.: Analysis and Simulation of Semiconductor Devices. Wien-New York: Springer. 1984.

2 Charge Transport in Semiconductors

2.1 Electron Dynamics

Electrons in a perfect crystal[1] can be described in terms of Bloch states, whose wave functions can be written as

$$\Psi_{nk}(\mathbf{r}) = u_{nk}(\mathbf{r})e^{i\mathbf{k}\cdot\mathbf{r}} \tag{2.1.1}$$

where n is a band index, $\hbar\mathbf{k}$ is the quasi momentum of the electron, and $u_{nk}(\mathbf{r})$ is a function of the space coordinate $\mathbf{r}$ with the periodicity of the crystal. For each band n, a value of the electron energy $\varepsilon_n(\mathbf{k})$ is associated to the electron state $\mathbf{k}$. The set of functions $\varepsilon_n(\mathbf{k})$ describes what is known as the band structure of the material. The knowledge of the band structure of a solid is the starting point for the study of any electronic property, and the theoretical determination of a band structure is still one of the major open problems of solid-state physics.

Considering increasing energy, semiconductors are characterized by a gap of forbidden energies (of width ε_g) between the last band whose states are occupied by electrons at zero temperature (valence band) and the first empty band (conduction band).

The region of the band structure of a semiconductor which is of interest when electron transport is studied is centered around the energy gap and extends some ε_g's above the minimum of the conduction band and below the top of the valence band. Electrons in the conduction band are simply mentioned as electrons while missing electrons in the valence band are referred to as holes.

At zero temperature, the valence band of a semiconductor is entirely occupied by electrons, and the conduction band is entirely empty. Thus, at $T = 0$ a semiconductor is actually an insulator.

If we now consider a temperature above zero, some electrons will leave the valence band and occupy some states of the conduction band. In such a case

---

[1] For a general treatment of solid state fundamentals, see, for example, the book by Ashcroft and Mermin [1].

the transport properties of the semiconductor are due to an equal number of electrons and holes.

If, however, some impurities create localized states, with energy within the energy gap, occupied at zero temperature, for $T > 0$ these electrons will tend to occupy the conduction band. Such impurities are called donors. On the other hand, if the states created by the impurities are empty at $T = 0$, at larger T they will tend to accept electrons from the valence band and create holes. Such impurities are called acceptors.

When, as it always happens, both acceptors and donors are present in the material, some compensation will occur, i.e. some electrons will leave donor states to occupy acceptor states, thus diminishing the number of electrons and holes that contribute to the conductivity.

The density of electrons and holes is controlled by the Fermi-Dirac statistics. If at a given temperature the electrical conduction is mainly due to electrons, the semiconductor is called extrinsic of n type; if the conduction is mainly due to holes, the semiconductor is said to be extrinsic of p type. If, finally, the number of carriers due to the presence of impurities is negligible with respect to the number of electrons that are thermally excited from the valence band to the conduction band, then the semiconductor is said to be intrinsic.

Around the maximum of the valence band and the minimum of the conduction band the function $\varepsilon(\mathbf{k})$ can be approximated with a second order function of $\mathbf{k}$ and for a high simmetry situation it reduces to

$$\varepsilon(k) = \frac{\hbar^2 k^2}{2m} \tag{2.1.2}$$

where the only difference with a free electron is the substitution of the actual electron mass m_0 with an effective mass m.

For simple analyses of transport problems, the approximation in Eq. (2.1.2) may be enough, but for most actual applications a better approximation of real band structures is to be considered. A description of the band structures of the semiconductors of major interest in electronics devices will be given in the next section.

Furthermore, in what follows we shall omit the band index n and we shall often use the word "electrons" for charge carriers, meaning holes as well, as is common practice in the transport literature.

It can be shown [1] that if an external electromagnetic field is applied to a crystal, electrons will change continuously their $\mathbf{k}$ state according to the law

$$\hbar\dot{\mathbf{k}} = -e\left(\mathbf{E} + \frac{1}{c}\mathbf{v} \times \mathbf{B}\right) \tag{2.1.3}$$

where $-e$ is the electron charge, $\mathbf{E}$ and $\mathbf{B}$ are the electric and magnetic fields, respectively, and

$$\mathbf{v} = \frac{1}{\hbar}\frac{\partial\varepsilon}{\partial\mathbf{k}} \tag{2.1.4}$$

is the group velocity of the electron, that reduces to

$$\mathbf{v} = \frac{\hbar\mathbf{k}}{m} \tag{2.1.5}$$

for a simple band given by Eq. (2.1.2).
As we shall see, in semiclassical transport electrons are treated, during free flights between scattering events, as classical point-like particles. For this approximation to be reasonable it is necessary to conceive a wave-packet with well-defined position and momentum. Such a condition must be confronted, however, with the uncertainty principle. If we require that the momentum uncertainty is much smaller than the average electron momentum $p = \hbar k$, and that, at the same time, the position uncertainty is much smaller than the mean free path $l = v\tau = (p/m)\tau$ (where τ is the mean time between collisions), we have

$$\Delta p \ll p, \qquad \Delta x \ll l. \tag{2.1.6}$$

Therefore, from the uncertainty principle,

$$\hbar \approx \Delta p \, \Delta x \ll pl = \frac{p^2}{m}\tau \sim K_B T \tau, \tag{2.1.7}$$

where K_B is the Boltzmann constant.
Thus the validity of the semiclassical approach is insured when

$$\tau \gg \frac{\hbar}{K_B T} \sim 10^{-13} - 10^{-14} \text{ sec}. \tag{2.1.8}$$

Collisions are often less frequent so that the above requirement is usually fulfilled. When electrons with very high energies (of the order of the eV) are considered, scattering times can approach the value in Eq. (2.1.8) and a quantum treatment of transport is necessary [2–8], where the wave nature of electron dynamics is fully accounted for.

2.2 Energy Bands

In this section we shall present a general model for the band structure which enables us to interpret the macroscopic properties of the semiconductors of interest in devices. Such a model will provide a description of the band structure for the whole class of cubic semiconductors with both diamond and zinc-blende symmetries.
The model consists of one conduction band, with three sets of minima, and three valence bands. The minima of the conduction band are located at the Γ point ($k = 0$), at the L points [$\mathbf{k} = (\pi/a_0, \pi/a_0, \pi/a_0)$, a_0 being the lattice parameter], and along the Δ lines ($\mathbf{k} = k, 0, 0$). The tops of the valence bands are located at Γ. Two of these bands are degenerate at this point, while the third one is split off by spin-orbit interaction. In transport analysis the

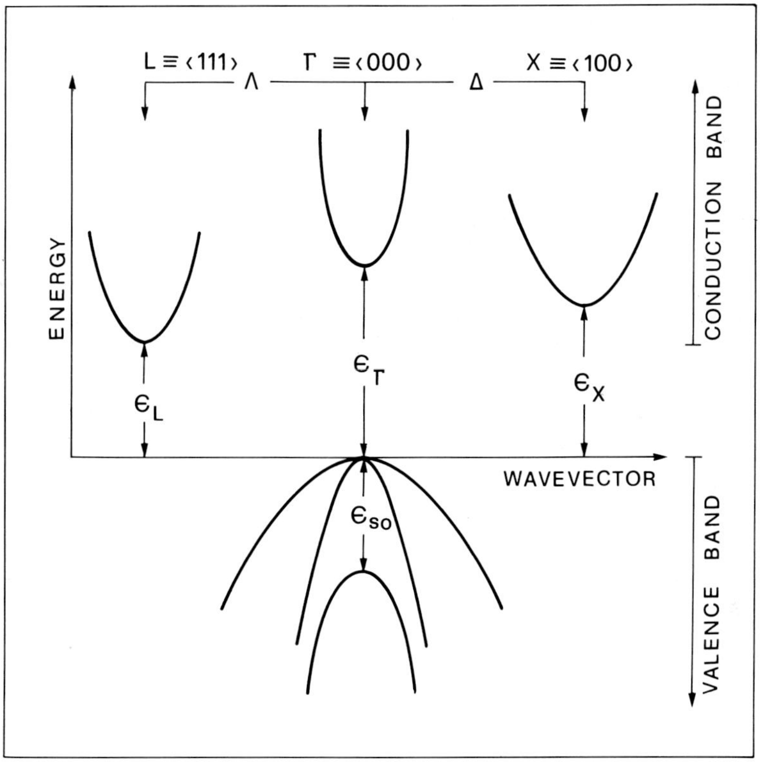

Fig. 2.1. Schematic representation of the band structure of a cubic model semiconductor

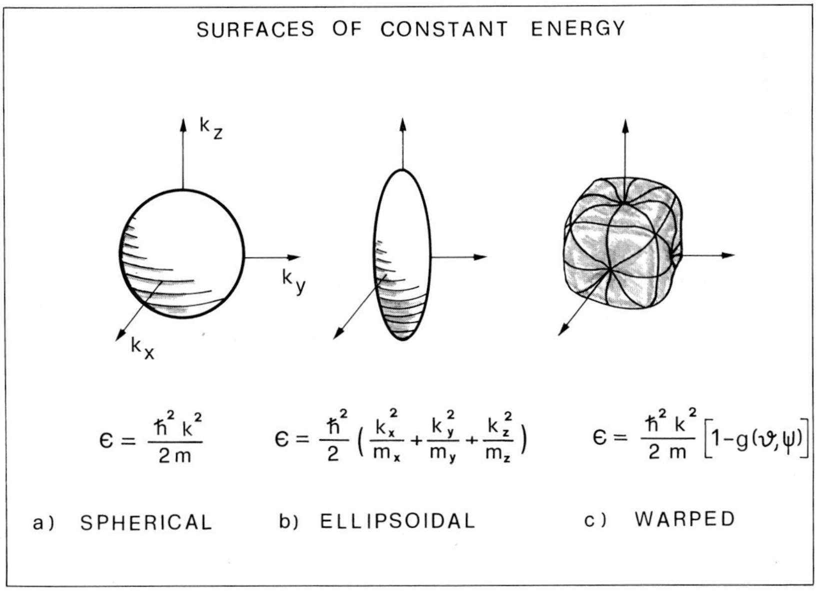

Fig. 2.2. Typical constant-energy surfaces for electrons and holes in cubic semiconductors

split-off valence band is usually neglected, owing to its low density of states and its energy separation.
The main features of the band structure in cubic semiconductors are summarized in Figs. 2.1 and 2.2.

2.2.1 *Relationship of Energy to Wavevector*

In the following we shall explicitly refer to electrons or holes when we consider $\mathbf{k}$ states belonging to the conduction or valence band, respectively.

a) Conduction Band

In the region around the minima of the conduction band, the function $\varepsilon(\mathbf{k})$ can be approximated by a quadratic function of $\mathbf{k}$ (parabolic bands):

$$\varepsilon(\mathbf{k}) = \sum_{i,j} \frac{\hbar^2}{2} \left(\frac{1}{m}\right)_{ij} k_i k_j, \tag{2.2.1}$$

where

$$\left(\frac{1}{m}\right)_{ij} = \frac{1}{\hbar^2} \left(\frac{\partial^2 \varepsilon}{\partial k_i \, \partial k_j}\right)_{\mathbf{k}=0} \tag{2.2.2}$$

is the inverse effective-mass tensor, and $\mathbf{k}$ is measured from the centers of the valleys. In the model considered here $\varepsilon(\mathbf{k})$ may assume, in particular, one of the following forms:

$$\varepsilon(\mathbf{k}) = \frac{\hbar^2 k^2}{2m}, \tag{2.2.3}$$

$$\varepsilon(\mathbf{k}) = \frac{\hbar^2}{2} \left[\frac{k_l^2}{m_l} + \frac{k_t^2}{m_t}\right], \tag{2.2.4}$$

Equation (2.2.3) (spherical case) represents a band with spherical equi-energetic surfaces with a single scalar effective mass m, and it is appropriate for the minimum of the conduction band located at Γ and for the maximum of the split-off valence band. This simple case, considered already in the previous section, is generally adopted as a model for any material when rough estimates of transport properties are sought.
Equation (2.2.4) (ellipsoidal case) represents a band with ellipsoidal equi-energetic surfaces, with a tensor effective mass. The ellipsoids have rotational symmetry around the crystallographic directions which contain the centers of the valleys. Here, k_l and k_t are the longitudinal and transverse components of $\mathbf{k}$ with respect to these directions; $1/m_l$ and $1/m_t$ are the longitudinal and transverse components, respectively, of the inverse effective-mass tensor. This case is appropriate for the minima of the conduction band located at L and along Δ; for symmetry reasons several equivalent valleys are present (many-valley model).

b) Valence Band

The valence bands of most cubic semiconductors are very similar to each other. The model we present here is obtained as an approximation of a pioneer $\mathbf{k}\cdot\mathbf{p}$ calculation of Kane [9] and well approximates (the larger the direct energy gap, the better the approximation) the energy-wave-vector relationship of the three bands around the center of the Brillouin zone, in terms of four band parameters characteristic of each material.
According to this model, the spherical, parabolic expression in Eq. (2.2.3) is appropriate also for the split-off valence band, while for the two degenerate valence bands the energy around the maxima is given by a more elaborate expression that does not allow the definition of an effective-mass tensor:

$$\varepsilon(\mathbf{k}) = ak^2[1 \mp g(\vartheta, \phi)]. \tag{2.2.5}$$

This equation represents bands with warped equienergetic surfaces (here $\mp$ refer to heavy and light holes, respectively). ϑ and ϕ are the polar and azimuthal angles of $\mathbf{k}$ with respect to crystallographic axes. $g(\vartheta, \phi)$ contains the angular dependence of the two degenerate bands, given by [10]

$$g(\vartheta, \phi) = [b^2 + c^2(\sin^4 \vartheta \cos^2 \phi \sin^2 \phi + \sin^2 \vartheta \cos^2 \vartheta)]^{1/2} \tag{2.2.6}$$

with

$$a = \frac{\hbar^2 |A|}{2m_0}, \qquad b = \frac{|B|}{|A|}, \qquad c = \frac{|C|}{|A|}, \tag{2.2.7}$$

where A, B and C are the inverse valence-band parameters [11].
The different shapes of the constant-energy surfaces for the three cases considered above are shown in Fig. 2.2.
The difficulties one encounters in dealing with three bands that include the full details of Eq. (2.2.6) for the two degenerate bands are usually overcome by neglecting the third band, because of its separation in energy and small density of states, and by introducing further simplifications in treating the remaining two bands (heavy and light holes). Among these simplified models we mention the following, in order of increasing sophistication [12, 13]:

(i) One single spherical, parabolic band with an effective mass related to some plausible average in $\mathbf{k}$ space of Eq. (2.2.6).
(ii) One single warped, parabolic band (heavy holes).
(iii) One single warped, non parabolic band (heavy holes).
(iv) Two spherical and parabolic bands.

2.2.2 *Effective Masses*

A simple parabolic band described by the energy-wave vector relationship in Eq. (2.2.3), is completely characterized by its effective mass m. Such a parameter plays a fundamental role in several physical problems of different nature: the acceleration of an electron in a crystal subject to an external force, the ratio of the electron momentum to its velocity, the electron density of states in energy space. When the band is not described by the simple Eq. (2.2.3), then the quantities that play the role of the effective mass in the different situations may be not the same. For this reason several types of effective mass have been defined in the literature, all of which coincide with m in the simple case.

a) Acceleration Effective Mass

If a wave packet is subject to a constant force $\mathbf{F}$, the i-th component of its acceleration is given by

$$a_i = \frac{dv_i}{dt} = \frac{d}{dt}\frac{1}{\hbar}\frac{\partial \varepsilon}{\partial k_i} = \sum_j \frac{1}{\hbar}\frac{\partial^2 \varepsilon}{\partial k_i\, \partial k_j}\dot{k}_j\,, \tag{2.2.8}$$

or

$$a_i = \sum_j \left(\frac{1}{m}\right)_{ij} F_j\,, \tag{2.2.9}$$

where

$$\left(\frac{1}{m}\right)_{ij} = \frac{1}{\hbar^2}\frac{\partial^2 \varepsilon}{\partial k_i\, \partial k_j} \tag{2.2.10}$$

is the inverse acceleration effective mass, or simply the inverse effective-mass tensor. The difference with respect to the situation of Eq. (2.2.2) is that here $(1/m)_{ij}$ is in general a function of $\mathbf{k}$. More precisely, Eq. (2.2.2) is the value of (2.2.10) at $k = 0$. It is clear that in the simple case of Eq. (2.2.3) the effective mass reduces to the scalar quantity m, while in the case of parabolic ellipsoidal bands given by Eq. (2.2.4), $(1/m)_{ij}$ is a constant tensor.

b) Conductivity Effective Mass

The ratio of the electron momentum to its velocity is usually called conductivity effective mass m_c:

$$\frac{\hbar \mathbf{k}}{m_c} = \frac{1}{\hbar}\frac{\partial \varepsilon}{\partial \mathbf{k}}\,. \tag{2.2.11}$$

In the simple case we have again $m_c = m$. For a parabolic, ellipsoidal band m_c^{-1} is the inverse effective-mass tensor. This indicates that $\mathbf{v}$ and $\mathbf{k}$ are not, in general, parallel in non-spherical bands. In this case a spherical approximation can be obtained by averaging over all possible directions. The resulting scalar conductivity inverse effective mass is [14]

$$\frac{1}{m_c} = \frac{1}{3}\left(\frac{1}{m_l}\right) + \frac{2}{3}\left(\frac{1}{m_t}\right). \tag{2.2.12}$$

For a cubic semiconductor, this is also the conductivity effective mass at any given direction, averaged over the valleys equivalent by symmetry.

c) Density-of-States Effective Mass

In the evaluation of the number of carriers in the conduction band, the effective mass plays again a role due to its influence on the electron density of state in energy space. In fact, the electron density n is given by [14]

$$n = \int f(\varepsilon) g(\varepsilon)\, d\varepsilon \tag{2.2.13}$$

where $f(\varepsilon)$ is the electron distribution function and $g(\varepsilon)$ is the density of states in energy space.

For a simple band, $g(\varepsilon)$ is given by

$$g(\varepsilon) = \frac{\sqrt{2}}{\pi^2 \hbar^3} m^{3/2} \varepsilon^{1/2} . \tag{2.2.14}$$

For non-degenerate statistics, with the above expression for $g(\varepsilon)$, n results in

$$n = N_c e^{-\varepsilon_c / K_B T} , \tag{2.2.15}$$

where ε_c is the energy of the bottom of the band with respect to the Fermi level, and

$$N_c = 2 \left(\frac{m K_B T}{2\pi\hbar^2} \right)^{3/2} . \tag{2.2.16}$$

For more complicated bands Eq. (2.2.13) still holds, but $g(\varepsilon)$ is not given by Eq. (2.2.14), and n has more elaborate expressions. The density-of-state effective mass m_d is defined in such a way that Eqs. (2.2.15) and (2.2.16) still hold with m_d in place of m. For a parabolic, ellipsoidal band m_d results in [14]

$$m_d = (m_l m_t^2)^{1/3} . \tag{2.2.17}$$

2.2.3 *Nonparabolicity*

For values of **k** far from the minima of the conduction band and/or from the maxima of the valence band, the energy deviates from the simple quadratic expressions seen above, and nonparabolicity occurs (see Fig. 2.3).

For the conduction band, a simple analytical way of introducing nonparabolicity is to consider an energy-wave vector relation of the type [15]

$$\varepsilon(1 + \alpha\varepsilon) = \gamma(\mathbf{k}) \tag{2.2.18}$$

or

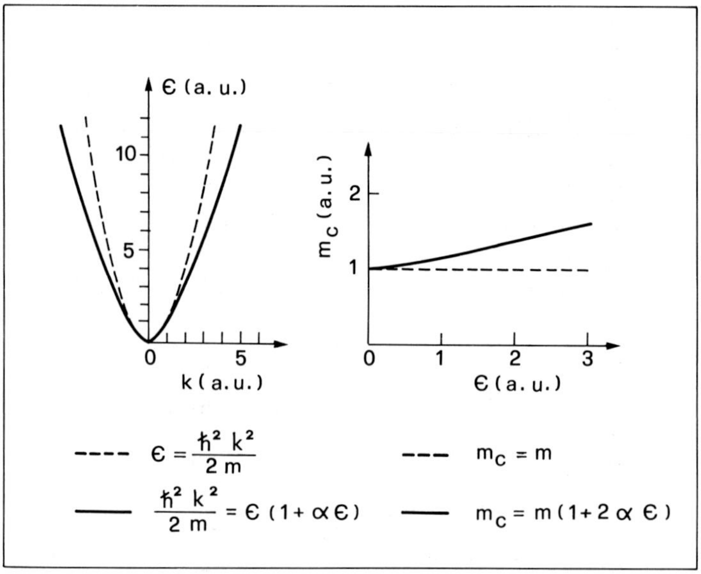

Fig. 2.3. Effect of non-parabolicity on a simple band diagram and corresponding effective mass. Arbitrary units are used with $\hbar^2/2m = 1$ and $\alpha = 0.1$

$$\varepsilon(\mathbf{k}) = \frac{-1 + \sqrt{1 + 4\alpha\gamma}}{2\alpha}, \tag{2.2.19}$$

where $\gamma(\mathbf{k})$ is given by one of the right-hand sides of Eqs. (2.2.3) and (2.2.4). α is a nonparabolicity parameter, which can be related to other band quantities [16–18]. Sometimes α is used as a fitting parameter in the interpretation of electron transport data.

For the valence band, nonparabolicity cannot be parameterized in a form like that of Eq. (2.2.18). In this case nonparabolicity has two main features [9]: (i) it is more pronounced along $\langle 110 \rangle$ and $\langle 111 \rangle$ directions for heavy and light holes, respectively; (ii) if ε_{so} is the split-off energy of the lowest valence band, nonparabolicity is mostly effective at energies near $\frac{1}{3}\varepsilon_{so}$. In the limits of $\varepsilon/\varepsilon_{so} \ll 1$ and $\varepsilon/\varepsilon_{so} \gg 1$ the bands are parabolic.

For a nonparabolic band of the type in Eq. (2.2.18), the velocity associated with a state $\mathbf{k}$ proves to be

$$\mathbf{v}(\mathbf{k}) = \frac{1}{\hbar}\frac{\partial \varepsilon}{\partial \mathbf{k}} = \frac{\hbar \mathbf{k}}{m(1 + 2\alpha\varepsilon)} \tag{2.2.20}$$

so that the conductivity effective mass is given by

$$m_c = m(1 + 2\alpha\varepsilon). \tag{2.2.21}$$

Nonparabolicity influences also the electron density of states. The temperature dependence of the density-of-states effective mass m_d is sketched in Fig. 2.4 for the case of electrons and holes [19]. From this figure it can be seen

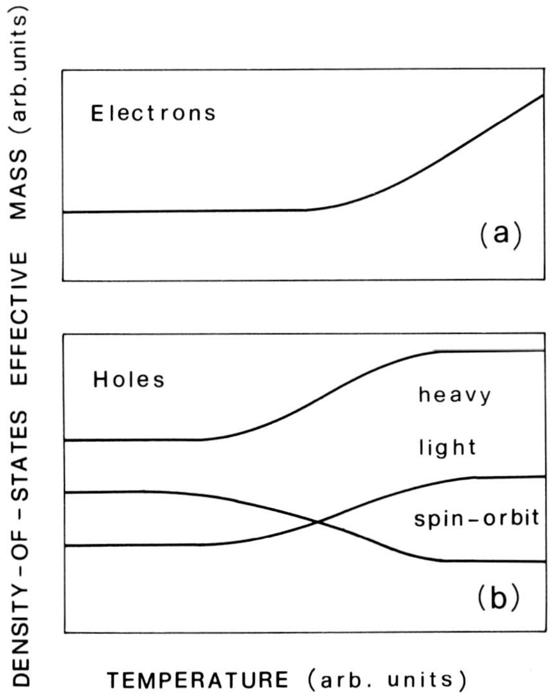

Fig. 2.4. Schematic representation of the temperature dependence of the density-of-state effective mass for electrons (*a*) and holes (*b*) [19]

how non-parabolicity effects lead to an unlimited increase of the effective mass in the case of electrons, while in the case of all three types of holes the increase (or decrease for the split-off band) of the effective mass is confined within a well-defined region of energy.

Taking into account what we have seen above, we may conclude that the effect of nonparabolicity is to reduce the velocity associated to a state **k** and to increase the density of states in energy. Since the scattering probabilities are proportional to the density of states, the second effect enhances the scattering rates. Thus nonparabolicity contributes to reduce the mobility of electrons at high temperatures and high fields, when they occupy high-energy states (see section 2.8). In some cases, for example, nonparabolicity could be enough to yield a negative differential mobility at low temperatures and high fields [20].

2.2.4 *Herring and Vogt Transformation*

When considering the ellipsoidal case of Eq. (2.2.4), in order to simplify analytical calculations, it is useful to introduce the Herring-Vogt transformation [21], which reduces the ellipsoidal equienergetic surfaces to spheres and is defined by

$$k_i^{*(m)} = T_{ij} k_j^{(m)}, \tag{2.2.22}$$

where $\mathbf{k}^{*(m)}$ is the transformed wave vector. For an electron in the mth valley the transformation matrix T_{ij} takes the form

$$T = \begin{vmatrix} \left[\dfrac{m_0}{m_t}\right]^{1/2} & 0 & 0 \\ 0 & \left[\dfrac{m_0}{m_t}\right]^{1/2} & 0 \\ 0 & 0 & \left[\dfrac{m_0}{m_l}\right]^{1/2} \end{vmatrix} \tag{2.2.23}$$

in the valley frame of reference, i.e., in the frame centered at the center of the valley, with the z axis along its symmetry axis. Consequently, the energy-wave-vector relationship in the starred space becomes of spherical type:

$$\varepsilon(\mathbf{k}) = \frac{\hbar^2}{2m_0} T_{ij} T_{il} k_j k_l = \frac{\hbar^2 k^{*2}}{2m_0}, \tag{2.2.24}$$

and the volume element $d\mathbf{k}$ is modified to $d\mathbf{k}^* = (m_0/m_d)^{3/2}\, d\mathbf{k}$, where m_d is the density-of-states effective mass given by Eq. (2.2.17).

To preserve vector equations, the transformation in Eq. (2.2.22) must be applied to other vector quantities, such as driving forces and phonon wave vector. Thus the equation of motion for an electron under the influence of an external force $\mathbf{F}$ becomes

$$\frac{d}{dt}(\hbar \mathbf{k}^*) = \mathbf{F}^*. \tag{2.2.25}$$

The electron velocity as a function of $\mathbf{k}^*$ is given by

$$v_i = \frac{\hbar}{m_0} T_{ij} k_j^*, \tag{2.2.26}$$

which is again generalized to the nonparabolic case by simply replacing m_0 with the expression $m_0(1 + 2\alpha\varepsilon)$.

2.2.5 *Actual Bands of Real Semiconductors*

In this section the main features of the band structure of the semiconductors of interest in microelectronic devices will be briefly described with regard to their use in charge transport calculations. The values of the parameters that describe conduction and valence bands for these materials are reported in Table 2.1, together with other parameters of interest.

The full band structures of the materials considered in this section are shown in Figs. 2.5–2.8.

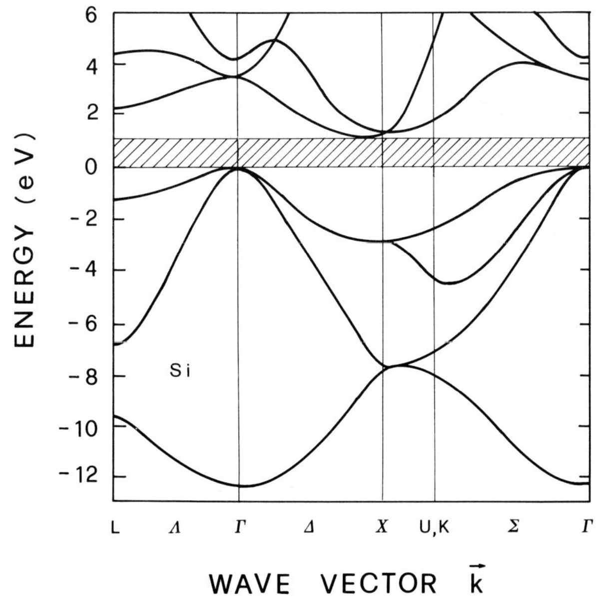

Fig. 2.5. Energy band structure of silicon [23]

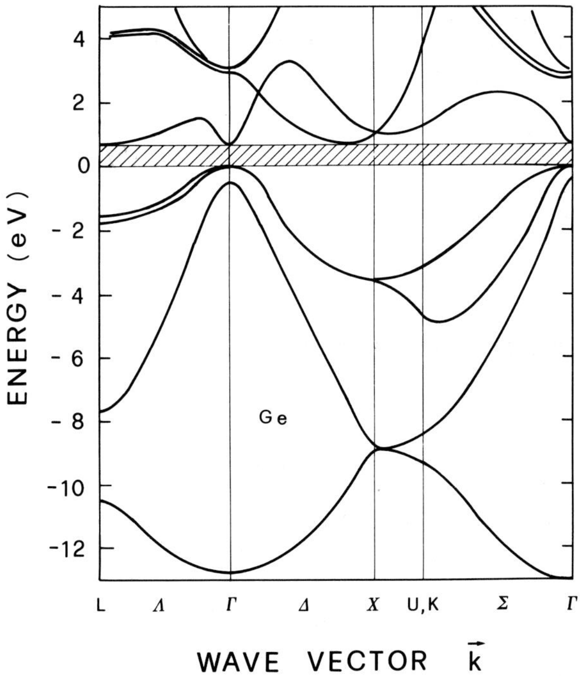

Fig. 2.6. Energy band structure of germanium [23]

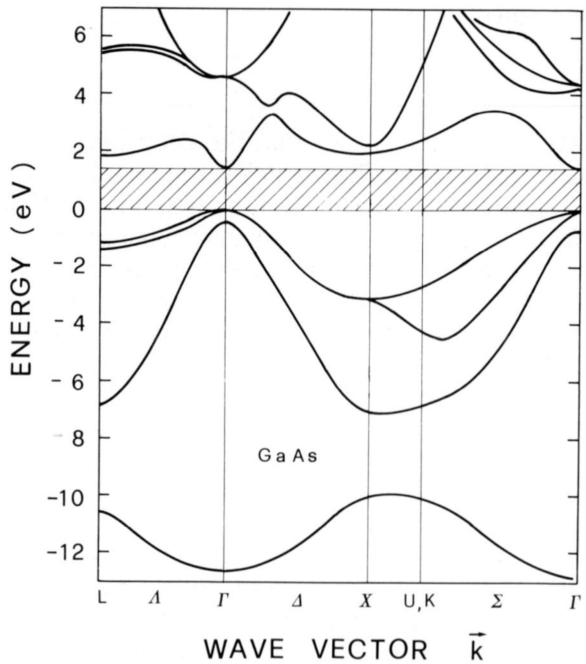

Fig. 2.7. Energy band structure of gallium arsenide [23]

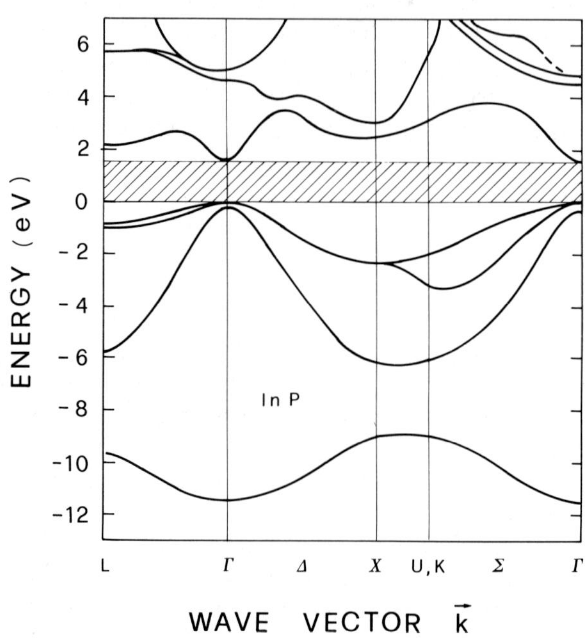

Fig. 2.8. Energy band structure of indium phosphide [23]

a) Conduction Bands

a.1) Silicon

The band structure of silicon is such that the electrons which contribute to charge transport, even at high electric fields, are those in the six equivalent ellipsoidal valleys along the ⟨100⟩ directions at about 0.85% of the Brillouin zone [24]. In fact, owing to their small density-of-states effective mass and distance in energy, the other secondary minima can be neglected.
The effect of band nonparabolicity is particularly important at high electric fields, when the electrons can reach energies far from the bottom of the conduction band [25].
The band structure of silicon is shown in Fig. 2.5.

a.2) Germanium

The absolute minima of the conduction band in germanium lie along the ⟨111⟩ directions at the L points. There are, therefore, four equivalent ellipsoidal valleys. Owing to their proximity in energy, the spherical upper valley at the center of the Brillouin zone and the six ellipsoidal upper valleys along the ⟨100⟩ directions can be populated by electrons at high electric fields, so that it is necessary to include them in high-field transport calculations.
As for the case of Si, nonparabolicity may influence the dynamical behavior of the electrons that populate the lowest bands, while for the bands associated with the upper minima this effect is expected to be much lower, due to the lower kinetic energies of the electrons that populate them. For this reason upper valleys are usually assumed to be parabolic in electron transport calculations.
The band structure of germanium is shown in Fig. 2.6.

a.3) Gallium Arsenide

The absolute minimum of the conduction band of GaAs (as well as of most III-V semiconductors) lies at the Γ point. Its nonparabolicity is quite important in transport since in polar materials the electron energy increases rapidly with the electric field. The assessment of the minima of upper valleys has given rise to some controversy in the recent past. The values currently accepted are given in Table 2.1. All valleys are normally populated at high electric fields. The population of the higher minima is at the origin of the negative differential mobility effect that will be described later in the chapter.
A very important characteristic of GaAs is its reduced value of the central-valley electron effective mass with respect to the average effective mass of silicon. The consequently higher electron mobility has triggered the interest

Table 2.1. *Band gaps and mass parameters of some cubic semiconductors* [22]

	ε_Γ (eV)	ε_L (eV)	ε_Δ (eV)	$\varepsilon_{SO}^{(a)}$ (eV)	m_l	m_t	m	$A(a)$	$B(a)$	$C(a)$	α $(\mathrm{eV})^{-1}$
Si	4.08	1.87	1.13	0.04	0.98	0.19	—	−4.22	−0.78	−4.80	0.5
Ge	0.89	0.76	0.96	0.29	1.64	0.082	—	−13.35	−8.50	−13.11	0.65
GaAs	1.42	1.71	1.90	0.34	—	—	0.067	−7.65	−4.82	−7.71	0.64
AlAs	2.95	2.67	2.16	0.28	—	—	2.0	−4.04	−1.56	−4.71	—
InP	1.26	2.0	2.3	0.13	—	—	0.080	−6.28	−4.16	−6.35	0.67

of the device community on GaAs as a suitable material for ultrafast devices. The values used in the literature for the average effective masses of the satellite, higher valleys are $m_L = 0.222$ and $m_X = 0.58$ [26]. The band structure of GaAs is shown in Fig. 2.7.

a.4) Indium Phosphide

The band structure of InP is very similar to the one of GaAs. Three main differences have, though, very important effects: the band gap and the Γ-valley effective mass are smaller, and the $\Gamma - L$ separation much higher in InP with respect to GaAs. Thus, intervalley transfer will be reduced in InP, and electron mobility increased.
The band structure of InSb is shown in Fig. 2.8.

b) Valence Bands

As we mentioned above, the valence bands of most cubic semiconductors are very similar to each other and reasonably well described by the model given above in this section. Table 2.1 reports the band parameters of the materials of interest in this book.

c) Alloys

When different materials have to be put into intimate contact to form heterostructures (which, as we shall see, are of great importance for semiconductor devices), lattice matching is a very important factor in determining the quality of the interfaces. Fig. 2.9 shows the correlation between energy

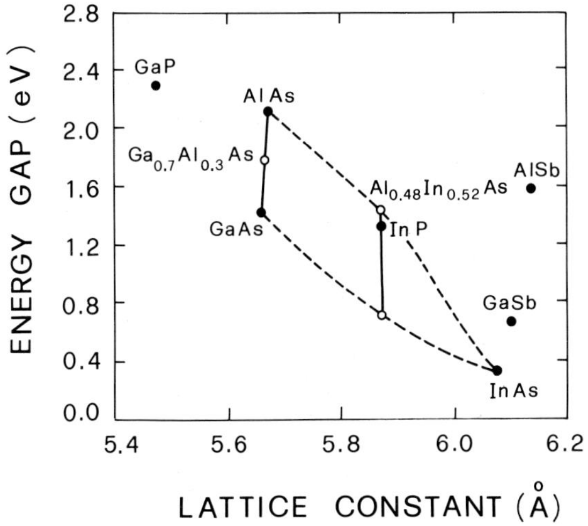

Fig. 2.9. Lattice constants and energy gaps of various compound semiconductors [27]

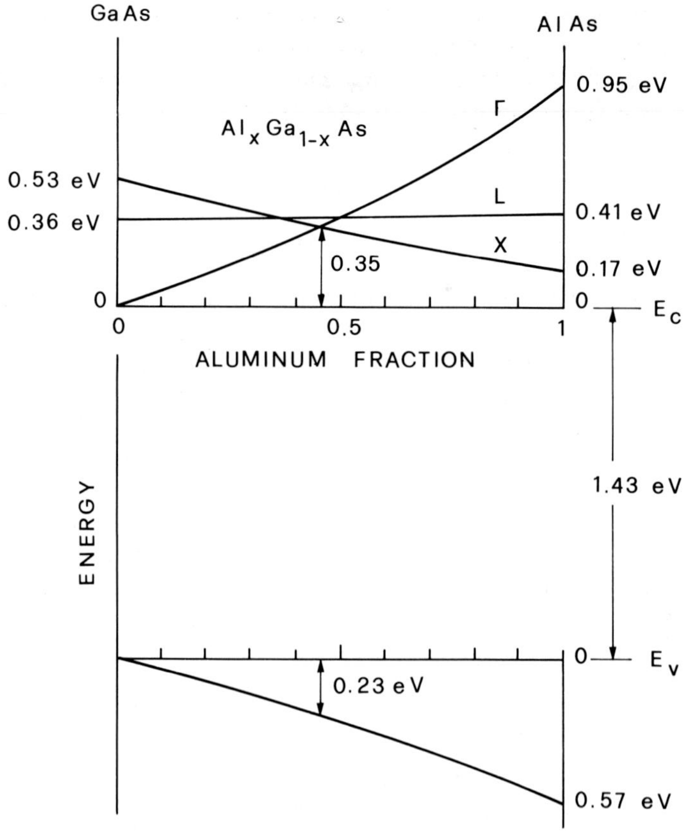

Fig. 2.10. Energy band extrema in the alloy $Al_xGa_{1-x}As$ as functions of the aluminum fraction x [28]

gap and lattice constant for a variety of III-V binary and ternary compound semiconductors.

It can be noticed that GaAs and AlAs have very similar lattice parameters. It is then possible to insert Al atoms into GaAs to form the alloy $Al_xGa_{1-x}As$ without any relevant changes (less than 1%) in the crystal structure. This allows the epitaxial growth of $Al_xGa_{1-x}As$ directly on GaAs. The dependence of the energy band extrema upon the Al content x in the alloy is presented in Fig. 2.10. The $x = 0$ value corresponds to GaAs ($\varepsilon_g = 1.43$ eV) and the $x = 1$ value to AlAs ($\varepsilon_g = 2.17$ eV). For $x < 0.45$, the compound is still direct-gap, with the minimum of the conduction band in Γ, while for $x > 0.45$ the minimum is along the [100] directions as in silicon. The importance of the band-gap dependence on Al content will clearly appear in the discussion of the two-dimensional electron gas in section 2.10.

2.3 Scattering Mechanisms

We now turn our attention to the scattering mechanisms that act on charge carriers in a host crystal. As usual in semiclassical transport, the dynamics of the electron interactions is assumed to be independent of the applied field, and the collisions are assumed to occur instantaneously.

All scattering calculations presented here will be carried out with a first-order perturbative approach; consequently only two-body interactions will be analyzed.

2.3.1 *Classification and Physical Discussion*

The electronic transitions of interest for charge transport in semiconductors can be classified as intravalley when the initial and final states lie in the same valley, or intervalley when they lie in different valleys. In the case of holes, transitions are correspondingly called intraband or interband.

The most important sources of scattering that determine these transitions in the bulk of homogeneous crystals are phonons, impurities, and other carriers

The interaction of phonons with charge carriers is due to the deformation of the otherwise perfect crystal produced by phonons through the deformation potential mechanism or through the electrostatic forces produced by the polarization waves that accompany the phonons. The first kind of interaction is typical of covalent semiconductors; the electrostatic interaction, typical of polar materials, is called piezoelectric interaction for acoustic phonons, and polar interaction for optical phonons. Often the attribute of piezoelectric or polar is given directly to the phonons involved. Thus we refer to piezoelectric phonons or polar optical phonons, as well as to intervalley phonons to indicate phonons that induce intervalley electron transitions.

Furthermore, impurities can be neutral or ionized. In the latter case the interaction is of Coulomb type, whereas in the former the interaction is of much shorter range and the overall effect of such impurities is in general much weaker.

Since the Coulomb cross section decreases rapidly at increasing momentum transfer Δk, owing to the large Δk involved in the intervalley transitions, the scattering from the Coulomb field of ionized impurities must, in general, be considered only for intravalley transitions.

Besides the above scatterings, other mechanisms should be considered, some of which might be of importance in special circumstances.

As a carrier moves at the surface of a definite region, or at the interface between two different regions, the corrugations of the potential profile at the interface can cause an additional scattering, usually called "surface roughness scattering" [29].

In disordered alloys, we may consider the fluctuations of composition as a source of scattering for electron eigenstates of the homogeneous system. This mechanism is usually called "alloy scattering" [30]. It has not received particular attention in connection with devices, even though it could have some effect in the heterostructures based on GaAlAs.

Finally, other phenomena involving trapping and detrapping of charges that can occur in semiconductors, such as impact ionization and generation-recombination, will be considered in this book, owing to their importance in device operation.

Before a formal treatment of the electron transitions due to the mechanisms indicated above, it is useful to present a physical discussion of the main features of each scattering process with particular attention to their influence on carrier transport.

a) Acoustic Scattering

As we shall see, the maximum energy transfer for an electronic interaction with acoustic phonons is, in general, much smaller than the electron energy, and thus very often acoustic scattering is treated as an elastic process.

If ohmic transport is investigated by analytical means, the energy distribution function is assumed to be the equilibrium Maxwellian (or Fermi) distribution, and no energy exchange of the electrons with the heat bath is explicitly required. If, on the other hand, a simulative approach is used, and nonlinear response is investigated, we need a mechanism that can exchange an infinitesimal amount of energy between the electrons and the crystal. To consider acoustic scattering as an elastic process is therefore, in general, not legitimate. When high temperatures or high fields are considered, the average electron energy is larger than the optical-phonon energy, and this kind of phonon can assume the task of exchanging energy between the electrons and the crystal.

b) Acoustic Scattering with Piezoelectric Interaction

The electrostatic nature of this mechanism leads to a scattering efficiency that decreases at increasing carrier energy; its importance is restricted to low temperature and low field situations, especially in high-purity materials. In dealing with hot-electron problems, where its influence is negligible, the piezoelectric scattering should always be considered together with the scattering due to deformation-potential interaction.

c) Optical-phonon Scatterig with Deformation-Potential Interaction

Usually the equivalent temperature of the optical phonons is assumed to be constant since the dispersion relation of such kind of phonons is quite flat for the q values involved in electronic intravalley transitions. As we shall see, this kind of scattering is isotropic. When realistic materials are considered, one should account for the selection rules that forbid this mechanism in particular band minima [31].

d) Optical-phonon Scattering with Polar Interaction

The electrostatic nature of the interaction is such that forward scattering is favored, so that this mechanism is strongly anisotropic. The treatment of this scattering is again simplified by the constancy of the phonon energy in the transition. At high electron energies the total scattering rate for polar optical scattering decreases with increasing energy, owing to the electrostatic nature of the interaction. This fact is responsible for the phenomenon called polar runaway [32, 33], which occurs at fields above a critical value E_{th}.

e) Intervalley Phonon Scattering

Electron transitions between states in two different equivalent valleys can be induced, as regards phonon scattering, by acoustic or optical modes. The phonon wave vector q involved in the transition is very close to the distance, in the Brillouin zone, between the minima of the initial and final valleys. Therefore, for a given branch of phonons, the energy involved in the transition is about constant, as for the case of optical intravalley scattering.

Polar or piezoelectric interaction is usually neglected for intervalley scattering, since the large momentum transfer involved in the transition makes its probability of occurrence very small.

In treating intervalley scattering, selection rules play a fundamental role. Transitions are forbidden or allowed depending on the positions of the valleys in the Brillouin zone and of the symmetry of the phonons involved [31]. One must consider, however, that the selection rules usually reported are valid for zero-order (in phonon wave vector) transitions and for electrons exactly at the minimum of the valleys, while thermal spread puts the electrons in an extended region around this point. This may explain why forbidden transitions appear to be present in transport phenomena [34].

When the electron energy is sufficiently high, electrons can scatter into valleys higher in energy than those corresponding to the absolute minima (upper valleys). The treatment of this type of scattering is almost identical to that for intervalley scattering between equivalent minima. The appropriate Δk in the Brillouin zone must of course be considered, together with the variation in the electronic kinetic energy due to the energy difference of the minima of the initial and the final valleys. Nonequivalent intervalley scattering is responsible for negative differential mobility at the origin of the Gunn effect in Ge [35, 20] and in most compound semiconductors [36].

f) Ionized Impurity Scattering

This type of collision is elastic in nature and therefore it cannot alone control the transport process in the presence of an external field. It must be accom-

panied by some other dissipating scattering mechanism if the proper energy distribution of electrons is to be derived from theory.
Owing to the electrostatic nature of the interaction, the efficiency of ionized impurity scattering decreases as the temperature of the crystal increases. This fact is accompanied by the increasing importance of phonon scattering. Furthermore, at any lattice temperature, as the field is increased, the effect of Coulomb scattering decreases when electrons are heated by the field.

g) Neutral Impurity Scattering

Neutral impurities have very small cross sections at normal concentrations (typically around 10^{16} cm^{-3}). Thus they influence the transport processes only at very low temperatures [37] and will be neglected in the present book.

h) Carrier-Carrier Scattering

Among the electron scattering mechanisms of interest in transport theory, carrier-carrier interaction is the most difficult to treat, because in this case the scattering probability depends on the distribution function, which is the unknown function to be determined. Thus carrier-carrier interaction makes the collision integral in the Boltzmann equation nonlinear.
We can focus briefly on the two most important types of carrier-carrier scattering, involving respectively electron-electron and electron-hole interactions.
The influence of electron-electron scattering on transport effects is, in general, quite small since these collisions do not change the total momentum nor the total energy of the colliding particles. It may be more effective on those particular quantities that strongly depend on the shape of the distribution function, such as the anisotropy of the drift velocity and the energy relaxation time [38–41].
An important effect due to *e-e* interaction can be obtained when fast electrons are injected into highly doped regions, characterized by large concentrations of thermal electrons. Very large exchange of energy and momentum can occur from the few fast carriers to the sea of ambient carriers. The situation is similar to that of the problem of the stopping power of a plasma when very energetic electrons are injected. Even in semiconductors, the collective excitation of the ambient electron gas generates an additional scattering mechanism, i.e. the electron-plasmon interaction [42].
The electron-hole scattering can be particularly important in the case of minority transport, that is, for instance, when electrons move in a *p*-type material in the presence of a large hole concentration. In such cases the transport characteristics can be drastically modified by *e-h* interaction.

i) Surface Scattering

As mentioned previously, this scattering mechanism [29] is due to the corrugation of interfaces and it becomes active when an electron travels in their vicinity. For this reason it is clear that the effect of such a mechanism is more pronounced when the transport phenomenon under investigation occurs in a narrow channel close to the interface. Furthermore, the effect is comparatively stronger at low temperatures when the phonon contribution is smaller.

Surface scattering is usually considered elastic, and simple classical models are often used that consider either specular or diffusive reflections of the electron at the interface.

j) Generation-Recombination and Impact Ionization

These two processes are quite important in devices since they change the number of active carriers in the conduction phenomenon.

Generation-recombination usually occurs under the effect of thermal phonons or of an electromagnetic radiation. In a generation process, by absorbing a quantum of energy, an electron enters the conduction band, leaving a lower state, while in the recombination the opposite process occurs. If the radiation is sufficiently energetic, the electron transition may occur from the valence band to the conduction band, and in this case two carriers, an electron and a hole, are created simultaneosly.

Electron-hole recombination usually occurs through a "recombination center", an impurity with an electron state deep in the energy gap where one of the two carriers first is trapped and then meets the partner necessary for recombination.

In the impact ionization process, a conducting carrier transfers energy to a localized carrier or to an electron in the valence band and promotes it to a conduction state. If in an impact ionization process, an electron leaves the valence band and occupies a state in the conduction band, two new carriers are generated.

Since impact ionization and generation-recombination mechanisms change the number of conducting carriers, they are important noise sources. In the case of impact ionization, if the newly created particle, under the effect of an external field, gains enough energy to impact-create other carriers, an avalanche process may occur and, in some cases, may lead to the destruction of the crystal (avalanche breakdown).

2.3.2 Fundamentals of Scattering

a) General Theory

In this section we shall review the general theory of electron transitions between different Bloch states in a crystal. The starting assumption is that the system can be separated into the electron of interest and the rest of the crystal. The state vector for the combined system can be written as the direct product

$$|\mathbf{k}, c\rangle = |\mathbf{k}\rangle|c\rangle \tag{2.3.1}$$

where $|\mathbf{k}\rangle$ and $|c\rangle$ represent the unperturbed states of the electron and of the crystal, respectively.

The expression for the transition probability per unit time from a state $|\mathbf{k}, c\rangle$ to a state $|\mathbf{k}', c'\rangle$ induced by a perturbation Hamiltonian H' is given, to first order, by the golden rule

$$P(\mathbf{k}, c; \mathbf{k}', c') = \frac{2\pi}{\hbar}|\langle\mathbf{k}', c'|H'|\mathbf{k}, c\rangle|^2\delta[\varepsilon(\mathbf{k}', c') - \varepsilon(\mathbf{k}, c)]. \tag{2.3.2}$$

where $\varepsilon(\mathbf{k}, c)$ represents the energy of the whole system. H' acts, in general, on the coordinate $\mathbf{r}$ of the electron and on the variables which describe the state $|c\rangle$ of the crystal, e.g., on the ion displacement $\mathbf{y}$ with respect to their equilibrium position and on the coordinates $\mathbf{r}'$ of the other electrons.

It is now convenient to expand H' in a Fourier series,

$$H'(\mathbf{r}, \mathbf{y}, \mathbf{r}') = \frac{(2\pi)^{3/2}}{V}\sum_{q}\mathscr{H}'(\mathbf{q}, \mathbf{y}, \mathbf{r}')\exp(i\mathbf{q}\cdot\mathbf{r}), \tag{2.3.3}$$

which gives to the matrix element in Eq. (2.3.2) the following form:

$$\langle\mathbf{k}', c'|H'|\mathbf{k}, c\rangle = \frac{(2\pi)^{3/2}}{V}\sum_{q}\langle c'|\mathscr{H}'(\mathbf{q}, \mathbf{y}, \mathbf{r}')|c\rangle\int d\mathbf{r}\,\psi_k^*(\mathbf{r})\exp(i\mathbf{q}\cdot\mathbf{r})\psi_{\mathbf{k}'}(\mathbf{r}), \tag{2.3.4}$$

where

$$\psi_{\mathbf{k}}(\mathbf{r}) = N^{-1/2}u_{\mathbf{k}}(\mathbf{r})\exp(i\mathbf{k}\cdot\mathbf{r}) \tag{2.3.5}$$

is the standard Bloch state; N is the number of unit cells in the crystal, and the factor $N^{-1/2}$ is introduced in order to have ψ_k normalized to one in the crystal with $u_k(\mathbf{r})$ normalized to one in the unit cell.

Let us now focus attention on the integral I on the right-hand side of Eq. (2.3.4). It can be calculated as a sum over the cells labelled by the vectors $\mathbf{R}$ of the direct lattice, and with the change of variable

$$\mathbf{r}'' = \mathbf{r} - \mathbf{R}, \tag{2.3.6}$$

it can be written as

$$I = \sum_{\mathbf{R}}\exp[i(\mathbf{k} - \mathbf{k}' + \mathbf{q})\cdot\mathbf{R}]\frac{1}{N}\int_{\text{cell}} d\mathbf{r}''u_{\mathbf{k}'}^*(\mathbf{r}'')u_{\mathbf{k}}(\mathbf{r}'') \times \exp[i(\mathbf{k} - \mathbf{k}' + \mathbf{q})\cdot\mathbf{r}'']. \tag{2.3.7}$$

By taking into account that $\mathbf{k} = \mathbf{k}' + \mathbf{q}$ is a vector of the reciprocal space and $\mathbf{R}$ a vector of the direct lattice, we have (see, for example, the textbook of Ziman [43]):

$$\sum_{\mathbf{R}} \exp[i(\mathbf{k} - \mathbf{k}' + \mathbf{q})\cdot\mathbf{R}] = \begin{cases} N & \text{if } \mathbf{k} - \mathbf{k}' + \mathbf{q} = \mathbf{G} \\ 0 & \text{otherwise} \end{cases} \tag{2.3.8}$$

where $\mathbf{G}$ is a vector of the reciprocal lattice. The terms of the sum in Eq. (2.3.4) which correspond to $\mathbf{G} = 0$ are called "normal" (or "N") terms, while when $\mathbf{G}$ is different from zero we speak of "umklapp" (or "U") terms.
By collecting the above results, we have

$$P(\mathbf{k}, c; \mathbf{k}', c') = \frac{(2\pi)^4}{\hbar V^2} \left| \sum_{\mathbf{q}} \langle c' | \mathscr{H}'(\mathbf{q}, \mathbf{y}) | c \rangle \right|^2 \mathscr{G}\, \delta[\varepsilon(\mathbf{k}', c') - \varepsilon(\mathbf{k}, c)] \tag{2.3.9}$$

where $\mathscr{G}$ is the overlap integral

$$\mathscr{G} = \left| \int_{\text{cell}} d\mathbf{r}\, u_{\mathbf{k}'}^*(\mathbf{r}) u_{\mathbf{k}}(\mathbf{r}) \exp(i\mathbf{G}\cdot\mathbf{r}) \right|^2 . \tag{2.3.10}$$

To proceed further, we must give explicit forms to $\mathscr{H}'$, which depend on the particular scattering mechanism considered. This will be done in the following section.

b) Overlap Factor

Expressions for the overlap factor $\mathscr{G}$ in Eq. (2.3.10) as a function of $\mathbf{k}$ and $\mathbf{k}'$ have been given in the literature for various cases. For electrons, an intravalley transition process is, in general, of "N" type, because the distance between $\mathbf{k}$ and $\mathbf{k}'$ is small compared to the dimensions of the Brillouin zone. For "N" processes, $\mathscr{G}$ is equal to unity for exact plane waves or for wave functions formed with pure s states. When lower symmetries are involved in the Bloch wave functions (as, for example, p states in the case of cubic semiconductors), an overlap factor less than unity is obtained, which depends mainly upon the angle θ between initial and final states $\mathbf{k}$ and $\mathbf{k}'$, measured from the center of the Brillouin zone.
In the case of a minimum located at Γ, $\mathscr{G}$ is strictly related to nonparabolicity, since both of them come from the presence of p terms in the electron wave functions. Fawcett et al. [16] gave for $\mathscr{G}$ the expression

$$\mathscr{G}(\mathbf{k}, \mathbf{k}') = \frac{[(1 + \alpha\varepsilon)^{1/2}(1 + \alpha\varepsilon')^{1/2} + \alpha(\varepsilon\varepsilon')^{1/2} \cos\theta]^2}{(1 + 2\alpha\varepsilon)(1 + 2\alpha\varepsilon')}, \tag{2.3.11}$$

where $\varepsilon' = \varepsilon(\mathbf{k}')$ is the energy of the electron in the state $\mathbf{k}'$.
In the many-valley model, for intervalley transitions the process can be of "U" type, because of the large values associated with $\mathbf{k}$ and $\mathbf{k}'$. For both intravalley and intervalley transitions, the angle θ between initial and final states depends mostly on the valleys involved in the transition, and $\mathscr{G}$ is thus

almost constant within each type of intravalley or intervalley scattering process [44]. The values for $\mathscr{G}$ in these cases may be included in the coupling constants.

For transitions of holes within heavy or light bands, Wiley [45] found the simple expression

$$\mathscr{G}(\mathbf{k}, \mathbf{k}') = \tfrac{1}{4}(1 + 3\cos^2\theta), \tag{2.3.12}$$

while for interband transitions he used

$$\mathscr{G}(\mathbf{k}, \mathbf{k}') = \tfrac{3}{4}\sin^2\theta. \tag{2.3.13}$$

2.4 Scattering Probabilities

The general theory reviewed in the previous section for electron scattering in crystals will now be applied to special cases of interest in device applications.

2.4.1 *Phonon Scattering, Deformation-Potential Interaction*

In the case of deformation-potential interaction, which is the one of interest in covalent semiconductors, the interaction Hamiltonian for a single band at Γ is written as [46]

$$H' = \mathscr{E}\frac{\partial \mathbf{y}}{\partial \mathbf{r}} \tag{2.4.1}$$

where $\partial\mathbf{y}/\partial\mathbf{r}$ is the deformation of the crystal due to the phonons, and $\mathscr{E}$ is a tensor that describes the shift of the electron band per unit deformation. In the approximation of a continuous medium, the ion displacement field $\mathbf{y}$ can be written in terms of phonon creation $a_q^\dagger$ and annihilation a_q operators as [46]

$$\mathbf{y} = \sum_q \left[\frac{\hbar}{2\rho V\omega_q}\right]^{1/2}(a_q + a_{-q}^\dagger)e^{i\mathbf{q}\cdot\mathbf{r}}\xi \tag{2.4.2}$$

where ρ is the density of the crystal, ω_q the phonon angular frequency, and ξ its polarization. By using Eq. (2.4.2) we obtain for Eq. (2.4.1) the expression

$$H' = \sum_q \left[\frac{\hbar}{2\rho V\omega_q}\right]^{1/2}(a_q + a_q^\dagger)e^{i\mathbf{q}\cdot\mathbf{r}}i\mathscr{E}_{lj}q_l\xi_j \tag{2.4.3}$$

where the sum is implied for the coordinate subscripts l and j. Equation (2.4.3) is an explicit form of the Fourier transform indicated in Eq. (2.3.3); a_q and $a_{-q}^\dagger$ appear in Eq. (2.4.3) in such a way that only two terms in the sum over $\mathbf{q}$ give a contribution to the matrix element; they correspond to states $|c'\rangle$ with the occupation number of one mode $\mathbf{q}$ changed by one unit with respect to $|c\rangle$. More precisely, they correspond to the emission and the absorption of a phonon with wave vector

$$\mathbf{q} = \mathbf{k}' - \mathbf{k} + \mathbf{G} \tag{2.4.4}$$

for absorption, or

$$\mathbf{q} = \mathbf{k} - \mathbf{k}' + \mathbf{G} \tag{2.4.5}$$

for emission. Eqs. (2.4.4) and (2.4.5) state the conservation law of crystal momentum.

For the two cases above we have

$$|\langle c'|a_q|c\rangle|^2 = N_q \tag{2.4.6}$$

and

$$|\langle c'|a_q^\dagger|c\rangle|^2 = N_q + 1 \tag{2.4.7}$$

for absorption and emission, respectively, where N_q is the phonon number in state $|c\rangle$.

By collecting the foregoing results and averaging over the possible phonon states, we obtain the following expression for the transition probability per unit time of an electron from state $\mathbf{k}$ to state $\mathbf{k}'$:

$$P(\mathbf{k}, \mathbf{k}') = \frac{\pi}{\rho V \omega_q} \begin{bmatrix} N_q \\ N_q + 1 \end{bmatrix} \mathscr{G} |\mathscr{E}_{ij} q_j \xi_i|^2 \, \delta[\varepsilon(\mathbf{k}') - \varepsilon(\mathbf{k}) \mp \hbar\omega_q], \tag{2.4.8}$$

where the upper and lower symbols refer to absorption and emission, respectively, and N_q stands now for the average phonon number.

This equation will be used in the following sections for the analysis of the various carrier transitions due to deformation potential phonons. In particular it will also be applied, in a slightly modified form [47, 48], to the case of optical modes, for which the above theory is not directly applicable [31, 49–52].

In what follows, the phonon number N_q will be assumed to be equal to its equilibrium value at the temperature of the crystal. There are situations where this assumption fails. For example, in the presence of large electrical or optical excitations, the phonon distribution can be driven out of equilibrium. Detailed analysis of this effect can be found in [53–55] and references therein.

a) Electron Intravalley Scattering. Acoustic Phonons

For a nondegenerate band at the center of the Brillouin zone of a cubic semiconductor, the deformation potential $\mathscr{E}$ of Eq. (2.4.8) is a second-rank tensor with cubic symmetry, which, as a rule, has a diagonal form with equal diagonal elements and therefore can be treated as a scalar quantity $\mathscr{E}_1$. The squared factor that appears in Eq. (2.4.8) reduces to $\mathscr{E}_1^2 q^2$ for longitudinal phonons, while it vanishes for transverse modes.

In this simple case, energy and momentum conservation imply

$$q = \mp 2\left[k \cos \theta - \frac{mu_l}{\hbar}\right], \tag{2.4.9}$$

where u_l is the longitudinal sound velocity; m the electron effective mass, θ the angle between $\mathbf{k}$ and $\mathbf{q}$, and upper and lower signs refer to absorption and emission, respectively. The maximum value of q is obtained for absorption with backward scattering, and is given by

$$q_{\max} = 2k + \frac{2mu_l}{\hbar}. \tag{2.4.10}$$

If the scattering mechanism were elastic, a backward collision would involve $q_{\max} = 2k$; thus the second term in Eq. (2.4.10) is the correction due to the energy of the phonon involved in the transition. The relative contribution of this term can be seen in Eq. (2.4.10) to be given by u_l/v, where v is velocity of the electron which, in general, is much larger than u_l. The maximum q involved in these transitions is therefore very close to $2k$; the corresponding maximum energy transfer is

$$\hbar q_{\max} u_l \approx 2\hbar k u_l = 2mvu_l, \tag{2.4.11}$$

which, again, is in general much smaller than the electron kinetic energy $mv^2/2$. For these reasons acoustic scattering is very often considered an elastic process.

As we have seen in section 2.3.1a, however, this approximation is not always legitimate. In what follows acoustic scattering will first be treated in elastic approximation and then with the correct energy exchange.

a.1) Elastic, Equipartition Approximation. Spherical, Parabolic Bands

When we deal with acoustic scattering in elastic approximation, the phonon population N_q is usually represented by the equipartition expression

$$N_q \simeq \frac{K_B T}{\hbar q u_l} - \frac{1}{2}. \tag{2.4.12}$$

This approximation is closely related to the elastic approximation discussed above. In fact, Eq. (2.4.12) is valid when $\hbar q u_l \ll K_B T$, i.e., when the thermal energy is much larger than the energy of the phonon involved in the transition. At very low temperatures and/or at very high fields, this condition may break down [56]. Furthermore, when a precise energy balance is to be obtained, exchange of electron energy with the heat bath via acoustic phonons must be taken into account, and in this case an exact expression for N_q or a good approximation of it must be used (see Appendix A).

The expression for the scattering probabilities per unit time from a state $\mathbf{k}$ to a state $\mathbf{k}'$ given by Eq. (2.4.8), in elastic and energy-equipartition approximations, becomes

$$P(\mathbf{k}, \mathbf{k}') = \frac{\pi q \mathscr{E}_1^2}{V u_l \rho}\left[\frac{K_B T}{\hbar q u_l} \mp \frac{1}{2}\right]\delta[\varepsilon(\mathbf{k}') - \varepsilon(\mathbf{k})] \tag{2.4.13}$$

where upper and lower signs refer to absorption and emission, and the overlap factor $\mathscr{G}$ has been taken as equal to one, in accord with the comments in Sec. 2.3.2b. Since in the elastic approximation no distinction is made between final states attained by means of absorption or emission processes, we can consider the sum of the two transition rates, given by

$$P(\mathbf{k}, \mathbf{k}') = \frac{2\pi K_B T}{\hbar V u_l^2 \rho} \mathscr{E}_1^2 \delta[\varepsilon(\mathbf{k}') - \varepsilon(\mathbf{k})]. \tag{2.4.14}$$

From this equation we conclude that, in the above approximations, acoustic scattering becomes isotropic.

Integration of Eq. (2.4.14) over all possible final states $\mathbf{k}'$ yields the integrated scattering probability per unit time which accounts for both absorption and emission processes:

$$P_{e,ac}(\varepsilon) = \frac{2^{1/2} m^{3/2} K_B T \mathscr{E}_1^2}{\pi \hbar^4 u_l^2 \rho} \varepsilon^{1/2}. \tag{2.4.15}$$

Its energy dependence, a simple proportionality $\varepsilon^{1/2}$, is shown in Fig. 2.11.

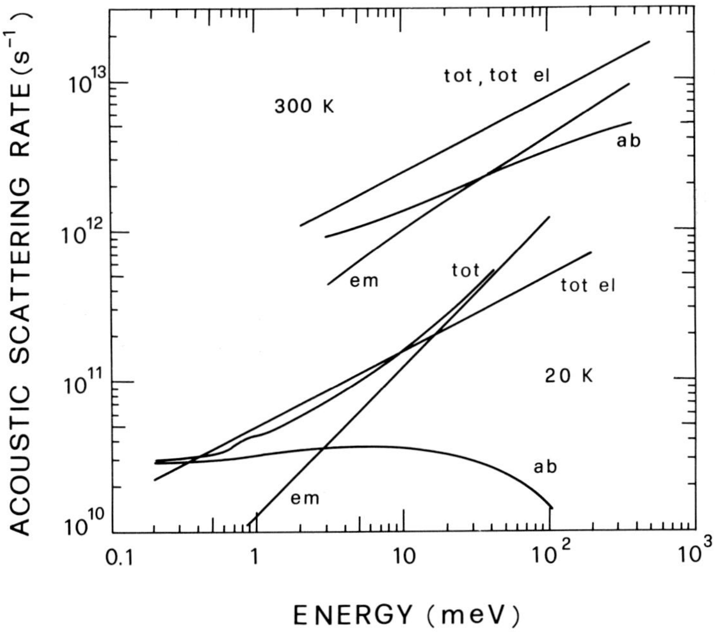

Fig. 2.11. Acoustic scattering rates at 300 K and 20 K for absorption (*ab*), emission (*em*), and total rate (*tot*) evaluated with the correct energy dissipation. The model used refers to electrons in Si with a parabolic band structure. For comparison, results obtained in the elastic approximation (*tot el*) are also shown [18]

a.2) Elastic, Equipartition Approximation. Ellipsoidal, Nonparabolic Bands

If we move from the simple spherical and parabolic case to more realistic band structures, many complications arise. For an ellipsoidal, nonparabolic model, the energy-wave vector relationship is given in section 2.2.3 and it is usually treated by means of the Herring-Vogt transformation.
In a model with valleys centered along $\langle 100 \rangle$ or $\langle 111 \rangle$ directions, for symmetry reasons we have two independent components Ξ_d and Ξ_u of the deformation-potential tensor [21], and interaction with acoustic phonons is allowed for transverse as well as longitudinal modes. In this case Eq. (2.4.14) becomes

$$P_l(\mathbf{k}, \mathbf{k}') = \frac{2\pi K_B T}{\hbar V u_l^2 \rho} (\Xi_d + \Xi_u \cos^2 \theta)^2 \delta[\varepsilon(\mathbf{k}') - \varepsilon(\mathbf{k})] \tag{2.4.16}$$

for longitudinal modes, and

$$P_t(\mathbf{k}, \mathbf{k}') = \frac{2\pi K_B T}{\hbar V u_t^2 \rho} (\Xi_u \sin \theta \cos \theta)^2 \delta[\varepsilon(\mathbf{k}') - \varepsilon(\mathbf{k})] \tag{2.4.17}$$

for transverse modes. Here u_t is the transverse sound velocity and θ is the angle between $\mathbf{q}$ and the longitudinal axis of the valley considered.
The effect of anisotropy is not large [57], so that it is in general neglected by replacing u_l and u_t with an average value $u = \frac{1}{3}(2u_t + u_l)$ and the expressions in brackets with mean values $\mathscr{E}_i^2$ over the angle θ. Thus a simple expression for the scattering probability per unit time is again used, of the type of Eq. (2.4.14):

$$P(\mathbf{k}, \mathbf{k}') = \frac{2\pi K_B T}{\hbar V u^2 \rho} \mathscr{E}_1^2 \delta[\varepsilon(\mathbf{k}^{*\prime}) - \varepsilon(\mathbf{k}^*)]. \tag{2.4.18}$$

Bands effects are carried on by the energy forms in the argument of the δ function. The last expression is again independent of the direction of $\mathbf{k}^*$ and $\mathbf{k}^{*\prime}$, so that the scattering is still isotropic in the starred space.
Integration over all possible final states $\mathbf{k}^{*\prime}$ yields the integrated scattering probability per unit time which accounts for both absorption and emission processes:

$$P_{e,ac}(\varepsilon) = \frac{2^{1/2} m_d^{3/2} K_B T \mathscr{E}_1^2}{\pi \hbar^4 u^2 \rho} \varepsilon^{1/2}(1 + 2\alpha\varepsilon)(1 + \alpha\varepsilon)^{1/2}. \tag{2.4.19}$$

From Eq. (2.4.19) it is also possible to determine the expression of $P_{e,ac}(\varepsilon)$ for "intermediate" models, i.e., spherical nonparabolic [$m_d = m$ and $u = u_l$] or ellipsoidal parabolic [$\alpha = 0$].

a.3) Inelastic Acoustic Scattering. Spherical, Parabolic Bands

In order to treat energy dissipation via acoustic phonons correctly, we must take into account the energy of the phonon involved in the energy balance of the collision:

$$\varepsilon(\mathbf{k}') = \varepsilon(\mathbf{k}) \pm \hbar q u_l\,. \tag{2.4.20}$$

In the notation that will be used throughout the book, upper and lower signs refer to absorption and emission, respectively.

In a simple spherical and parabolic band, only longitudinal modes contribute to the scattering, and energy and momentum conservation imply [see Eq. (2.4.9)]

$$\cos\theta = \pm\frac{q}{2k} + \frac{m u_l}{\hbar k}\,. \tag{2.4.21}$$

The condition $-1 \leqslant \cos \leqslant 1$ determines the range of phonon wave vectors that may be involved in a collision with an electron in state $\mathbf{k}$.

The results are reported in Table 2.2, where, instead of q, use has been made of the more convenient dimensionless variable $x = (\hbar q u_l)/(K_B T)$, and where $\varepsilon_u = m u_l^2/2$ is the kinetic energy of an electron with velocity equal to the longitudinal sound velocity.

In order to be consistent with the correct treatment of energy exchange in acoustic-phonon scattering, the exact expression of the phonon number N_q must be included in the calculation. Thus the transition probability of Eq. (2.4.8) for the case of interest becomes

$$P(\mathbf{k}, \mathbf{k}') = \frac{\pi q \mathscr{E}_1^2}{V\rho u_l}\begin{bmatrix} N_q \\ N_q + 1\end{bmatrix}\delta[\varepsilon(\mathbf{k}') - \varepsilon(\mathbf{k}) \mp \hbar q u_l] \tag{2.4.22}$$

Table 2.2. *Limits of integration for acoustic-phonon scattering*

Spherical Parabolic Bands		
Absorption	Emission	
$x_{1,a} = C(\varepsilon_u^{1/2} - \varepsilon^{1/2})$ $x_{2,a} = C(\varepsilon_u^{1/2} + \varepsilon^{1/2})$	absent	$\varepsilon < \varepsilon_u$
$x_{1,a} = 0$ $x_{2,a} = C(\varepsilon^{1/2} + \varepsilon_u^{1/2})$	$x_{1,e} = 0$ $x_{2,e} = C(\varepsilon^{1/2} - \varepsilon_u^{1/2})$	$\varepsilon > \varepsilon_u$

Ellipsoidal Non-Parabolic Bands		
Absorption	Emission	
$x_{1,a} = C^*(\alpha)\{\sqrt{\varepsilon_s^*(1+2\alpha\varepsilon)} - \sqrt{\gamma}\}$ $x_{2,a} = C^*(\alpha)\{\sqrt{\varepsilon_s^*(1+2\alpha\varepsilon)} + \sqrt{\gamma}\}$	absent	$\gamma < \dfrac{\varepsilon_s^*}{1-4\alpha\varepsilon_s^*}$
$x_{1,a} = 0$ $x_{2,a} = C^*(\alpha)\{\sqrt{\gamma} + \sqrt{\varepsilon_s^*(1+2\alpha\varepsilon)}\}$	$x_{1,e} = 0$ $x_{2,e} = C^*(\alpha)\{\sqrt{\gamma} - \sqrt{\varepsilon_s^*(1+2\alpha\varepsilon)}\}$	$\gamma > \dfrac{\varepsilon_s^*}{1-4\alpha\varepsilon_s^*}$
$C = \dfrac{4\varepsilon_u^{1/2}}{K_B T}$	$C^*(\alpha) = \dfrac{4\varepsilon_s^{*\,1/2}}{K_B T(1-4\alpha\varepsilon_s^*)}$	

To perform the integration over all possible final states, it is convenient to consider integration over **q**, and to use polar coordinates with the polar axis along **k**. The δ function of energy conservation can be used to integrate over the polar angle θ between **q** and **k**, which leads to

$$P(k, q)\, dq = \frac{m\mathscr{E}_1^2}{4\pi\rho u_l \hbar^2 k}\begin{bmatrix} N_q \\ N_q + 1 \end{bmatrix} q^2\, dq. \tag{2.4.23}$$

Final integration over q leads to the integrated scattering probability per unit time as a function of energy:

$$P_{e,ac}(\varepsilon) = \frac{m^{1/2}(K_B T)^3 \mathscr{E}_1^2}{2^{5/2}\pi\hbar^4 u_l^4 \rho}\varepsilon^{-1/2}\begin{Bmatrix} F_1(x_{2,a}) - F_1(x_{1,a}) \\ G_1(x_{2,e}) - G_1(x_{1,e}) \end{Bmatrix}, \tag{2.4.24}$$

where

$$F_1(x) = \int_0^x N_q(x')x'^2\, dx', \tag{2.4.25}$$

$$G_1(x) = \int_0^x [N_q(x') + 1]x'^2\, dx', \tag{2.4.26}$$

relate to absorption and emission processes, respectively, and $x_{1,a}$, $x_{2,a}$, $x_{1,e}$, and $x_{2,e}$ are those given in Table 2.2. The numerical evaluation of the functions $F_1(x)$ and $G_1(x)$ is reported in Appendix A.
In Fig. 2.11, the integrated acoustic scattering probabilities, given by Eq. (2.4.24), are shown and compared with the results of the elastic approximation.

a.4) Inelastic Acoustic Scattering. Ellipsoidal, Nonparabolic Bands

When all details of ellipsoidal many-valley electron bands and acoustic energy relaxation are considered, the transition probability $P(\mathbf{k}, \mathbf{k}')$ has the form given in Eqs. (2.4.16) and (2.4.17), where, in addition, the argument of the δ function contains the energy of the phonon, as in Eq. (2.4.22), and nonparabolicity is accounted for by assuming, for the energy-wave vector relationship, the expression in Eq. (2.2.18).
We shall again keep the isotropic approximation for the deformation-potential coupling and, in applying the Herring-Vogt transformation for **k** and **q** vectors, approximate the magnitude of **q** as

$$q = q^*\left[\frac{m_l}{m_0}\cos^2\theta^* + \frac{m_t}{m_0}\sin^2\theta^*\right]^{1/2} \approx q^*\left[\frac{m_d}{m_0}\right]^{1/2}, \tag{2.4.27}$$

where θ^* is the angle between $\mathbf{k}^*$ and the principal axis of the valley.
The requirement that the argument of the δ vanishes yields

$$\cos\theta = \mp\frac{q^*}{2k^*}(1 - 4\alpha\varepsilon_u^*) + \left[\frac{\varepsilon_u^*}{\gamma(\varepsilon)}\right]^{1/2}(1 + 2\alpha\varepsilon), \tag{2.4.28}$$

where θ is the angle between $\mathbf{q}^*$ and $\mathbf{k}^*$, and $\varepsilon_u^* = m_d u^2/2$

As in the standard procedure, the condition $-1 \leqslant \cos\theta \leqslant 1$ yields the limits of variability of q^*. These limits are conveniently expressed in terms of the dimensionless variable $x = (\hbar q^* u/K_B T)(m_d/m_0)^{1/2}$ and are given in Table 2.2 for the case, verified in our cases, that $4\alpha\varepsilon_u^* < 1$. By using polar coordinates as for the previous case, we obtain, by integration over the angle variables

$$P(\varepsilon, x)\, dx = \frac{m_d^{1/2}(K_B T)^3 \mathscr{E}_1^2}{2^{5/2}\pi\hbar^4 u^4 \rho}\gamma^{-1/2}(\varepsilon)\begin{bmatrix} N_q(x) \\ N_q(x)+1 \end{bmatrix}(1 + 2\alpha\varepsilon \pm 2\alpha K_B T x)x^2\, dx. \tag{2.4.29}$$

Final integration over x leads to the integrated scattering probability per unit time as a function of energy,

$$P_{e,ac}(\varepsilon) = \frac{m_d^{1/2}(K_B T)^3 \mathscr{E}_1^2}{2^{5/2}\pi\hbar^4 u^4 \rho}\gamma^{-1/2}(\varepsilon) \times \begin{Bmatrix} (1+2\alpha\varepsilon)[F_1(x_{2,a})-F_1(x_{1,a})]+2\alpha K_B T[F_2(x_{2,a})-F_2(x_{1,a})] \\ (1+2\alpha\varepsilon)[G_1(x_{2,e})-G_1(x_{1,e})]-2\alpha K_B T[G_2(x_{2,e})-G_2(x_{1'e})] \end{Bmatrix}, \tag{2.4.30}$$

where

$$F_2(x) = \int_0^x N_q(x')x'^3\, dx', \tag{2.4.31}$$

$$G_2(x) = \int_0^x [N_q(x')+1]x'^3\, dx'. \tag{2.4.32}$$

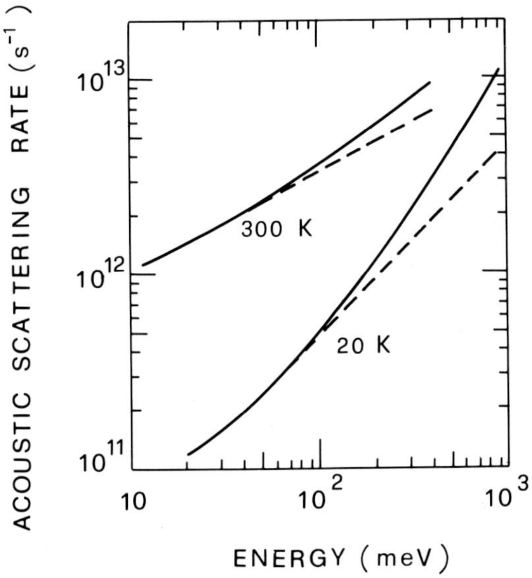

Fig. 2.12. Acoustic scattering rates with correct energy dissipation at 300 K and 20 K for a non-parabolic band (continuous line) compared to the case of a parabolic band (dashed line) [18]

The numerical evaluation of the functions $F_2(x)$ and $G_2(x)$ is reported in Appendix A.
The intermediate case of spherical nonparabolic bands can be dealt with by substituting m for m_d in Eq. (2.4.30). The nonspherical parabolic result is recovered by putting $\alpha = 0$ in Eq. (2.4.30). In Fig. 2.12 the integrated acoustic probabilities given by Eq. (2.4.30) are shown and compared with the results of the parabolic case.

b) Electron Intravalley Scattering. Optical Phonons

For optical modes, the perturbation Hamiltonian to the lowest value of approximation is assumed to be proportional to the atomic displacement [47, 48, 50] and not to its derivative, as in the case of acoustic vibrations. The continuous-medium approximation adopted for long-wavelength acoustic modes would be inconsistent for optical phonons. Nevertheless, the final result of the correct theory is the same, for the simplest case, which is obtained by assuming a wavelength equal to the lattice constant, to represent the fact that atoms belonging to the same unit cell vibrate in opposite directions. The scattering probability can then be written, starting from Eq. (2.4.8), by replacing $|\mathscr{E}_{ij}q_j\xi_i|^2$ with a squared optical coupling constant $(D_tK)^2$ which can also include the $\mathscr{G}$ factor, as indicated in section 2.3.2b. The energy associated with optical phonons in intravalley transitions can be assumed constant, given by $\hbar\omega_{op} = K_B\theta_{op}$, where θ_{op} is the equivalent temperature of the phonon, since the dispersion curve of such a phonon is quite flat for the q values involved in electronic intravalley transitions. For the same reason N_q becomes q independent ($N_q = N_{op}$). The resulting scattering probability per unit time is

$$P(\mathbf{k}, \mathbf{k}') = \frac{\pi(D_tK)^2}{\rho V\omega_{op}} \begin{bmatrix} N_{op} \\ N_{op} + 1 \end{bmatrix} \delta[\varepsilon(\mathbf{k}') - \varepsilon(\mathbf{k}) \mp \hbar\omega_{op}]. \qquad (2.4.33)$$

No angular dependence upon the direction of $\mathbf{k}'$ is present, and the scattering is therefore isotropic.

b.1) Spherical, Parabolic Bands

For spherical and parabolic bands, integration of Eq. (2.4.33) over $\mathbf{k}'$ yields, with standard calculations, the integrated scattering probability per unit time

$$P_{e,op}(\varepsilon) = \frac{(D_tK)^2 m^{3/2}}{2^{1/2}\pi\hbar^3\rho\omega_{op}} \begin{bmatrix} N_{op} \\ N_{op} + 1 \end{bmatrix} (\varepsilon \pm \hbar\omega_{op})^{1/2}. \qquad (2.4.34)$$

The probability of emission is obviously zero when $\varepsilon < \hbar\omega_{op}$, since the carrier does not have enough energy to emit the phonon.

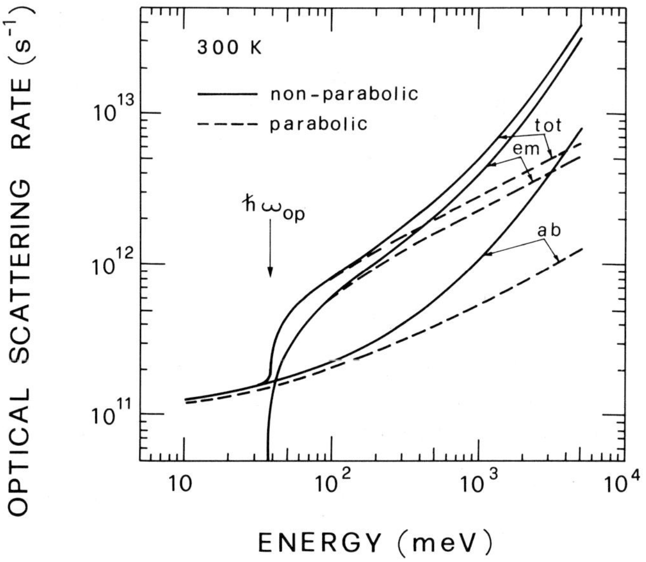

Fig. 2.13. Optical-phonon scattering rates as functions of energy. The model used refers to electrons in Ge with parabolic (dashed curves) and nonparabolic (solid curves) band structures. *ab* refers to absorption, *em* to emission, and *tot* to the total rate

Figure 2.13 shows the optical-phonon scattering rate as a function of energy, given in Eq. (2.4.34). A simple proportionality to $\varepsilon^{1/2}$, as for acoustic modes, is attained when the carrier energy is much larger than the phonon energy, so that the scattering becomes approximately elastic.

b.2) Ellipsoidal, Nonparabolic Bands

In dealing with ellipsoidal bands it is convenient to apply the Herring-Vogt transformation, as for acoustic phonons. The transition probability can be derived from the general expression in Eq. (2.4.8), as indicated above for simple bands. Since the transition probability is independent of $\mathbf{q}$, approximation in Eq. (2.4.27) is not necessary. The resulting transition probability is the same as in Eq. (2.4.33), with $\mathbf{k}^*$ and $\mathbf{k}^{*\prime}$; and so the integrated scattering probability per unit time is the same as in Eq. (2.4.34), with m replaced by m_d. When nonparabolicity is included, integration over $\mathbf{k}^*$ yields an extra factor $[1 + 2\alpha(\varepsilon \pm \hbar\omega_{op})]$, so that the integrated scattering probability per unit time for optical phonons is given by

$$P_{e,op}(\varepsilon) = \frac{(D_t K)^2 m_d^{3/2}}{2^{1/2}\pi\hbar^3\rho\omega_{op}} \begin{bmatrix} N_{op} \\ N_{op}+1 \end{bmatrix} \gamma^{1/2}(\varepsilon \pm \hbar\omega_{op})[1 + 2\alpha(\varepsilon \pm \hbar\omega_{op})], \tag{2.4.35}$$

where γ is defined in Eq. (2.4.5).

In Fig. 2.13 the integrated optical intravalley scattering probabilities for the parabolic and nonparabolic cases are shown as a function of energy.

c) Electron Intervalley Scattering

Electron transitions between states in two different equivalent valleys can be induced by both acoustic and optical-mode phonon scattering.
The phonon wave vector **q** involved in a transition remains very close to the distance between the minima of the initial and final valleys, even for high-energy electrons. Consequently, given these two valleys, $\mathbf{\Delta k}$ is almost constant, and, for a given branch of phonons, the energy $\hbar\omega_i$ involved in the transition is also about constant, as in the case of optical intravalley scattering. Thus intervalley scattering is usually treated, formally, in the same way as intravalley scattering by optical phonons with a deformation-potential interaction [47, 57].
The squared coupling constant $(D_t K)_i^2$ depends upon the kinds of valleys (initial and final) and the branch of phonons involved in the transition, and it may include, as usual, an overlap factor.
When then electron energy is sufficiently high, electrons can scatter into valleys higher in energy than the lowest-energy valleys. In this case the appropriate $\mathbf{\Delta k}$ in the Brillouin zone must of course be considered together with the variation of the electron kinetic energy due to the energy difference between minima of the initial and final valleys.
The integrated scattering probability per unit time for intervalley transitions

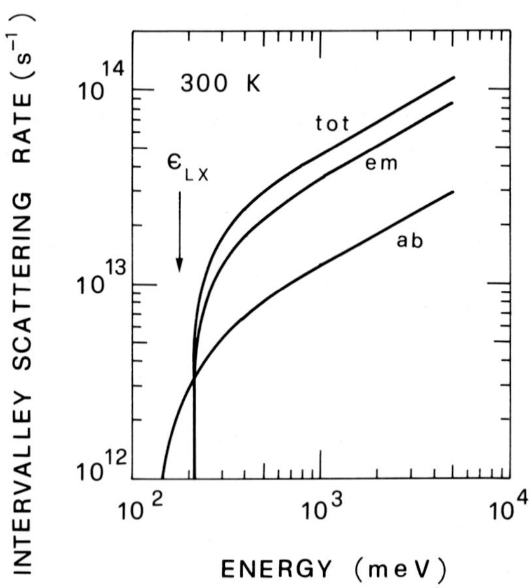

Fig. 2.14. Scattering rates for intervalley transitions to non-equivalent valleys as functions of energy. The model used refers to electrons in Ge with parabolic upper bands [18]

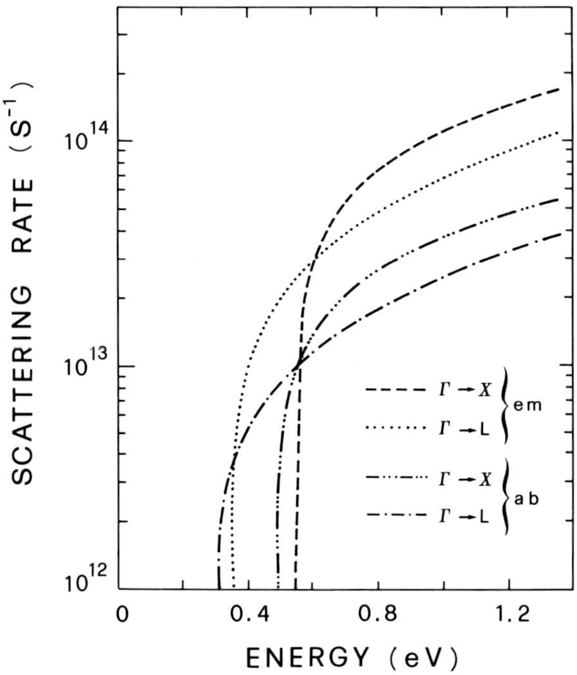

Fig. 2.15. Energy dependence of the intervalley scattering rate in the Γ valley of GaAs at 300 K [78]

due to phonons, in spherical and parabolic valleys, is therefore given by

$$P_{e,i}(\varepsilon) = \frac{(D_t K)_i^2 m^{3/2} Z_f}{2^{1/2} \pi \rho \hbar^3 \omega_i} \begin{bmatrix} N_i \\ N_i + 1 \end{bmatrix} (\varepsilon \pm \hbar\omega_i - \Delta\varepsilon_{fi})^{1/2}, \tag{2.4.36}$$

where Z_f is the number of possible final equivalent valleys for the type of intervalley scattering under consideration, N_i is the number of phonons involved in the transition, and $\Delta\varepsilon_{fi}$ is the difference between the energies of the bottoms of the final and initial valleys.

When ellipsoidal and/or nonparabolic valleys are considered, the same corrections must be introduced relative to m_d and to the extra factor $[1 + 2\alpha(\varepsilon \pm \hbar\omega_i - \Delta\varepsilon_{fi})]$ as in the case of optical intravalley phonon scattering.

Figures 2.14 and 2.15 show the integrated scattering rate for electron intervalley transitions to nonequivalent valleys in Ge and GaAs, respectively. For equivalent valleys, when $\Delta\varepsilon_{fi} = 0$, the behavior of $P_{e,i}(\varepsilon)$ is the same as that of $P_{e,op}(\varepsilon)$ (see Fig. 2.13).

d) Hole Intraband Scattering. Acoustic Phonons

Owing to the complexity of the valence-band structure, with its peculiarities of degeneracy and warping, the description of acoustic-phonon interaction requires three deformation-potential parameters [48, 58, 59]. Furthermore, the characteristic p-like symmetry of hole wave functions introduces an overlap factor which, in a good approximation, assumes the forms given in Eqs. (2.3.12) and (2.3.13).

To overcome analytical difficulties, at the same time keeping the physics, a single coupling constant, which for convenience of notation will be called $\mathscr{E}_1^0$, is used in place of the original three parameters, and a single warped parabolic band, the heavy one, will be considered in the following, with the notation given by Eqs. (2.2.5) and (2.2.6).

When a warped, parabolic band is considered, the hole intraband scattering with acoustic phonons closely follows the electron case treated in subsection a.1 above. The transition probability in Eq. (2.4.8) for the case of interest becomes

$$P(\mathbf{k}, \mathbf{k}') = \frac{\pi q \mathscr{E}_1^{0\,2}}{V \rho u} \begin{bmatrix} N_q \\ N_q + 1 \end{bmatrix} \frac{1}{4}(1 + 3\cos^2\theta)\delta[\varepsilon(\mathbf{k}') - \varepsilon(\mathbf{k}) \mp \hbar q u], \tag{2.4.37}$$

where θ, the angle between $\mathbf{k}$ and $\mathbf{k}'$, is expressed in terms of initial ϑ, ψ and final ϑ', ψ' angles as

$$\cos\theta = \cos\vartheta\cos\vartheta' + \sin\vartheta\sin\vartheta'\cos(\psi - \psi'). \tag{2.4.38}$$

To perform integration in Eq. (2.4.37), we can replace q to a good approximation with $2^{1/2}k(1 - \cos\theta)^{1/2}$, and by integrating over k' we find

$$P(\mathbf{k}, \vartheta', \psi') = \frac{\mathscr{E}_1^{0\,2}}{2^{1/2}\pi^2 \rho u} \begin{bmatrix} N_q \\ N_q + 1 \end{bmatrix} (1 + 3\cos^2\theta) k \frac{\{(1 - \cos\theta)[\varepsilon(\mathbf{k}) \pm \varepsilon_{ph}]\}^{1/2}}{\{\alpha[1 - g(\vartheta', \psi')]\}^{3/2}}, \tag{2.4.39}$$

where $\varepsilon_{ph} = \hbar q u$ is the energy of the phonon. This expression will be useful later on for the determination of the state $\mathbf{k}'$ after scattering.

In order to obtain the integrated scattering probabilities, we carry out the integration over the solid angle Ω' in the "nearly elastic approximation" (i.e. $(\varepsilon \pm \varepsilon_{ph})^{1/2} \simeq \varepsilon^{1/2}[1 \pm \varepsilon_{ph}/(2\varepsilon)]$) and in "small warped approximation" (i.e. $a[1 - g(\vartheta', \psi')] = \hbar^2/(2m_h)$) by taking $\mathbf{k}$ as polar axis for the angular integration [60]. The former approximation is sufficiently accurate down to temperatures of the order of 10 K; the latter approximation is quite reasonable, since it substitutes the density-of-state effective mass for a warping-dependent effective mass [60].

By introducing the dimensionless variable $x = (2m_h^{1/2}u\varepsilon^{1/2})/K_B T)$, we obtain the integrated scattering probability per unit time for acoustic absorption and emission processes

$$P_{h,ac}(\varepsilon) = \frac{\mathscr{E}_1^{0\,2} m_h^{1/2} (K_B T)^3}{2^{9/2}\pi\rho u^4 \hbar^4} \varepsilon^{-1/2} \begin{Bmatrix} F_3(x) + (K_B T/\varepsilon) F_4(x) \\ G_3(x) - (K_B T/\varepsilon) G_4(x) \end{Bmatrix} \tag{2.4.40}$$

with

$$F_3(x) = \int_0^{\sqrt{2x}} \frac{z^2[1 + 3(1 - z^2/x^2)^2]}{\exp(z) - 1} dz, \tag{2.4.41}$$

$$F_4(x) = \int_0^{\sqrt{2x}} \frac{z^3[1 + 3(1 - z^2/x^2)^2]}{\exp(z) - 1} dz, \tag{2.4.42}$$

$$G_3(x) = \int_0^{\sqrt{2x}} \frac{z^2[1 + 3(1 - z^2/x^2)^2]}{1 - \exp(-z)} dz, \tag{2.4.43}$$

$$G_4(x) = \int_0^{\sqrt{2x}} \frac{z^3[1 + 3(1 - z^2/x^2)^2]}{1 - \exp(-z)} dz. \tag{2.4.44}$$

The numerical evaluation of the functions $F_3(x)$, $F_4(x)$, $G_3(x)$ and $G_4(x)$ is reported in Appendix A.
The energy dependence of $P_{h,ac}(\varepsilon)$ is analogous to that of Eq. (2.4.24) for electrons reported in Fig. 2.11.

e) Hole Intraband Scattering. Optical Phonons

Optical scattering for the case of holes is described in terms of one deformation-potential parameter and has been found to be isotropic even when the symmetry of the hole wave function is taken into account [48, 61]. In analogy with the preceding section, a single warped band will be considered. The transition probability of Eq. (2.4.8) for the case of interest becomes

$$P(\mathbf{k}, \mathbf{k}') = \frac{3\pi d_0^2}{2\rho V \omega_{op} a_0^2} \begin{bmatrix} N_{op} \\ N_{op} + 1 \end{bmatrix} \delta[\varepsilon(\mathbf{k}') - \varepsilon(\mathbf{k}) \mp \hbar\omega_{op}], \tag{2.4.45}$$

where d_0 is the optical deformation potential. Again the probability of emission is obviously zero when $\varepsilon < \hbar\omega_{op}$. By integrating Eq. (2.4.45) over k' we find

$$P(\mathbf{k}, \vartheta', \psi') = \frac{3d_0^2}{2^5\pi^2 \rho \omega_0 a_0^2} \begin{bmatrix} N_{op} \\ N_{op} + 1 \end{bmatrix} \frac{(\varepsilon \pm \hbar\omega_{op})^{1/2}}{a[1 - g(\vartheta', \psi')]^{3/2}}. \tag{2.4.46}$$

Integration of Eq. (2.4.46) over the solid angle Ω' yields the integrated scattering probability per unit time

$$P_{h,op}(\varepsilon) = \frac{3d_0^2 m_h^{3/2}}{2^{3/2}\pi\hbar^3 \rho \omega_{op}} \begin{bmatrix} N_{op} \\ N_{op} + 1 \end{bmatrix} (\varepsilon \pm \hbar\omega_{op})^{1/2}, \tag{2.4.47}$$

with

$$m_h^{3/2} = \frac{m_0^{3/2}}{4\pi |A|^{3/2}} \int_0^{2\pi} d\psi' \int_0^{\pi} \sin\vartheta' [1 - g(\vartheta', \psi')]^{-3/2} d\vartheta'. \tag{2.4.48}$$

The angular integral in Eq. (2.4.48) can be performed numerically.
The energy dependence of $P_{h,op}(\varepsilon)$ is analogous to that of Eq. (2.4.34) reported in Fig. 2.13.

f) Hole Interband Scattering

When a two-band model, with heavy and light holes, is considered, interband scattering should be accounted for. For the case of two spherical and parabolic bands, the scattering probabilities for different microscopic mechanisms have been calculated by Costato and Reggiani [62]. For the case of acoustic-phonon scattering, inclusion of energy dissipation and exact phonon population has been considered by Bosi et al. [13]. Owing to the complexity of the calculations involved and to their rather limited interest in practical cases, we shall not treat this subject further, but refer the reader to the appropriate literature.

g) Selection Rules

Intravalley optical and intervalley phonons are subject to selection rules when initial and final **k** states are along high-symmetry directions (see, for example, the textbook of Bir and Pikus [31]. In dealing with cubic semiconductors, intravalley optical-phonon interaction is forbidden to zero order in the phonon wave vector for electrons at the Γ point (e.g., in GaAs) and along ⟨100⟩ directions (e.g., in Si and diamond), while it is allowed along ⟨111⟩ directions (e.g., in Ge) [47, 17]. Similarly, zero-order transitions are allowed for the valence band at Γ [47, 48].
Concerning intervalley scattering, group-theoretical analysis [63–67] shows that in the case of Si and diamond longitudinal-optical modes assist g scattering (between parallel valleys) and longitudinal-acoustic as well as transverse-optical modes assist f scattering (between perpendicular valleys). In the case of Ge, longitudinal-acoustic and longitudinal-optical modes assist intervalley scattering.
For GaAs, transitions from Γ to L valleys are assisted by longitudinal modes, both acoustic and optical. It is worth noting that, in practice, initial and final states do not exactly coincide with high-symmetry points, and consequently the selection rules need not be strictly fulfilled, as is confirmed in magnetophonon experiments [68]. However, from continuity, it is reasonable to expect that the forbidden transitions will remain weak when compared with allowed processes.
Ferry [69] has calculated the matrix element for optical and/or intervalley scattering to first order in the phonon wave vector; this may become significant when the zero-order transition is forbidden by symmetry.

2.4.2 Phonon Scattering, Electrostatic Interaction

In polar materials, in addition to the deformation-potential scattering, another electron-phonon interaction exists due to the electrostatic polarization associated to the atom vibrations. As we have seen in section 2.3.1, this type of interaction has different names, and it is actually treated in different ways [70], when due to acoustic modes (where the deformation of the cristal produces a piezoelectric field) or to optical modes (where the polarization associated to the two atoms vibrating with opposite phases plays the dominant role). In both cases the interaction Hamiltonian can be written as

$$H' = \int \rho(\mathbf{r})\phi(\mathbf{r})\, d\mathbf{r}, \tag{2.4.49}$$

where $\rho(\mathbf{r})$ is the charge density of the electron and $\phi(\mathbf{r})$ is the electric potential associated with the polarization $\mathbf{P}(\mathbf{r})$ of the cristal. However, in the case of acoustic modes the polarization is proportional to the acoustic strain, while for optical modes it is assumed proportional to the optical displacement of the atoms.

In both cases, the electrostatic nature of the mechanism is such that the efficiency of the scattering mechanism decreases at increasing electron energies, that is at increasing temperatures and/or applied fields.

a) Acoustic, Piezoelectric Phonons

A plane acoustic wave propagating in a piezoelectric material produces an electrostatic field given by [70]

$$E_i = -p_{ijk} e_{jk}/\kappa, \tag{2.4.50}$$

where p_{ijk} is the appropriate component of the piezoelectric tensor, e_{jk} the strain tensor, and κ the dielectric constant. When the potential energy associated with such a field is considered in the perturbation Hamiltonian in Eq. (2.4.49), and the standard machinery for the application of the golden rule is used, the following expression for the transition rate from state $\mathbf{k}$ to state $\mathbf{k}'$ is obtained [70]:

$$P(\mathbf{k}, \mathbf{k}') = \frac{4\pi^2 e^2 p^2}{V q u \kappa^2 \rho} \begin{bmatrix} N_q \\ N_q + 1 \end{bmatrix} \left(\frac{q^2}{q^2 + \beta^2} \right)^2 \mathscr{G} \delta[\varepsilon(\mathbf{k}') - \varepsilon(\mathbf{k}) \mp \hbar q u], \tag{2.4.51}$$

where upper and lower signs refer to absorption and emission, u is an average sound velocity, p an appropriate piezoelectric constant. β is the inverse screening length, which, for the non-degenerate equilibrium case, is given by the Debye formula:

$$\beta = \left(\frac{4\pi^2 e^2 n}{\kappa K_B T} \right)^{1/2}, \tag{2.4.52}$$

where n is the free-carrier concentration.

We now consider, as usually done in dealing with piezoelectric scattering, a simple parabolic band, elastic and equipartition approximation, and an

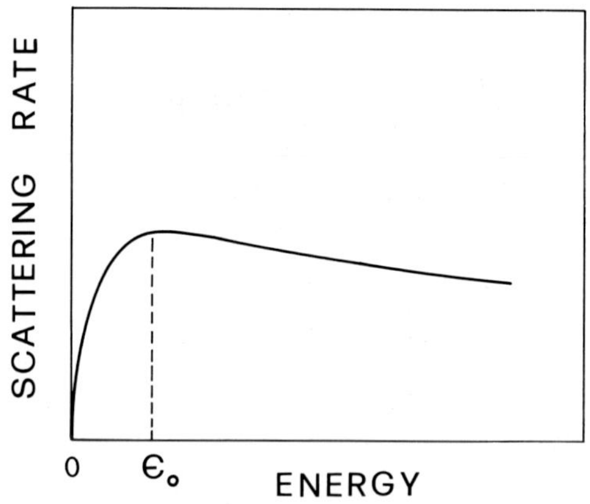

Fig. 2.16. Schematic representation of the energy dependence of the electron scattering rate for piezoelectric phonons ($\varepsilon_0 = \hbar^2 \beta^2/8m$) [70]

overlap factor equal to unity. Since in the elastic approximation no distinction is made between final states attained by means of absorption or emission processes, we can consider the sum of the two transition rates, given by

$$P(\mathbf{k}, \mathbf{k}') = \frac{8\pi^2 e^2 p^2 K_B T}{\hbar V q^2 u^2 \rho \kappa^2} \left(\frac{q^2}{q^2 + \beta^2} \right)^2 \delta[\varepsilon(\mathbf{k}') - \varepsilon(\mathbf{k})]. \qquad (2.4.53)$$

Integration of Eq. (2.4.53) over all possible final states $\mathbf{k}'$ yields the integrated scattering probability per unit time which accounts for both absorption and emission processes:

$$P_{e,pz}(\varepsilon) = \frac{\sqrt{m} e^2 p^2 K_B T}{\sqrt{8} \kappa^2 \hbar^2 \rho u^2} \varepsilon^{-1/2} \left(\log\left(1 + \frac{8m\varepsilon}{\hbar^2 \beta^2}\right) - \frac{1}{1 + (\hbar^2 \beta^2/8m\varepsilon)} \right). \qquad (2.4.54)$$

Its energy dependence is shown qualitatively in Fig. 2.16. For vanishing screening ($\beta = 0$) this total scattering rate would diverge.

b) Polar Optical Phonons

As two atoms vibrate along the polar axis in a longitudinal optical mode, they create an oscillating electric dipole and hence an alternating electric field and potential that can scatter the carriers. The interaction Hamiltonian has been derived by Froelich [71]. The transition rates are again calculated with the Fermi golden rule. For polar optical transitions we have [16]

$$P(\mathbf{k}, \mathbf{k}') = \frac{2\pi}{\hbar} B_0(\mathbf{k}, \mathbf{k}') \begin{bmatrix} N_q \\ N_q + 1 \end{bmatrix} \delta(\varepsilon_{\mathbf{k}'} - \varepsilon_{\mathbf{k}} \mp \hbar\omega_{op}), \qquad (2.4.55)$$

where

$$B_0(\mathbf{k}, \mathbf{k}') = \frac{2\pi e^2 \hbar\omega_{op}}{Vq^2}\left(\frac{1}{\kappa_\infty} - \frac{1}{\kappa_o}\right)\mathscr{G}(\mathbf{k}, \mathbf{k}'), \tag{2.4.56}$$

and $\mathscr{G}$ is the overlap factor given in Eq. (2.3.11).
The total scattering rate is calculated as

$$P_{e,po}(\varepsilon) = \frac{e^2\sqrt{m\omega_{op}}}{\sqrt{2}\hbar}\left(\frac{1}{\kappa_\infty} - \frac{1}{\kappa_o}\right)\frac{1 + 2\alpha\varepsilon'}{\gamma^{1/2}(\varepsilon)} F_o(\varepsilon, \varepsilon')\begin{bmatrix} N_q \\ N_q + 1 \end{bmatrix} \tag{2.4.57}$$

where $\gamma(\varepsilon) = \varepsilon(1 + \alpha\varepsilon) = \hbar^2 k^2/2m$, as seen in section 2.2.3, $\varepsilon' = \varepsilon \pm \hbar\omega_{op}$ is the energy of the electron after the transition, and

$$F_o(\varepsilon, \varepsilon') = C^{-1}\left\{A \ln\left|\frac{\gamma^{1/2}(\varepsilon) + \gamma^{1/2}(\varepsilon')}{\gamma^{1/2}(\varepsilon) - \gamma^{1/2}(\varepsilon')}\right| + B\right\},$$

$$A = \{2(1 + 2\alpha\varepsilon)(1 + \alpha\varepsilon') + \alpha[\gamma(\varepsilon) + \gamma(\varepsilon')]\}^2,$$

$$B = -2\alpha\gamma^{1/2}(\varepsilon)\gamma^{1/2}(\varepsilon')\{4(1 + \alpha\varepsilon)(1 + \alpha\varepsilon') + \alpha[\gamma(\varepsilon) + \gamma(\varepsilon')]\},$$

$$C = 4(1 + \alpha\varepsilon)(1 + \alpha\varepsilon')(1 + 2\alpha\varepsilon)(1 + 2\alpha\varepsilon').$$

To recover the result for non-parabolic bands, without taking account of the admixture of p-functions, we let $\alpha \to 0$ in A, B and C; then $A = C = 4$ and $B = 0$.
$P_{e,po}$ is zero for emission if the electron energy is less than the phonon energy. Quite often, an "effective field strength" defined as

$$E_o = \frac{me\hbar\omega_{op}}{h^2}\left(\frac{1}{\kappa_\infty} - \frac{1}{\kappa_o}\right)$$

is used in the total scattering rate. It has the value of $E_o = 5.95$ kV/cm for the Γ-valley of GaAs.
From Eq. (2.4.55) we obtain the angular dependence of the scattering

$$P(\theta)\, d\theta \sim \frac{(\sqrt{\gamma\gamma'} + \alpha\varepsilon\varepsilon' \cos\theta)^2}{[\gamma + \gamma' - 2\sqrt{\gamma\gamma'}\cos\theta]} \sin\theta\, d\theta. \tag{2.4.58}$$

where $\gamma = \gamma(\varepsilon)$, $\gamma' = \gamma(\varepsilon')$ and θ is the angle between $\mathbf{k}$ and $\mathbf{k}'$.
In Fig. 2.17 the energy dependence of the polar optical scattering rate in the Γ-valley of GaAs at 300 K is shown. Fig. 2.18 shows the angular dependence of the scattering probability for electron energies of 0.1 eV and 0.4 eV. It is seen that, due to the $1/q^2$ term in the magnitude of the matrix element, the mechanism is strongest for small q, or $\mathbf{k}' \approx \mathbf{k}$. The scattering funnels the carriers into the electric field direction, creating a strong spike in momentum space. It is also evident from Fig. 2.17 that the scattering efficiency decreases with increasing energy.
If screening effects are to be considered, a factor $[q^2/(q^2 + \beta^2)]^2$ must be introduced in the scattering rate from state $\mathbf{k}$ to $\mathbf{k}' = \mathbf{k} \pm \mathbf{q}$.

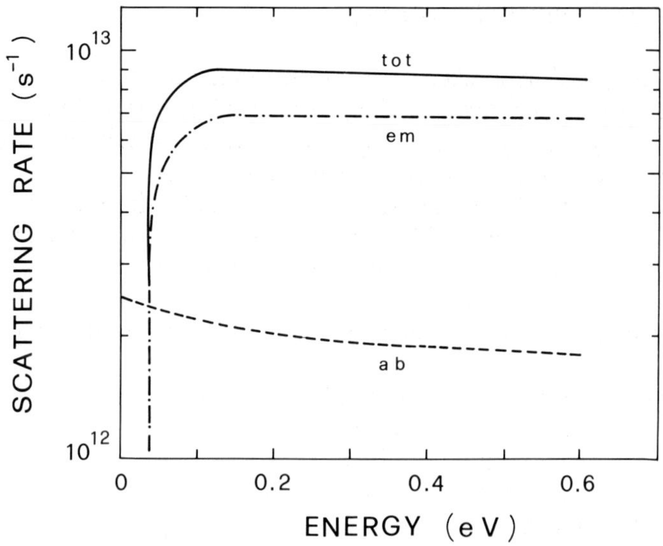

Fig. 2.17. Energy dependence of the polar optical scattering rate for the Γ valley of GaAs at 300 K [78]

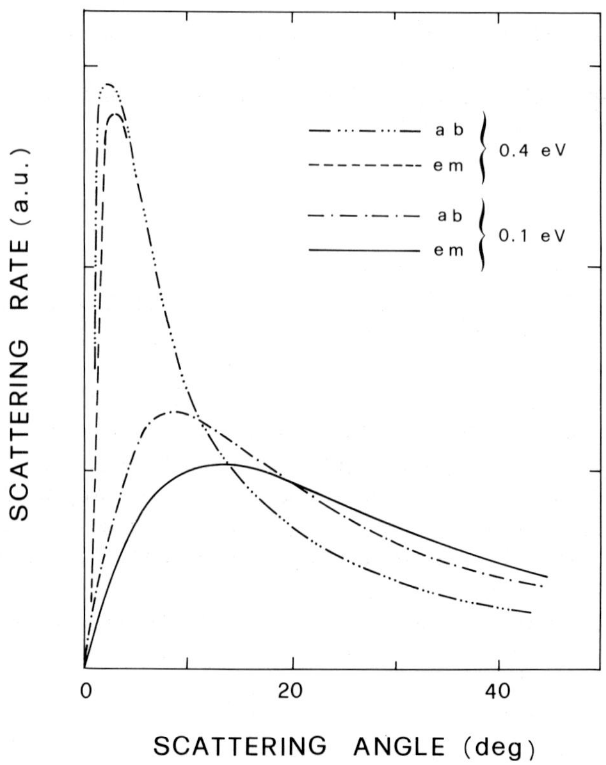

Fig. 2.18. Angular distribution of polar-optical scattering probability for electrons in the Γ valley of GaAs at 300 K with energy 0.1 and 0.4 eV [78]

2.4.3 Ionized Impurity Scattering

This type of collision is elastic in nature and, therefore, it cannot alone control the transport process in the presence of an external field; it must be accompanied by some dissipative scattering mechanism if the proper energy distribution of electrons is to be derived from theory.

For an ionized impurity, the scattering source can be simplified as a screened Coulomb potential. The problem was initially dealt with by two different formulations: the Conwell and Weisskopf approach (CW) [72], and the Brooks and Herring approach (BH) [73]. Since then several authors have presented refinements of the theory of ionized impurity scattering. A detailed analysis can be found in [74].

The BH and CW approaches differ in the model used to screen the potential of the ion, and both of them use the Born approximation, consistent with the scattering theory outlined in section 2.3.2.

In the BH approach, a simple exponential screening factor is introduced, so that the potential scattering is given by

$$\mathscr{V}(r) = \frac{Ze^2}{\kappa r} \exp(-\beta r), \tag{2.4.59}$$

where β^{-1} is the screening length, κ the dielectric constant of the material, and Z the number of charge units of the impurity.

In the CW approach, a bare potential is assumed, cut off at the mean distance b between impurities:

$$b = \left[\frac{3}{4\pi n_I}\right]^{1/3} \tag{2.4.60}$$

where n_I is the impurity concentration.

It is useful to unify the calculations by using the potential given in Eq. (2.4.59) also for the CW approach, taking for this case $\beta = 0$, and a maximum impact parameter b, or a minimum scattering angle θ_m.

When a high degree of compensation is present, very few free carriers are available to screen the ionized impurities, which, however, effectively screen each other, so that the CW approach seems to be more convenient for this condition. When, in the opposite case, each ionized impurity contributes one free carrier, n_I can be used instead of n in the calculation of the screening length. Figure 2.19 shows, for this case, the space dependence of the scattering potential in the two approaches.

The Fourier transform of the screened Coulomb potential is given by

$$\begin{aligned} \mathscr{H}'(q) &= \frac{1}{(2\pi)^{3/2}} \int d\mathbf{r}\, \mathscr{V}(\mathbf{r}) \exp(-i\mathbf{q}\cdot\mathbf{r}) \\ &= \frac{2^{1/2} Ze^2}{\pi^{1/2}\kappa} \frac{1}{\beta^2 + q^2}. \end{aligned} \tag{2.4.61}$$

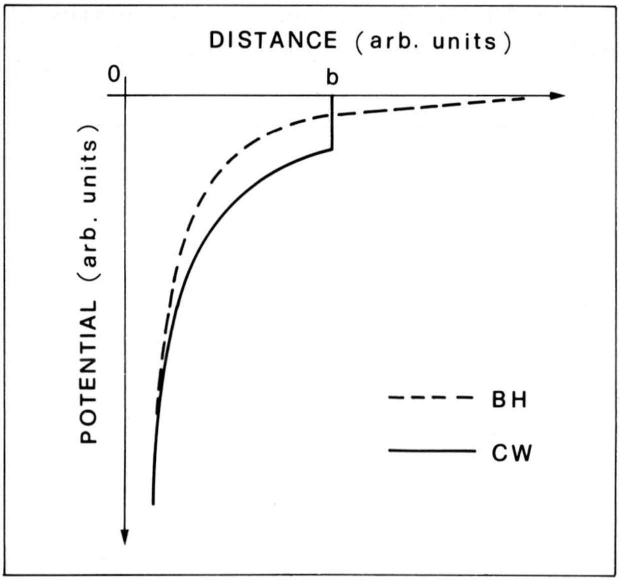

Fig. 2.19. Space dependence of the ionized-impurity potential in the Brooks-Herring and Conwell-Weisskopf approaches

This expression inserted in Eq. (2.3.2) and multiplied by the number of ionized impurities in the crystal yields the scattering probability per unit time:

$$P(\mathbf{k}, \mathbf{k}') = \frac{2^5\pi^3 Z^2 n_I e^4}{\hbar V \kappa^2} \mathscr{G} \frac{1}{(\beta^2 + q^2)^2} \delta[\varepsilon(\mathbf{k}') - \varepsilon(\mathbf{k})]. \tag{2.4.62}$$

a) Electrons

a.1) Spherical, Parabolic Bands

The integrated scattering probability is obtained from Eq. (2.4.62) by taking $\mathscr{G} = 1$, multiplying by the density of states $V/(2\pi)^3$, and integrating over the possible states $\mathbf{k}'$. In this simple band model we thus obtained

$$P(\mathbf{k}) = \frac{2^3\pi Z^2 n_I e^4 mk}{\hbar^3 \kappa^2} \int_{-1}^{\cos\theta_m} \frac{d\xi}{[\beta^2 + 2k^2(1-\xi)]^2}. \tag{2.4.63}$$

Final integration over $\xi = \cos\theta$ gives the integrated scattering probability per unit time

$$P_{e,I}(\varepsilon) = \frac{2^{3/2}\pi Z^2 n_I e^4 m^{1/2}}{\hbar^2 \kappa^2} \varepsilon^{-1/2} \left[\frac{1}{\beta^2 + \frac{4m\varepsilon}{\hbar^2}(1-\cos\theta_m)} - \frac{1}{\beta^2 + \frac{8m}{\hbar^2}\varepsilon} \right]. \tag{2.4.64}$$

Expression (2.4.64) can be written in more physical terms by introducing the maximum impact parameter b and by using the following notation:

$$\varepsilon_\beta = \frac{\hbar^2 \beta^2}{2m}, \tag{2.4.65}$$

$$\varepsilon_b = \frac{e^2}{2\kappa b}, \tag{2.4.66}$$

$$a_b = \frac{\varepsilon_b^2}{\varepsilon^2 + \varepsilon_b^2}. \tag{2.4.67}$$

From the classical relation between b and θ_m

$$\mathrm{tg}\left[\frac{\theta_m}{2}\right] = \frac{\varepsilon_b}{\varepsilon}, \tag{2.4.68}$$

Eq. (2.4.64) becomes

$$P_{e,I}(\varepsilon) = n_I \sigma v \frac{Z^2}{\left[1 + \dfrac{\varepsilon_\beta}{4\varepsilon}\right]\left[1 + \dfrac{\varepsilon_\beta}{4a_b\varepsilon}\right]} \tag{2.4.69}$$

where $\sigma = \pi b^2$ is the geometric cross section and v the electron velocity. In the CW approach $\beta = 0$, thus $\varepsilon_\beta = 0$, and Eq. (2.4.69) reduces to

$$P_{e,I}^{CW}(\varepsilon) = n_I \sigma v Z^2 = \pi n_I Z^2 b^2 (2/m)^{1/2} \varepsilon^{1/2}. \tag{2.4.70}$$

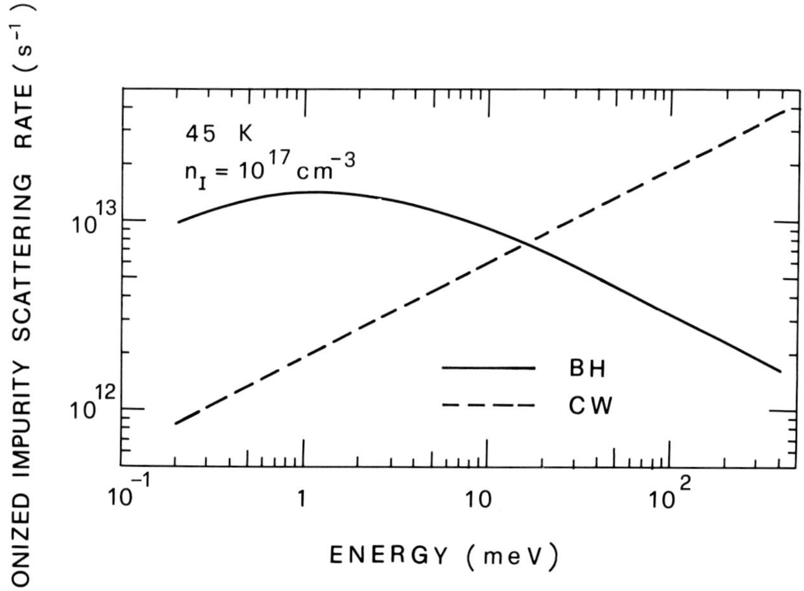

Fig. 2.20. Scattering rates for ionized-impurity scattering as functions of energy. The model used refers to electrons in Si with a parabolic band

In the *BH* approach $b^{-1} = 0$, and we have, from Eq. (2.4.69)

$$P_{e,I}^{BH} = \frac{2^{5/2}\pi n_I Z^2 e^4}{\kappa^2 \varepsilon_\beta^2 m^{1/2}} \frac{\varepsilon^{1/2}}{(1 + 4\varepsilon/\varepsilon_\beta)} . \tag{2.4.71}$$

Figure 2.20 shows the dependence upon energy of the integrated scattering probability per unit time for ionized impurities in the two different approaches.

In evaluating the importance of impurity scattering, Eqs. (2.4.70) and (2.4.71) must be set against the fact that, at high energies, Coulomb scattering is strongly peaked along the forward direction, so that a large number of scattering events may result in a small effect on the electron path. In this respect the *BH* formula of Eq. (2.4.70) seems to be more appropriate, since the integrated scattering probability itself decreases at sufficiently high energies.

a.2) Ellipsoidal, Nonparabolic Bands

As for the other scattering mechanisms, when ellipsoidal and/or nonparabolic valleys are taken into account, the Herring-Vogt transformation is applied, and nonparabolicity must be considered in the integration over the magnitude of $\mathbf{k}^{*\prime}$, which is performed by means of the δ function. The calculations are straightforward, and the results can be given in a form which shows the generalizations of Eqs. (2.4.70) and (2.4.71):

$$P_{e,I}^{CW}(\varepsilon) = \pi n_I Z^2 b^2 (2/m_d)^{1/2} \frac{\varepsilon^2(1 + 2\alpha\varepsilon)}{\gamma^{3/2}(\varepsilon)} , \tag{2.4.72}$$

$$P_{e,I}^{BH}(\varepsilon) = \frac{2^{5/2}\pi n_I Z^2 e^4}{\kappa^2 \varepsilon_\beta^{1/2} m_d^{1/2}} \gamma^{1/2}(\varepsilon) \frac{1 + 2\alpha\varepsilon}{1 + 4\dfrac{\gamma(\varepsilon)}{\varepsilon'_\beta}} , \tag{2.4.73}$$

for the *CW* and *BH* approaches, respectively, where

$$\varepsilon'_\beta = \frac{\hbar^2 \beta^2}{2m_d} . \tag{2.4.74}$$

b) Holes

b.1) Spherical, Parabolic Bands

The scattering probability given in Eq. (2.4.61) can be applied also to the case of holes for a single spherical and parabolic band accounting for the overlap factor as given in Eq. (2.3.12) [62]. By using the *BH* approach, we obtain the ionized-impurity scattering probability per unit time for holes

$$P(\mathbf{k}, \mathbf{k}') = \frac{2^5\pi^3 Z^2 n_I e^4}{\hbar V \kappa^2} \frac{\frac{1}{4}(1 + 3\cos^2\theta)}{(\beta^2 + q^2)^2} \times \delta[\varepsilon(\mathbf{k}') - \varepsilon(\mathbf{k})], \qquad (2.4.75)$$

where θ is the angle between $\mathbf{k}$ and $\mathbf{k}'$.
A first integration of Eq. (2.4.75) yields

$$P(\mathbf{k}, \theta) = \frac{2^{1/2} Z^2 n_I e^4}{\kappa^2 m^{1/2}} \varepsilon^{-3/2} \frac{1}{4} \frac{1 + 3\cos^2\theta}{[2(1 - \cos\theta) + \varepsilon_\beta/\varepsilon]^2}. \qquad (2.4.76)$$

Final integration of Eq. (2.4.76) gives the scattering rate

$$P_{h,I}^{BH}(\varepsilon) = \frac{2^{5/2}\pi Z^2 n_I e^4}{\kappa^2 \varepsilon_\beta^2 m^{1/2}} \frac{\varepsilon^{1/2}}{(1 + 4\varepsilon/\varepsilon_\beta)} F_{ov} \qquad (2.4.77)$$

with

$$F_{ov} = \frac{1}{4}\left[1 + \frac{3(1 + 4\varepsilon/\varepsilon_\beta)(1 + 2\varepsilon/\varepsilon_\beta)^2}{4(\varepsilon/\varepsilon_\beta)^2}\left[\frac{1}{(1 + 4\varepsilon/\varepsilon_\beta)} + \frac{1}{(1 + 2\varepsilon/\varepsilon_\beta)^2} + \frac{\ln[1/(1 + 4\varepsilon/\varepsilon_\beta)]}{2(1 + 2\varepsilon/\varepsilon_\beta)\varepsilon/\varepsilon_\beta}\right]\right]. \qquad (2.4.78)$$

Note that the inclusion of the overlap has the net effect of decreasing the efficiency of the impurity scattering mechanism.

b.2) Warped Band

Band warping can be included [75] by using in the Eq. (2.4.62) $\varepsilon(\mathbf{k})$ as given by Eqs. (2.2.5) and (2.2.6). By integration over k' we obtain

$$P(\mathbf{k}, \vartheta', \psi') = \frac{2Z^2 n_I e^4}{\hbar \kappa^2}[a(1 - g)]^{1/2}\varepsilon^{-3/2} \times \frac{\frac{1}{4}(1 + 3\cos^2\theta)(1 - g)^{3/2}}{(1 - g')^{3/2}\left[1 + \frac{1 - g}{1 - g'} - 2\frac{(1 - g)^{1/2}}{(1 - g')^{1/2}}\cos\theta + \beta^2 a(1 - g)/\varepsilon\right]^2}, \qquad (2.4.79)$$

where g and g' refer to initial (ϑ, ψ) and final (ϑ', ψ') angles taken with respect to crystallographic axes. Integration of Eq. (2.4.79) over the solid angle Ω' yields the integrated scattering probability. As follows from Eq. (2.4.79), warping introduces a dependence of the integrated scattering probability on the initial orientation of $\mathbf{k}$. Provided the warping is not very pronounced, Eq. (2.4.77) can still be used with an appropriate average value for the effective mass m_h [see Eq. (2.4.48)]: this is accomplished by setting $g = g'$ and $a(1 - g) = \hbar^2/(2m_h)$ in Eq. (2.4.79).

2.4.4 Carrier-Carrier Scattering

The interaction among carriers in semiconductors can be analyzed starting from the electronic Hamiltonian

$$H_e = \sum_i \frac{p_i^2}{2m} + \frac{e^2}{2\kappa} \sum_{i,j i \neq j} \frac{1}{|\mathbf{r}_i - \mathbf{r}_j|} \tag{2.4.80}$$

The first term on the right-hand side represents the free-electron energy, and the second one the bare Coulomb interaction among electrons. By following the procedure outlined by Bohm and Pines [76] it is possible to rewrite H_e in the random phase approximation (RPA) as the sum of four different contributions:

$$H_e = H_e^{(k)} + H_e^{(Sc)} + H_e^{(pl)} + H_e^{(e-pl)} \tag{2.4.81}$$

which can be explicitly written (in the occupation number representation) as [77]:

$$H_e^{(k)} = \sum_{\mathbf{k}} \frac{\hbar^2 k^2}{2m} c_{\mathbf{k}}^+ c_{\mathbf{k}}, \tag{2.4.82}$$

$$H_e^{(Sc)} = \sum_{\mathbf{k} > \mathbf{q}_c} \frac{2\pi e^2}{V\kappa k^2} \sum_{\lambda\mu} c_{\mathbf{k}_\lambda + \mathbf{k}}^+ c_{\mathbf{k}_\mu - \mathbf{k}}^+ c_{\mathbf{k}_\mu} c_{\mathbf{k}_\lambda}, \tag{2.4.83}$$

$$H_e^{(pl)} = \sum_{\mathbf{k}} \hbar\omega_p (a_{\mathbf{k}}^+ a_{\mathbf{k}} + \tfrac{1}{2}), \tag{2.4.84}$$

$$H_e^{(e-pl)} = \sum_{\mathbf{k} < \mathbf{q}_c} \left(\frac{\pi e^2 \hbar^2}{2V\kappa\omega_p m^2 k^2} \right)^{1/2} \times \sum_{\mathbf{k}'} (2\mathbf{k} \cdot \mathbf{k}' + k^2)(a_{\mathbf{k}} c_{\mathbf{k}'+\mathbf{k}}^+ c_{\mathbf{k}}^+ + a_{-\mathbf{k}} c_{\mathbf{k}'-\mathbf{k}}^+ c_{\mathbf{k}'}), \tag{2.4.85}$$

The form of Eq. (2.4.81) is obtained after several transformations of the original hamiltonian, which involve a Fourier transform of the interparticle term and the introduction of additional canonical conjugate variables. Details can be found in the reference.

The four terms in Eq. (2.4.81) have a precise physical meaning. $H_e^{(k)}$ represents the free electron kinetic term, with $c_{\mathbf{k}}^+$ and $c_{\mathbf{k}}$ respectively creation and annihilation electron operators. The second term $H_e^{(Sc)}$ (Eq. 2.4.83) describes a screened Coulomb interaction in which the momentum $\mathbf{k}$ is transferred from the μth to the λth electron (annihilation of the two electrons with momentum $\mathbf{k}_\mu$ and $\mathbf{k}_\lambda$, and creation of two electrons with momentum $\mathbf{k}_\lambda + \mathbf{k}$ and $\mathbf{k}_\mu - \mathbf{k}$).

The screening parameters are given by the cut-off wave vector q_c which is introduced arbitrarily to separate long range and short range components into the Fourier expansion of the Coulombic term of Eq. (2.4.80). The physical significance of q_c will be discussed below. The term $H_e^{(pl)}$ represents the energy of a gas of interacting bosons, which are here identified with the

plasmons, i.e., with the quantized oscillations of the electron gas. In Eq. (2.4.84) $a_{\mathbf{k}}^{+}$ and $a_{\mathbf{k}}$ are respectively the creation and annihilation operators of a plasmon of wave vector **k**, and $\hbar\omega_p$ is the energy of the plasmon. The last term, given by Eq. (2.4.85), describes the electron-plasmon interaction, through a process in which a momentum **k** is transferred to an electron with absorption of a plasmon **k** or emission of a plasmon $-\mathbf{k}$. Processes involving more than one plasmon are neglected in the RPA.
According to the previous analysis, two main contributions to the carrier-carrier scattering can be identified:

— the individual carrier-carrier interaction via a screened Coulomb potential of the form

$$V(r) = \frac{e^2}{\kappa r} e^{-\beta_c r} \tag{2.4.86}$$

which accounts for two-body short-range interaction;
— the electron-plasmon interaction, which accounts for the collective long-range behaviour of the electron gas.

In semiconductors, the plasmon energy at reasonable density can be of the same order of magnitude as the characteristic phonon energies (for example in GaAs at a concentration of 5×10^{17} cm^{-3} the plasmon energy $\hbar\omega_p$ is equal to 30 meV). A determination of the value of the cut-off wave vector q_c can be obtained from an independent analysis of the complex wave vector and frequency-dependent dielectric function [42, 78]. It is found that it is reasonable for most cases to assume q_c equal to the inverse Debye screening length given by (2.4.52).

a) Screened Electron-Electron Interaction

Using the screened potential of Eq. (2.4.86) with $q_c = \beta$ (where β is inverse Debye screening length), the transition probability of two electrons from the wave-vectors **k** and $\mathbf{k}_0$ to $\mathbf{k}'$ and $\mathbf{k}_0'$ is obtained (using the Fermi Golden rule) as:

$$S_{\mathbf{k}_0,\mathbf{k}\to\mathbf{k}_0',\mathbf{k}'} = \frac{2\pi}{\hbar}|M|^2 f_{\mathbf{k}_0} f_{\mathbf{k}}(1 - f_{\mathbf{k}_0'})(1 - f_{\mathbf{k}'})\delta(\varepsilon_{\mathbf{k}_0'} + \varepsilon_{\mathbf{k}'} - \varepsilon_{\mathbf{k}_0} - \varepsilon_{\mathbf{k}}), \tag{2.4.87}$$

where f's and ε's are the occupation probabilities and the energy at each wave vector, respectively. A parabolic energy-momentum dispersion is used. The transition matrix element is given by

$$M = \langle \mathbf{k}_0, \mathbf{k}|V(r)|\mathbf{k}_0', \mathbf{k}'\rangle = \frac{4\pi e^2}{V\kappa} \frac{\delta_{\mathbf{k}_0+\mathbf{k},\mathbf{k}_0'+\mathbf{k}'}}{|\mathbf{k}' - \mathbf{k}_0|^2 + \beta^2}, \tag{2.4.88}$$

where V represents the volume of the crystal and the electron wave functions are approximated by simple plane waves. The matrix element describes a two-body collision, where the total momentum is conserved, as indicated by the Kroenecker δ-function. The total scattering rate $\Gamma_{ee}(\mathbf{k}_0)$ for electron-electron interaction is obtained from (2.4.87) by summing over $\mathbf{k}$, $\mathbf{k}'$ and $\mathbf{k}_0'$. For non-degenerate situations ($f_{\mathbf{k}} = f_{\mathbf{k}'} = 0$) we have [78]:

$$\Gamma_{e-e}(\mathbf{k}_0) = \frac{me^4}{\hbar^3 V\kappa^2} \sum_{\mathbf{k}} f_{\mathbf{k}} \frac{|\mathbf{k} - \mathbf{k}_0|}{\beta^2(|\mathbf{k} - \mathbf{k}_0|^2 + \beta^2)} . \tag{2.4.89}$$

The choice of the wave vector $\mathbf{k}$ determines the value of the wave vector difference $\mathbf{g} = \mathbf{k} - \mathbf{k}_0$ of the two electrons before the collision. After the collision the relative wave vectors will be changed to $\mathbf{g}' = \mathbf{k}' - \mathbf{k}_0'$. Since the scattering conserves the magnitude of $\mathbf{g}$, the new vector $\mathbf{g}'$ can be determined from the angular distribution probability $P(\theta)\, d\theta$ with θ equal to the angle between $\mathbf{g}$ and $\mathbf{g}'$. The angular probability $P(\theta)\, d\theta$, determined from the differential scattering probability $S(\mathbf{g}, \mathbf{g}')$ is given by

$$P(\theta)\, d\theta = \frac{\sin\theta\, d\theta}{g^2 \sin^2(\theta/2) + \beta^2} , \tag{2.4.90}$$

with $g = |\mathbf{g}|$.

b) Electron-Plasmon Interaction

As was shown earlier, the splitting of the hamiltonian of an electron gas gives a term describing the interaction of the electrons with the quasiparticles associated with collective behaviour of the electron gas, i.e. plasmons. The interaction hamiltonian for the electron-plasmon scattering is given by

$$H_{e-pl} = \sum_{\mathbf{q}<\mathbf{q}_c} \left(\frac{\pi e^2 \hbar^3}{2V\kappa\omega_p m^2 q^2} \right)^{1/2} \sum_{\mathbf{k}} (2\mathbf{k}\cdot\mathbf{q} + q^2) \cdot (a_{\mathbf{q}} c^+_{\mathbf{k}+\mathbf{q}} c_{\mathbf{k}} + a_{-\mathbf{q}} c^+_{\mathbf{k}-\mathbf{q}} c_{\mathbf{k}}) . \tag{2.4.91}$$

The squared matrix element can be easily calculated for plane waves and the following expression is obtained [77]

$$|M|^2 = \frac{\pi e^2 \hbar^3}{2V\kappa\omega_p q^2 m^2} (2\mathbf{k}\cdot\mathbf{q} + q^2)^2 [N_q \delta(\varepsilon_{k+q} - \varepsilon_k - \hbar\omega_p) + (N_q + 1)\delta(\varepsilon_{k-q} - \varepsilon_k + \hbar\omega_p)] \tag{2.4.92}$$

Here, N_q is the Bose-Einstein equilibrium distribution population for the plasmon population and $\mathbf{q}$ is the wave vector of the plasmon involved in the interaction. Equation (2.4.92) describes a process in which an electron with wave vector $\mathbf{k}$ absorbs or emits a plasmon $\mathbf{q}$, with an energy exchange equal to $\hbar\omega_p(q)$. If the small-q limit is taken, the energy involved in the transition is just $\hbar\omega_p$, where ω_p is the plasma frequency. The upper limit q_c in (2.4.92) defines the maximum wave vector for which plasmon exists as an independent excitation of the electron gas. Its value can be determined from the analysis of the complex dielectric function. The final closed form for Γ_{e-pl} obtained from Fermi golden rule is fairly complex, since different regions are defined by the requirements of energy and momentum conservation [78]:

$$\Gamma_{e-pl}(k) = \frac{e^2\omega_p m^*}{\pi\kappa\hbar k}\left[N_q \ln\left(\frac{q_a^+}{q_a^-}\right) + (N_q + 1)\ln\left(\frac{q_e^+}{q_e^-}\right)u(\varepsilon_k - \hbar\omega_p)\right], \tag{2.4.93}$$

where $u(\varepsilon_k - \hbar\omega_p)$, the usual step function, is zero for $\varepsilon_k < \hbar\omega_p$ and is unity for $\varepsilon_k > \hbar\omega_p$. The terms in the argument of the logarithmic functions are given by:

$$q_a^- = k\left[\left(1 + \frac{\hbar\omega_p}{\varepsilon_k}\right)^{1/2} - 1\right],$$

$$q_e^- = k\left[1 - \left(1 - \frac{\hbar\omega_p}{\varepsilon_k}\right)^{1/2}\right],$$

$$q_a^+ = \min\left\{q_c, k\left[\left(1 + \frac{\hbar\omega_p}{\varepsilon_k}\right)^{1/2} - 1\right]\right\},$$

$$q_e^+ = \min\left\{q_c, k\left[\left(1 - \frac{\hbar\omega_p}{\varepsilon_k}\right)^{1/2} + 1\right]\right\},$$

where min{ } indicates the minimum between the two quantities enclosed in the brackets and the suffix a and e indicate respectively absorption and emission. If the wave vector k_p associated with the plasma frequency is introduced (with $\hbar^2 k_p^2/2m = \hbar\omega_p$), Eq. (2.4.93) can be rewritten in the following form: for the emission process with $k > k_p$,

$$k_p > q_c;\ \Gamma_{e-pl}(k) = \begin{cases} C(N_q + 1)\ln\dfrac{q_c/k}{1 - (1 - \hbar\omega_k/\varepsilon_k)^{1/2}}, & k > \dfrac{q_c^2 + k_p^2}{2q_c}, \\[2ex] 0, & k_p < k < \dfrac{q_c^2 + k_p^2}{2q_c}, \end{cases} \tag{2.4.94a}$$

$$k_p < q_c;\ \Gamma_{e-pl}(k) = \begin{cases} C(N_q + 1)\ln\dfrac{1 + (1 - \hbar\omega_p/\varepsilon_k)^{1/2}}{1 - (1 - \hbar\omega_p/\varepsilon_k)^{1/2}}, & k_p < k < \dfrac{q_c^2 + k_p^2}{2q_c}, \\[2ex] C(N_q + 1)\ln\dfrac{q_c/k}{1 - (1 - \hbar\omega_p/\varepsilon_k)^{1/2}}, & k > \dfrac{q_c^2 + k_p^2}{2q_c}. \end{cases}$$

For the absorption,

$$k_p > q_c;\ \Gamma_{e-pl}(k) = \begin{cases} CN_q \ln \dfrac{q_c/k}{-1 + (1 + \hbar\omega_p/\varepsilon_k)^{1/2}}, & k > \dfrac{-q_c^2 + k_p^2}{2q_c}, \\ 0, & k < \dfrac{-q_c^2 + k_p^2}{2q_c}, \end{cases}$$

$$k_p < q_c;\ \Gamma_{e-pl}(k) = \begin{cases} CN_q \ln \dfrac{1 + (1 + \hbar\omega_p/\varepsilon_k)^{1/2}}{-1 + (1 + \hbar\omega_p/\varepsilon_k)^{1/2}}, & k < \dfrac{q_c^2 - k_p^2}{2q_c}, \\ CN_q \ln \dfrac{q_c/k}{-1 + (1 + \hbar\omega_p/\varepsilon_k)^{1/2}}, & k < \dfrac{q_c^2 - k_p^2}{2q_c}. \end{cases} \tag{2.4.94b}$$

Here,

$$C = \frac{me^2\omega_p}{\kappa\hbar^2 k}.$$

These expressions are similar, but more complete, than the ones derived by Kumenov and Perel [79].

In the limit of small enough q_c, the condition $k_p > q_c$ is easily satisfied. Taking q_c for the non-degenerate case as $\alpha\beta$, with $\alpha < 1$, the condition $(k > (q_c^2 + k_p^2)/2q_c)$ for the emission process reduces to

$$\varepsilon_k > \frac{2K_BT}{\alpha^2}.$$

For $\alpha = 0.5$ we recover the result found by Davidov [80] for plasmon losses in ionized plasmas: only electrons with energy higher than twice the thermal energy can emit plasmons.

If the electron energy ε_k is much higher than the plasma energy $\hbar\omega_p$, the term $1 - [1 - (\hbar\omega_p/\varepsilon_k)]^{1/2} = \frac{1}{2}\hbar\omega_p$, and the scattering rate for plasmon emission is

$$\Gamma_{e-pl}^{(em)} = \frac{m^* e^2\omega_p}{\kappa\hbar k}(N_q + 1)\ln\left(\frac{\hbar k}{m^*}\frac{q_c}{\omega_p}\right). \tag{2.4.95}$$

From the expression for the square matrix element $|M|^2$, and the fact that $2\mathbf{k}\cdot\mathbf{p} + q^2 = (2m/\hbar)\omega_p$, it follows that the electron-plasmon differential scattering probability $S(\mathbf{k}, \mathbf{k}_0)$ is of the same form as the one obtained for the polar optical phonons (at least concerning the q dependence). The angular dependence of the scattering is also of the same type, and it is given for parabolic bands by

$$P(\beta)\, d\beta = \frac{\sin\beta\, d\beta}{\varepsilon + \varepsilon' - 2\sqrt{\varepsilon\varepsilon'}\cos\theta}, \tag{2.4.96}$$

where ε and ε' are the electron energies before and after the scattering.

c) Plasmon-Phonon Coupling

In polar materials, such as GaAs, the interaction between electrons and lattice vibration introduces novel features in the properties of the material. When electrons are present in the conduction band, the screening effects modify both the optical mode frequency and the electron-phonon interaction. For non-degenerate conditions, Ehrenreich [81] found that the optical-phonon frequency was raised for wavelengths longer than the thermal wavelength of the electrons. If carriers are degenerate, the frequencies of the charge density fluctuations are comparable to the optical frequencies. The longitudinal phonon is no longer a normal mode, since the longitudinal electric field which accompany such phonons must couple strongly with the charge-density fluctuations of the electron gas. The coupling of the phonons with the plasmons leads to important changes in the nature of the collective modes. Varga [82] showed that, in the RPA, the polarizabilities of electrons and ions are additive. Thus, the total dielectric function for the coupled system is given by

$$\varepsilon_T(q, \omega) = \varepsilon_e(q, \omega) + \varepsilon_L(q, \omega) - 1, \tag{2.4.97}$$

where ε_e is the electron dielectric function. $\varepsilon_L(q, \omega)$ is the lattice contribution given by

$$\varepsilon_L(q, \omega) = \kappa_\infty \frac{\omega_{lo}^2 - \omega_{to}^2}{\omega^2 - \omega_{to}^2} \tag{2.4.98}$$

where ω_{lo} and ω_{to} are the frequencies of the longitudinal and transverse optical phonon, respectively. The dispersion of the optical phonon modes has been neglected because, at the concentration of interest here, the Fermi wave vector is a few percent of the Brillouin-zone width. For this reason, the dielectric function of the lattice alone is actually only frequency dependent. The transverse + longitudinal optical frequencies are connected by the Lyddane-Sachs-Teller relation

$$\omega_{lo}^2 = (\kappa_0/\kappa_\infty)\omega_{to}^2$$

κ_0 being the static dielectric constant, and equal to 10.3 in GaAs. In the long-wavelength limit, Eq. (2.4.97) can be written as

$$\varepsilon_T(q, \omega) = \kappa_\infty + \kappa_\infty \frac{\omega_{lo}^2 - \omega_{to}^2}{\omega^2 - \omega_{to}^2} - \kappa_\infty \left(\frac{\omega_p}{\omega}\right)^2, \tag{2.4.99}$$

where the $q \to 0$ limit of the electron dielectric function

$$\varepsilon_e(q, \omega) = \kappa_\infty \left(1 - \frac{\omega_p^2}{\omega^2}\right) \tag{2.4.100}$$

has been used. The zeros of Eq. (2.4.98) give the dispersion curve for the coupled modes. At $q = 0$ the coupled modes, which are described by two branches, denoted ω_+ and ω_-, are given by

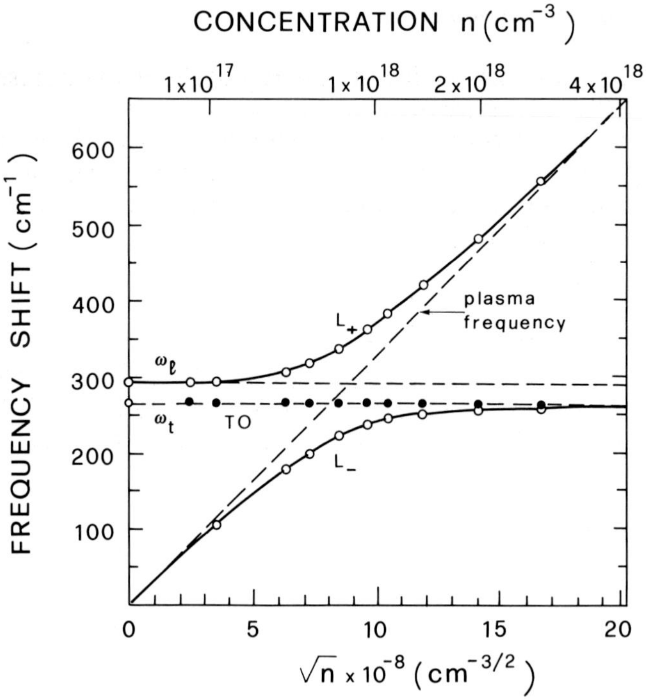

Fig. 2.21. Frequency shift of the Raman-scattered light in GaAs at 300 K as a function of electron concentration [78]. Circles refer to data in [83]

$$\omega_{\pm}^2(q=0) = \tfrac{1}{2}[\omega_p^2 + \omega_{lo}^2] \pm [(\omega_p^2 + \omega_{lo}^2)^2 - 4\omega_p^2\omega_{to}^2]^{1/2}. \quad (2.4.101)$$

Figure 2.21 shows a comparison of Eq. (2.4.101) (solid line) with the data of Mooradian and Wright [83] obtained with a Raman-scattering experiment (points). The agreement is excellent. For a small concentration, the lowest root starts out as a plasmon of frequency $(ne^2/\kappa_0 m)^{1/2}$, at higher concentration it becomes more phonon-like. In the limit of large density, it is the longitudinal optical phonon mode of the crystal. The frequency is ω_{to} and not ω_{lo}, as it would be expected, because the charge carriers have shielded out the ionic field, which, in a pure crystal, raise ω_{lo} compared to ω_{to}. Conversely, the root ω_+ starts out at the LO phonon and turns into a plasmon (now at frequency ω_p) as n increases. The coupled modes are a mixture of collective electrons and ion motion.

The analysis of the total dielectric function shows very interesting properties of polar semiconductors in the presence of high electron densities. In particular, it has been shown [82, 84] that the scattering strength of the coupled plasmon-phonon modes is proportional to $\mathrm{Im}(1/\varepsilon_T)$. In the limit of very low density the unscreened Froehlich scattering lifetime is recovered from the analysis of $\mathrm{Im}(1/\varepsilon_T)$. For very high carrier concentration the highly screened LO phonon limit [85] is obtained.

d) Electron-Hole Interaction

For the electron-hole interaction, we can repeat the analysis presented at the beginning of the present section.
In particular, the two body coulomb interaction can still be represented by a screened potential. If electrons and holes are characterized by two different temperatures (see section 2.7), respectively T_e and T_h, the inverse screening length β is given by [86]:

$$\beta^2 = \frac{4\pi n e^2}{\kappa K_B}\left[\frac{1}{T_e} + \frac{1}{T_h}\right]. \tag{2.4.102}$$

In the calculation of the scattering rates using Fermi golden rule, the same procedure seen before for the electron-electron scattering is followed, introducing here the relative wave vector $\mathbf{g}$ and the reduced mass μ defined by

$$\mathbf{g} = 2\mu\left(\frac{\mathbf{k}_0}{m_e} - \frac{\mathbf{k}}{m_h}\right) \tag{2.4.103}$$

and

$$\mu = \frac{m_e m_h}{m_e + m_h}, \tag{2.4.104}$$

where m_e and m_h are respectively the electron and hole effective mass.
The total scattering rate for electron-hole interaction is finally given by [87]

$$\Gamma_{e-h}(\mathbf{k}_0) = \frac{2\mu e^4}{\hbar^3 \kappa^2 V} \sum_{\mathbf{k}} f_{\mathbf{k}} \frac{g}{\beta^2[g^2 + \beta^2]}. \tag{2.4.105}$$

More insight can be gained by rewriting Eq. (2.4.105) in integral form for both $\Gamma_{e-h}(\mathbf{k}_0)$ and $\Gamma_{h-e}(\mathbf{k}_0)$

$$\Gamma_{e-h}(\mathbf{k}_0) = \frac{2p\mu e^4}{\kappa^2 \hbar^3} \int d\mathbf{k}\, f_h(\mathbf{k}) \frac{2\mu\left|\frac{\mathbf{k}_0}{m_e} - \frac{\mathbf{k}}{m_h}\right|}{\beta^2\left[4\mu^2\left|\frac{\mathbf{k}_0}{m_e} - \frac{\mathbf{k}}{m_h}\right|^2 + \beta^2\right]}, \tag{2.4.106}$$

$$\Gamma_{h-e}(\mathbf{k}_0) = \frac{2n\mu e^4}{\kappa^2 \hbar^3} \int d\mathbf{k}\, f_e(\mathbf{k}) \frac{2\mu\left|\frac{\mathbf{k}_0}{m_h} - \frac{\mathbf{k}}{m_e}\right|}{\beta^2\left[4\mu^2\left|\frac{\mathbf{k}_0}{m_h} - \frac{\mathbf{k}}{m_e}\right|^2 + \beta^2\right]}, \tag{2.4.107}$$

where n and p are the electron and hole concentrations, respectively. It is obvious that generally $\Gamma_{e-h} \neq \Gamma'_{h-e}$, because the electron distribution function can be different from that of holes. Furthermore, the concentrations of electrons might in general be different from that of holes.

2.5 Transport Equation

The transport theory used in almost all treatments of hot-electron effects is semiclassical in nature. In fact, charge carriers are considered to move according to classical mechanics between two successive scattering events (see section 2.1), while the scattering cross-sections, in the imperfect host crystal, are derived from the quantum theory of scattering (see section 2.2). The fundamental quantity in classical transport theory is the carrier distribution function $f(\mathbf{r}, \mathbf{k}, t)$, defined as proportional to the density of the electrons in the six-dimensional space $(\mathbf{r}, \mathbf{k})$, where $\mathbf{r}$ indicates the position in the crystal. The normalization constant can be chosen in such a way that

$$2/(2\pi)^3 \int d\mathbf{k} \int d\mathbf{r}\, f(\mathbf{r}, \mathbf{k}, t) = N, \tag{2.5.1}$$

where N is the number of the electrons in the crystal. Since spin interactions are usually not involved in transport phenomena in the presence of the electric fields alone, the spin variables are not explicitly considered.

The equation that describes transport phenomena in the semiclassical approach is the Boltzmann equation [88]

$$\partial f/\partial t + \mathbf{v}\cdot\nabla_r f + \dot{\mathbf{k}}\cdot\nabla_k f = \partial f/\partial t|_{\text{coll}}, \tag{2.5.2}$$

where the right-hand side is the rate of change of f due to collisions.

The Boltzmann equation can be derived from the Liouville theorem. An elementary derivation can be given considering a small volume in phase-space and counting the number of particles moving in and out of the volume in the unit time.

Once the distribution function is known, all quantities of interest, such as carrier drift velocity, mean energy, and diffusion coefficient, can be obtained as functions of the applied field, temperature, and/or carrier concentration gradient.

As can be seen from inspection of the Boltzmann equation and Eqs. (2.1.3) and (2.1.4), the basic knowledge required for the analysis of any transport phenomenon is concerned with the band structure $\varepsilon(\mathbf{k})$ of the specific material under consideration.

Statistical mechanics tells us that the equilibrium distribution function f_0 is the Fermi-Dirac distribution. For $\varepsilon - \varepsilon_F \gg K_B T$ (where ε_F is the Fermi level) f_0 can be approximated by the Maxwell-Boltzmann distribution

$$f_{MB} = C \exp(-\varepsilon/K_B T); \tag{2.5.3}$$

C is here a normalization constant.

The collision term on the right-hand side of Eq. (2.5.2) is due to the fact that the imperfections of the ideal lattice, including phonons, can induce transitions between different Bloch states. Let $P(\mathbf{k}, \mathbf{k}')$ be the probability per unit time of an electron transition from a state $\mathbf{k}$ into an empty state $\mathbf{k}'$, induced by the lattice imperfections (we are assuming here that no spin flip can be caused by the transitions). The collision term can be expressed as the difference between the electrons scattered in and out of **k**:

$$\left(\frac{\partial f}{\partial t}\right)_{\text{coll}} = \frac{V}{(2\pi)^3} \int [f(\mathbf{k}')P(\mathbf{k}'\mathbf{k})(1 - f(\mathbf{k})) - f(\mathbf{k})P(\mathbf{k}\mathbf{k}')(1 - f(\mathbf{k}'))]\, d\mathbf{k}' . \tag{2.5.4}$$

The $(1 - f)$ coefficients account for the Pauli exclusion principle; in the approximation of Eq. (2.5.3) these factors do not contribute since it is always assumed that $f \ll 1$. By substituting Eq. (2.5.4) into the Boltzmann equation, an integral-differential equation is obtained. The complexity of such an equation strongly depends on the models used for the scattering mechanisms and for the band structure.

It is hard to find solutions to the Boltzmann equation. Even in the linear response case, with simple scattering mechanisms, approximations are needed. Very efficient numerical techniques have been developed, that are able to provide definite and accurate descriptions of the transport properties of a given material once an adequate theoretical model of its microscopic properties is given.

From the analytical point of view, transport phenomena in nonlinear regimes would be completely described by solving the Boltzmann equation without linearizing it with respect to the external force. This is a formidable mathematical problem whose analytical solution has resisted repeated attacks for several decades; it is still an unsolved problem today. However, the analytical techniques applied to extremely simple semiconductor models have the merit of providing physical insight into the problem of nonlinear transport by introducing simple concepts such as energy and momentum relaxation times and electron temperature. On the other hand, to get detailed analytical results, it is necessary to perform such drastic approximations that it is no longer clear whether the features of interest in the results are due to the microscopic model or to the mathematical approximations.

A big step forward in the solution of the Boltzmann equation has been achieved by the introduction of numerical techniques. The two most important, which date from 1966 (the Kyoto Semiconductor Conference), are the iterative technique and the Monte Carlo method.

The iterative technique [89–91] yields a solution of the Boltzmann equation by means of an iterative procedure and processes the whole distribution function at each step of the procedure. For this reason it can be useful when we deal with physical phenomena that depend on details of the distribution function.

The Monte Carlo method [16], [68], [35], [92–93] is by far the most popular of the two numerical techniques. This method, as applied to the study of stationary transport properties in semiconductors, will be treated at length in the following chapters.

The Monte Carlo technique, as a direct simulation of the dynamics of charge carriers inside the crystal, enables us to extract any required physical information while the solution of the transport equation is being built up. Furthermore, it enables us to simulate particular physical situations unattainable in experiments and even to investigate into nonexisting materials, in order to emphasize special features of the phenomenon under study. This use of Monte Carlo makes it similar to an experimental technique: the simulated "experiment" can in fact be compared with analytically formulated theories.

2.6 Linear Response and the Relaxation Time Approximation

The complexity of Boltzmann equation is drastically reduced if we are only interested in the linear response to the external electric field **E**. The distribution function can then be approximated by

$$f(\mathbf{k}) \approx f_0(\mathbf{k}) + f_1(\mathbf{k}), \tag{2.6.1}$$

where $f_1(\mathbf{k})$ is linear in **E**. The Boltzmann equation can then be linearized in the external field by using Eq. (2.6.1) for the distribution function and discarding all terms proportional to E^2 and higher powers of E.
A very useful simplification of the Boltzmann equation is obtained by the so called relaxation-time approximation. There, the collision integral is approximated by a much simpler expression given by:

$$\left(\frac{\partial f}{\partial t}\right)_{\text{coll}} = -\frac{f - f_0}{\tau(\varepsilon)}, \tag{2.6.2}$$

where τ is the relaxation time. τ is assumed to depend only on the energy. Under linear conditions, such assumption is justified for many materials of interest. Eq. (2.6.2) describes a relaxation phenomenon: if the electron distribution is initially disturbed and, at $t = 0$, the perturbation is removed, the exponential relaxation to equilibrium is controlled by the characteristic time τ as follows:

$$f_1(t) = f_1(0)e^{-t/\tau}. \tag{2.6.3}$$

The collision integral can be rigorously expressed in the relaxation time form only in two cases: velocity-randomizing collisions and elastic collisions (see below).
In linear response theory, the carrier distribution function in energy is not changed with respect to the equilibrium one, and τ represents only the momentum relaxation time. In order to discuss non-linear phenomena it is necessary to introduce, when possible, an additional energy relaxation time [57].
On the basis of simple calculations, it is possible to show that the Ohmic mobility in the relaxation time approximation is given by

$$\mu = \frac{-e}{m}\frac{\langle \tau\varepsilon \rangle}{\langle \varepsilon \rangle} \tag{2.6.4}$$

where $\langle\ \rangle$ indicates the average over the equilibrium distribution function. Such a result is valid in this form for simple spherical and parabolic bands. For more complicated bands, m must be substituted by the conductivity effective mass (see section 2.2.2 b).

a) Velocity-randomizing Collisions

Velocity randomizing collisions are defined by the following property:

$$P(\mathbf{k}, \mathbf{k}') = P(\mathbf{k}, \mathbf{k}^*), \tag{2.6.5}$$

where $\mathbf{k}^*$ describes an electron state with the opposite velocity of the state $\mathbf{k}'$. When we are dealing with a spherically symmetric band Eq. (2.6.5) reduces to

$$P(\mathbf{k}, \mathbf{k}') = P(\mathbf{k}, -\mathbf{k}'). \tag{2.6.6}$$

In this case, at each scattering event in average all the electron momentum is lost, and a relaxation time can be defined, given by

$$\frac{1}{\tau(\varepsilon)} = \frac{V}{(2\pi)^3} \int P(\mathbf{k}, \mathbf{k}')\{1 - f_0(\mathbf{k}')\}\, d\mathbf{k}' \tag{2.6.7}$$

For non-degenerate semiconductors ($f_0 \ll 1$) Eq. (2.6.7) reduces to the total scattering rate:

$$\frac{1}{\tau(\varepsilon)} = \frac{V}{(2\pi)^3} \int P(\mathbf{k}, \mathbf{k}')\, d\mathbf{k}' \tag{2.6.8}$$

b) Elastic Collisions

In the absence of magnetic fields a relaxation time can be defined also for the case of elastic collisions, for which $P(\mathbf{k}, \mathbf{k}')$ is different from zero only if $\varepsilon(k) = \varepsilon(k')$. If, in addition, we have also spherical symmetry P depends only on the angle θ between $\mathbf{k}$ and $\mathbf{k}'$. For this case

$$\frac{1}{\tau(\varepsilon)} = \frac{V}{(2\pi)^3} \int P(k, \theta)(1 - \cos\theta)\, d\mathbf{k}'. \tag{2.6.9}$$

The difference between the relaxation times in Eq. (2.6.8) (velocity-randomizing collisions) and Eq. (2.6.9) (elastic collisions) can be easily understood if we consider that in the first case the electrons lose, in average, all their momentum at each collision, while in the second case the fraction of lost momentum at each scattering process must be weighted by the factor $(1 - \cos\theta)$. If elastic and velocity-randomizing collisions are considered, the average of $\cos\theta$ in Eq. (2.6.9) is obviously zero, and Eq. (2.6.9) reduces, as expected, to Eq. (2.6.8). In order to further verify the consistency of this picture, we can notice that Eq. (2.6.7) reduces to Eq. (2.6.8) also for degenerate statistics in the case of elastic collisions.

2.6.1 Relaxation Times for the Various Scattering Mechanisms

a) Acoustic Phonons, Deformation Potential Interaction

We have seen in section 2.4 that acoustic scattering is in general treated in the elastic and equiparation approximation. Consequently an acustic relaxation time can be defined, which results to be given by

$$\frac{1}{\tau_{ac}} = \frac{(2m)^{3/2} K_B T \mathscr{E}_i^2 \varepsilon^{1/2}}{2\pi\rho u^2 \hbar^4}. \tag{2.6.10}$$

If this expression is inserted in Eq. (2.6.4) the following mobility results:

$$\mu_{ac} = \frac{2^{3/2}\pi^{1/2}\hbar^4 e\rho u_s^2}{3m(mK_B T)^{3/2}\mathscr{E}_i^2} \propto T^{-3/2}. \tag{2.6.11}$$

The efficiency of this mechanism increases with temperature, and the mobility decreases, at increasing T, as $T^{-3/2}$. This kind of scattering is important for covalent semiconductors. It controls the carrier mobility together with ionized impurity scattering at low temperatures ($T \leqslant 200$ K). At high temperatures it must be accompanied by optical phonon scattering.

b) Optical and Intervalley Phonons, Deformation Potential Interaction

The energy exchange due to these types of scattering is of the order of the ambient temperature. Consequently the scattering cannot be considered elastic. However the transition probability is independent of the direction of electron state after scattering so that the collisions are velocity randomizing and a relaxation time can be defined which is given by

$$\frac{1}{\tau_{op}} = \frac{(D_t K)^2 m^{3/2}}{\sqrt{2}\pi\hbar^3 \rho\omega_0}[N_{op}(\varepsilon + \hbar\omega_{op})^{1/2} + (N_{op} + 1)(\varepsilon - \hbar\omega_{op})^{1/2}], \tag{2.6.12}$$

where N_{op} is the phonon population at frequency ω_{op}. This relaxation time does not yield a simple expression for the mobility in general. At high temperatures ($K_B T \gg \hbar\omega_{op}$) a mobility results which decreases as $T^{-5/2}$.

c) Acoustic Phonons, Piezoelectric Interaction

If the acoustic collisions are considered in elastic and equipartition approximation, a relaxation time can be defined and it is given by

$$\frac{1}{\tau_{pz}} = \frac{\sqrt{2mK_BT}}{\hbar^2 \rho u^2}\left(\frac{p}{\kappa}\right)^2 \varepsilon^{-1/2}, \qquad (2.6.13)$$

where the screening effect has been neglected. A mobility results that decreases at increasing temperature as $T^{-1/2}$.

d) Polar Optical Phonons

In this case collisions are neither elastic nor velocity randomizing, so that it is not possible to define a relaxation time. Usually an analytical expression obtained by Ehrenreich [94] is used for the mobility due to polar-optical phonons. This theoretical result is quite complicated and it contains a function tabulated in the original paper.

e) Ionized Impurities

The scattering is elastic, and it is possible to define a relaxation time, given by

$$\frac{1}{\tau_I} = \frac{\pi}{\sqrt{2m}} \frac{e^4 n_I}{\kappa^2} \varepsilon^{-3/2} L(\varepsilon), \qquad (2.6.14)$$

where $L(\varepsilon)$ is slowly varying with energy and is rendered by

$$L(\varepsilon) = \ln(1 + \xi) - \frac{\xi}{(1 + \xi)} \qquad (2.6.15)$$

with

$$\xi = \frac{8m\varepsilon}{\hbar^2 \beta^2}. \qquad (2.6.16)$$

The relaxation time is an increasing function of energy since the facter the electron the lesser is the interaction with the Coulomb center. Collisions with ionized impurities are therefore more effective for low-energy electrons, so that this mechanism becomes dominant, in semiconductors, at low temperatures, unless we deal with highly pure samples.
If we apply Eq. (2.6.4) we obtain the following expression for mobility in the presence of ionized impurities:

$$\mu_I = \frac{2^{1/2}}{\pi^{3/2}} \frac{\kappa^2 (K_B T)^{3/2}}{e^3 m^{1/2} n_I} L(3KT). \qquad (2.6.17)$$

These results are obtained with the Brooks and Herring approach. If the

Conwell and Weisskopf approach is used, the only difference with respect to the results seen above is an $L(\varepsilon)$ given by

$$L_{CW}(\varepsilon) = \ln\left[1 + \left(\frac{\kappa\varepsilon}{e^2 n_I^{1/3}}\right)^2\right]. \tag{2.6.18}$$

The most important feature of the mobility from ionized impurities is that it increases with temperature as $T^{3/2}$:

$$\mu_I \propto T^{3/2}. \tag{2.6.19}$$

2.6.2 *Carrier Mobilities in Various Materials*

a) Matthiessen Rule

Usually electron mobility in semiconductor materials is determined by the effect of several scattering mechanisms. The transition probability per unit time from a state $\mathbf{k}$ to a state $\mathbf{k}'$ is, in this case, obtained by summing different probabilities, each of them corresponding to an interaction process. If it is possible to define a relaxation time for each interaction, it can be easily seen from the definition of τ that a relaxation time exists also for all the scattering mechanisms combined, and it is given by

$$\frac{1}{\tau} = \frac{1}{\tau_1} + \frac{1}{\tau_2} + \cdots. \tag{2.6.20}$$

Eq. (2.6.20) does not straightforwardly imply Matthiessen rule:

$$\frac{1}{\mu} = \frac{1}{\mu_1} + \frac{1}{\mu_2} + \cdots, \tag{2.6.21}$$

but it allows us to suppose that this rule is, in general, well verfied. The Matthiessen rule is very useful, since it gives an estimate of the charge mobility for a given material when the single mobilities due to the various scattering processes are known.

b) Electron Mobilities

The temperature dependence of electron mobility in Si is shown schematically in Figs. 2.22 and 2.23 for various doping densities [95, 96]. At low temperature, impurity scattering dominates, and the mobility shows a strong reduction as the impurity concentration increases. At higher temperatures, phonons become more important, and the mobility tends to high temperature values which are independent of doping. In the energy range where phonon interaction is dominant, a deviation of the $(T^{-3/2})$ behaviour (typical

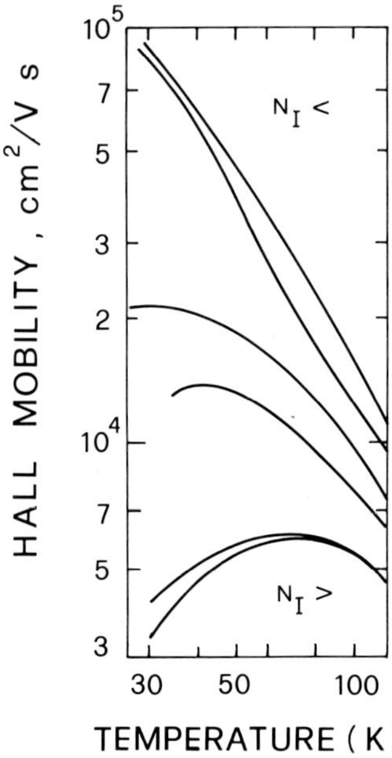

Fig. 2.22. Electron mobility in Si as a function of temperature for different impurity concentrations, increasing towards the bottom of the figure [95]

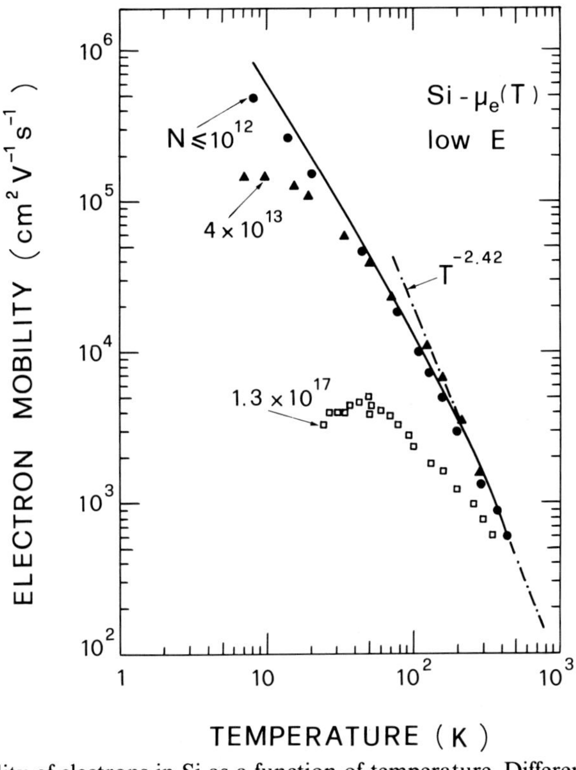

Fig. 2.23. Mobility of electrons in Si as a function of temperature. Different points refer to various experiments; the solid curve indicates results of MC calculations where only lattice scattering is considered; the dot-dashed curve gives a $T^{-2.42}$ dependence around room temperature. [After Jacoboni, C., et al.: Solid-State Electronics 20, 77 (1977)]

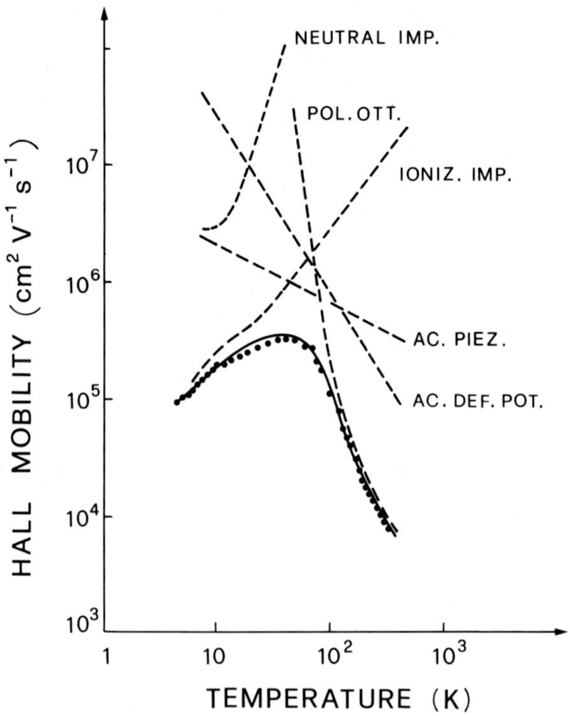

Fig. 2.24. Electron mobility in GaAs as a function of temperature. Dots indicate experimental results; dashed lines indicate theoretical curves obtained by considering sigle scattering mechanisms; the continuous line gives the theoretical mobility obtained when all mechanisms are active [97]

of acoustic phonons) is observed caused by the onset of optical and intervalley modes.

In polar semiconductors at room temperature, the ohmic mobility is mostly controlled by polar optical phonons. At lower temperatures, acoustic phonons (both with deformation potential and piezoelectric interaction) and ionized impurities become increasingly important. Figure 2.24 shows the electron mobility of GaAs as a function of temperature, together with the contribution of the individual scatterers [97]

c) Hole Mobilities

The hole mobility is in general influenced by the same scattering mechanisms as the electron one. Due to the higher effective masses, the hole mobility is usually lower than the electron mobility. Figures 2.25 and 2.26 illustrate the calculated and measured temperature behaviour of holes in Si and GaAs, respectively [98, 99].

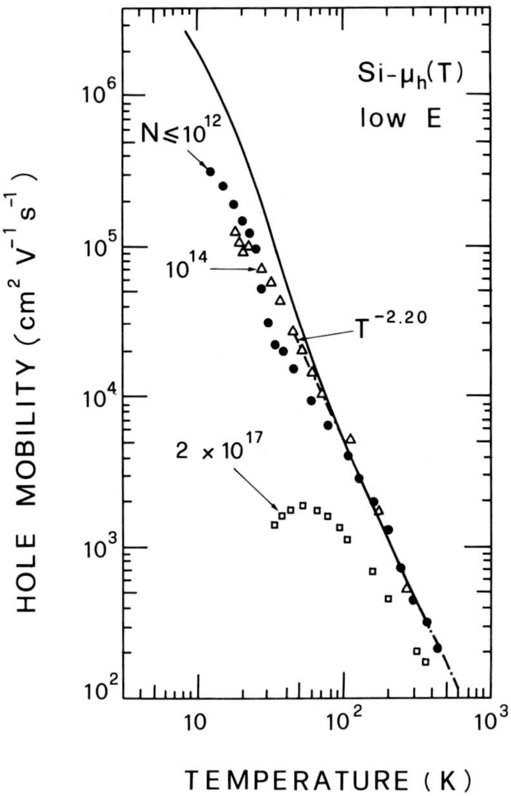

Fig. 2.25. Experimental (points) and theoretical (line) mobility of holes in Si as a function of temperature [98]

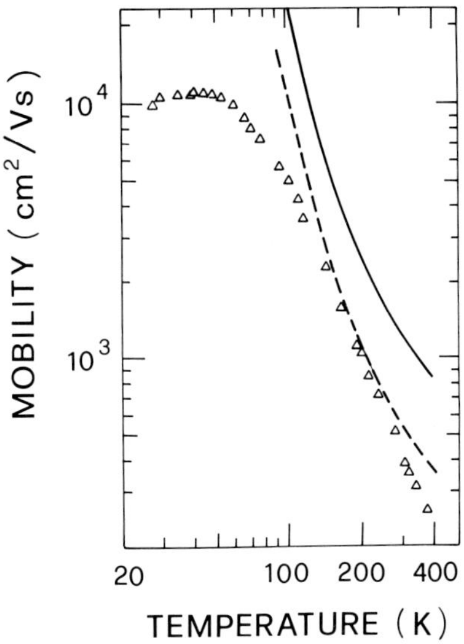

Fig. 2.26. Hole mobility in GaAs as a function of temperature. Experimental results (triangles) are compared with theoretical results obtained neglecting (dashed line) or considering (continuous line) the overlap integral in the scattering probabilities [99]

2.7 Diffusion, Noise, and Velocity Autocorrelation Function

2.7.1 *Basic Macroscopic Equations of Diffusion*

If the electron gas inside a crystal has any nonuniform distribution in space, the phenomenon of diffusion occurs, tending to make the concentration uniform through the spreading out of the carriers.

Diffusion is described, at a phenomenological level, by Fick's first equation:

$$j_i = -eD_{ij}[\partial n(\mathbf{r})/\partial x_j], \tag{2.7.1}$$

where $\mathbf{r}$ is the space position with components x_i, $n(\mathbf{r})$ and $\mathbf{j}$ the particle density and current density, respectively, and D_{ij} the diffusion coefficient tensor ($i, j = 1, 2, 3$); the sum over repeated indices is implied. The diffusion coefficient reduces to a scalar quantity (i.e. to a diagonal tensor with equal diagonal elements) by symmetry for isotropic or cubic systems. This is the frequent case of cubic semiconductor crystals with low applied electric fields. At high fields the direction of the applied field breaks the symmetry and D is no longer a scalar quantity.

If an electric field $\mathbf{E}$ that would produce, in the absence of diffusion, a drift velocity $\mathbf{v}_d$ is also present, and if diffusion and drift do not influence each other, then we can substitute for Eq. (2.7.1),

$$j_i = -e\{n(\mathbf{r})v_{d_i} - D_{ij}[\partial n(\mathbf{r})/\partial x_j]\}. \tag{2.7.2}$$

By combining this equation with the continuity equation, we obtain the diffusion equation

$$\partial n/\partial t = -v_{d_i}(\partial n/\partial x_i) + D_{ij}(\partial^2 n/\partial x_i\,\partial x_j), \tag{2.7.3}$$

where v_d and D_{ij} have been assumed to be space independent. If $\mathbf{E}$ is applied along a high-symmetry direction of a cubic crystal (say, e.g. $\langle 111\rangle$ and $\langle 100\rangle$), then D_{ij} reduces to a diagonal form, with a longitudinal component D_l and two transverse components D_t.

For a simple one-dimensional example Eq. (2.7.3) reduces to

$$\frac{\partial n(x, t)}{\partial t} = D\frac{\partial^2 n}{\partial x^2}. \tag{2.7.4}$$

If the initial conditions are such that all particles at $t = 0$ are in $x = 0$, the solution of Eq. (2.7.4) is

$$n(x, t) = \frac{N}{\sqrt{4\pi Dt}}\exp(-x^2/4Dt), \tag{2.7.5}$$

whose behaviour is shown in Fig. 2.27.

If only one particle is considered which is in $x = 0$ at $= 0$, then Eq. (2.7.5) is to be interpreted as the probability density to find the particle in x at t.

If a drift v_d is also present, Eq. (2.7.4) becomes

$$\frac{\partial n}{\partial t} = -v_d\frac{\partial n}{\partial x} + D\frac{\partial^2 n}{\partial x^2}, \tag{2.7.6}$$

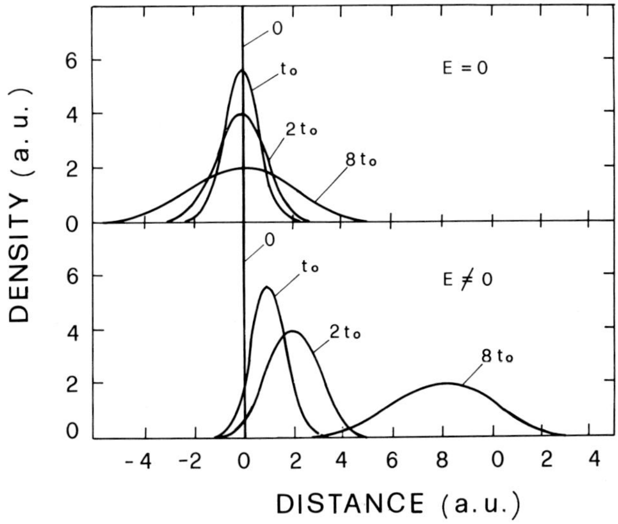

Fig. 2.27. Diffusion of a cloud of particules without and with a driving force

with solution

$$n(x,\, t) = \frac{N}{\sqrt{4\pi D t}}\, e^{(z - v_d t)^2/4Dt}\,, \tag{2.7.7}$$

whose behaviour is again shown in Fig. 2.27.

For vanishing small electric-field strengths, diffusivity D and mobility μ are field independent and verify the Einstein relation:

$$D = \mu K_B T/e\,. \tag{2.7.8}$$

At high fields the Einstein relation fails, and the study of diffusion is generally performed through Eq. (2.7.3) by introducing a field-dependent diffusion coefficient $D_{ij}(\mathbf{E})$. As a matter of fact, a rigorous derivation of Eq. (2.7.3) from the Boltzmann equation with $D = D(\mathbf{E})$ can be performed only by assuming small concentration gradients, no carrier-carrier interaction, and times longer than both the transient-transport time (see Section 2.9) and the time necessary for setting up the correct space-velocity correlations at the basis of the diffusion phenomena (see below).

For the sake of simplicity let us assume again that $n(\mathbf{r})$ is a function of only one coordinate x parallel to the direction of $\mathbf{v}_d$, and let us define the mth moment M_m of the distribution $n(x)$ as

$$M_m = \langle x^m \rangle = N^{-1} \int x^m n(x)\, dx \tag{2.7.9}$$

(N being the total number of particles). Using Eq. (2.7.3), we obtain [93, 100]

$$(d/dt)\langle x \rangle = v_d\,, \tag{2.7.10}$$

$$(d/dt)\langle (x - \langle x \rangle)^2 \rangle = 2D_l\,, \tag{2.7.11}$$

where D_l is the diagonal component of D along $\mathbf{v}_d$.

From Eq. (2.7.11) another expression for D can be obtained by differentiation:

$$D = \langle \delta x(t)\, \delta v(t) \rangle , \tag{2.7.12}$$

where $\delta x(t) = x(t) - \langle x \rangle$ and $\delta v(t) = v(t) - \langle v \rangle$ are the fluctuations of the particle position and velocity with respect to their ensemble averages. Equation (2.7.12) gives a clear physical insight into the diffusion phenomenon: the spreading of a carrier ensemble comes from the particle space-velocity correlations that arise during the time evolution of the system. Particles with a velocity higher than average, having accumulated this positive fluctuation in the near past, are, in general, in the front of the carrier space distribution and, similarly, the slower particles are in the rear of the distribution. Thus particles at the front will tend to run away, and particles at the rear will tend to be left behind, which causes a further spreading of the distribution.

It will be seen in next chapter that equation (2.7.7) can be used to determine the diffusion coefficient by means of a Monte Carlo simulation.

2.7.2 Diffusion, Autocorrelation Function, and Noise

The diffusion phenomenon is strictly related to noise. This is so because both are essentially due to the stochastic velocity fluctuations $\delta v(t)$ over the drift value v_d. The mathematical quantity that describes the common origin of diffusion and noise is the autocorrelation function of velocity fluctuations, which carries the information on how large these fluctuations are and how they decay in time.

If we consider, for simplicity, a one-dimensional situation, the autocorrelation function of velocity fluctuations is defined as

$$C(t) = \langle \delta v(t')\, \delta v(t' + t) \rangle , \tag{2.7.13}$$

where the angle brackets indicate ensemble average, that in steady-state conditions is independent of t'. $C(t)$ is related to the diffusion D through [88]

$$D = \int_0^\infty C(t)\, dt . \tag{2.7.14}$$

Thus D can be determined, from the evaluation of $C(t)$, by Eq. (2.7.14). It is worth noting that the calculation of $C(t)$ may be of interest in itself because it yields physical information on the time evolution of the dynamic system under investigation.

Another important relation can be established between the diffusion coefficient and the noise spectrum of velocity fluctuations, defined by

$$S_v(\omega) = \lim_{T \to \infty} T^{-1} \left\langle \left| \int_0^T \delta v(t) e^{i\omega t}\, dt \right|^2 \right\rangle . \tag{2.7.15}$$

The Wiener-Kintchine theorem connects the autocorrelation function $C(t)$

and $S_v(\omega)$ [101]:

$$S_v(\omega) = 2\int_0^\infty C(t)e^{i\omega t}\,dt. \tag{2.7.16}$$

At vanishing frequency, Eqs. (2.7.14) and (2.7.15) yield

$$D = \tfrac{1}{2}S_v(0). \tag{2.7.17}$$

This equation can be generalized to frequency dependent diffusion, as will be seen later in this chapter.

2.7.3 *Electron Lifetime and Diffusion Length*

Let us consider a charge carrier that is located in $x = 0$ at $t = 0$. Let $1/\tau_r$ be the probability per unit time for its recombination, i.e. dt/τ_r is the probability of this carrier to recombine during the infinitesimal time interval dt. Let finally $P(t)$ be the probability that it did not recombine at time t. From these definitions it follows immediately that

$$P(t + dt) = P(t)(1 - dt/\tau_r), \tag{2.7.18}$$

and therefore

$$P(t) = e^{-t/\tau_r}, \tag{2.7.19}$$

where we have taken into account that $P(0) = 1$. The probability that recombination occurs during an interval of time dt around t is then given by

$$P(t)\,dt/\tau_r = \exp(-t/\tau_r)\,dt/\tau_r. \tag{2.7.20}$$

Thus the lifetime of the carrier is

$$\langle t\rangle = \int_0^\infty P(t)\frac{t}{\tau_r}\,dt = \int_0^\infty \frac{t}{\tau_r}e^{-t/\tau_r}\,dt = \tau_r. \tag{2.7.21}$$

The probability $p(x)\,dx$ that the particle covers, before recombination, a distance between x and $x + dx$ (for simplicity we treat again a one-dimensional case) is given by the probability that it lives for a time between t and $t + dt$, times the probability that it covers a distance between x and $x + dx$, integrated over all possible t. Therefore the square mean distance covered before recombination is

$$\langle x^2\rangle = \int_{-\infty}^\infty x^2 p(x)\,dx = 2D\tau_r \tag{2.7.22}$$

or

$$x_{rms} = \sqrt{2}L_d, \qquad L_d = \sqrt{D\tau_r}; \tag{2.7.23}$$

L_d is called diffusion length.

2.8 Hot Electrons

The name 'hot electron', given to the problem of non-linear conduction in semiconductors, is due to the fact, pointed out in some fundamental works by Froehlich [32], that non-linear phenomena are associated with a rise of the electron mean energy above its thermal value $\frac{3}{2}K_B T$. For many years since then physicists have been speaking of an electron temperature T_e higher than the lattice temperature T. It is now known that this idea does not correctly describe reality since the electronic distribution function very often is not such as to allow an unambiguous definition of an electron temperature; that is, it is not of a Maxwellian type. It must be recognized, however, that the concept of an electron temperature has greatly helped the understanding of non-linear transport and it still gives a useful terminology for a heuristic investigation of these problems.

Nonlinear transport deals with problems that arise when a sufficiently strong electric field **E** is applied to a material so that the current density deviates from the linear response.

Carrier heating and its relation to nonlinearity can be easily understood in a simple model semiconductor (spherical and parabolic bands). The presence of the field inside the crystal modifies the free paths of the electrons between two successive scattering events by adding to the velocity a component in the direction of the field:

$$\mathbf{v}^{(i)} = \mathbf{v}_0^{(i)} - (e\mathbf{E}/m)t^{(i)}, \tag{2.8.1}$$

where $\mathbf{v}^{(i)}$ is the instantaneous velocity of the ith electron, $t^{(i)}$ is the time elapsed after the last scattering event of the ith electron, and $\mathbf{v}_0^{(i)}$ is the initial value of $\mathbf{v}^{(i)}$ at $t^{(i)} = 0$.

The drift velocity of the electron gas is obtained by averaging $\mathbf{v}^{(i)}(t)$ over all electrons at a given time t, corresponding to a different value of $t^{(i)}$ for each electron. If there is no accumulation of the effect of the electric field on the initial velocities $\mathbf{v}_0^{(i)}$ of the successive free-electron paths, all these initial velocities reproduce the electron distribution function in the absence of the field. The mean value of the electron energy differs from its zero-field value only by the amount gained during each single flight. As long as this energy is negligible with respect to the electron energy, since the electron scattering probabilities are functions of the electron energies, the distribution of the $t^{(i)}$ at any given time t is independent of E. The drift velocity is therefore linear in the electric field, according to Ohm's law.

On the other hand, if the field is sufficiently high, the energy gained by the electrons during each single flight is not negligible. The electrons cannot fully dissipate this energy at each scattering event, and the effect of the field accumulates on the initial velocities $\mathbf{v}_0^{(i)}$. The distribution of the $t^{(i)}$ also depends on the field, and the linearity of Ohm's law is lost in the average of Eq. (2.8.1). The mean energy of the carriers then increases, and it would continue to increase up to the breakdown of the sample were it not for scattering mechanisms whose dissipation capacity increases at increasing

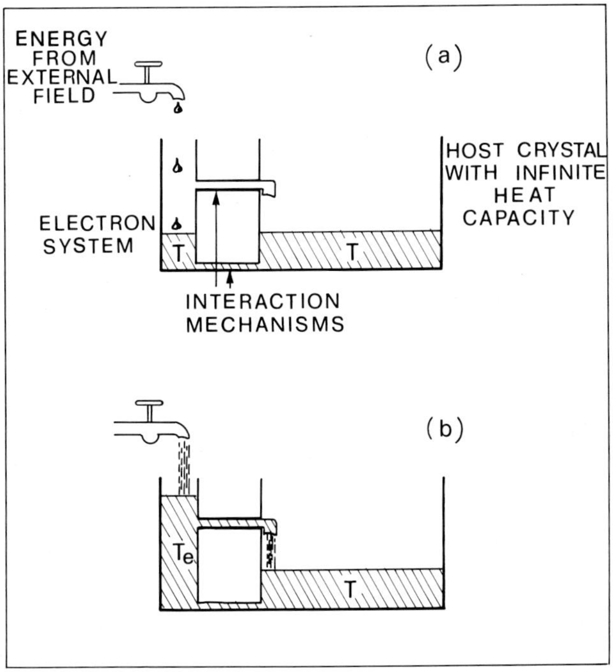

Fig. 2.28. Hydraulic analog of the concept of hot electrons: *a* low-field case; *b* high-field case (see text)

electron energies. These scattering mechanisms establish a new stationary state in which average electron energy is higher than at equilibrium. This is the origin of the expression "hot-electron effects", applied to nonlinear transport phenomena in semiconductors. Fig. 2.28 shows a hydraulic analog of the concept of hot electrons. Energy is transferred from the external electric field, to the electron system, to the host crystal. In the low-field case (Fig. 2.28a) some interaction mechanisms are capable of maintaining the temperature of the electron system equal to that of the thermal bath. At high fields (Fig. 2.28b) the energy supplied to the electrons is higher, and a new stationary state is attained by an increased efficiency of the scattering mechanisms already active at low fields and, sometimes, by the onset of new mechanisms.

A general picture of hot-electron effects in pure semiconductors (see Fig. 2.29) can be given by dividing all semiconductors into two categories, depending on whether the most effective scattering mechanisms are due to electrostatic fields (compound materials) or to a deformation potential (elemental materials).

We begin by recalling that, as the external field increases above the Ohmic region, the mean energy of the electrons in steady-state conditions increases. For elemental materials the scattering mechanisms become more effective as the electron energy increases, so that the mobility $\mu(E)$ decreases as the

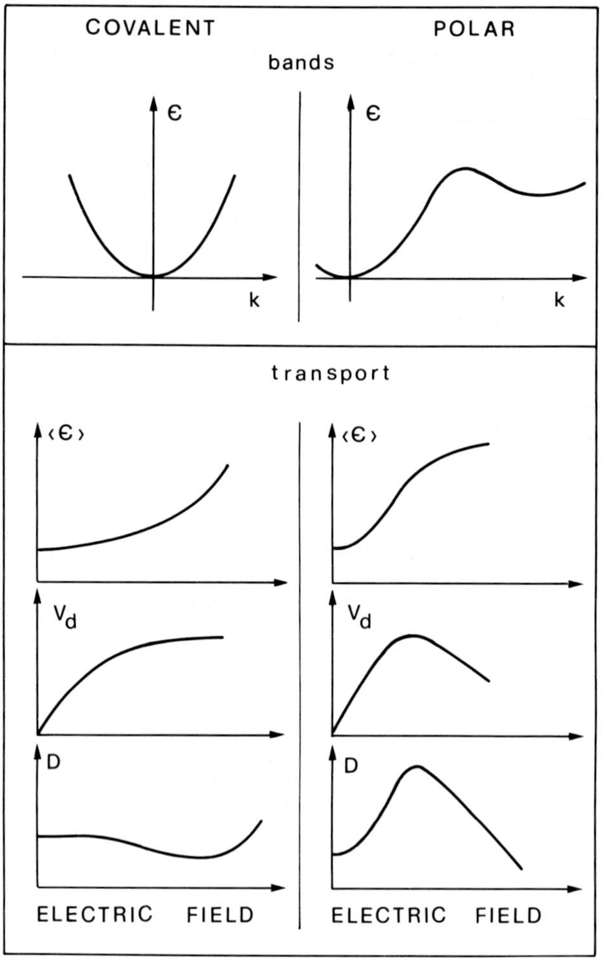

Fig. 2.29. Simple physical picture of hot-electron effects in semiconductors (see text)

electric field E is increased. The steady state is attained with a small energy rise above thermal equilibrium, and the electron mean energy increases slowly with field. Eventually, upper valleys are reached; this fact, however, does not occur in Si since the upper valleys are so high in energy that breakdown occurs first.

The experimental velocity-field curves for n-Si measured with time-of-flight technique are shown in Fig. 2.30.

For compound materials the efficiency of the scattering mechanisms does not increase with energy. Thus the electron mean energy increases very rapidly to reach a steady-state condition as the applied field increases, whereas the mobility does not decrease. Above the threshold field E_{th} for polar runaway, the mean energy would increase indefinitely, giving rise to dielectric breakdown. However, if upper valleys are present with energy

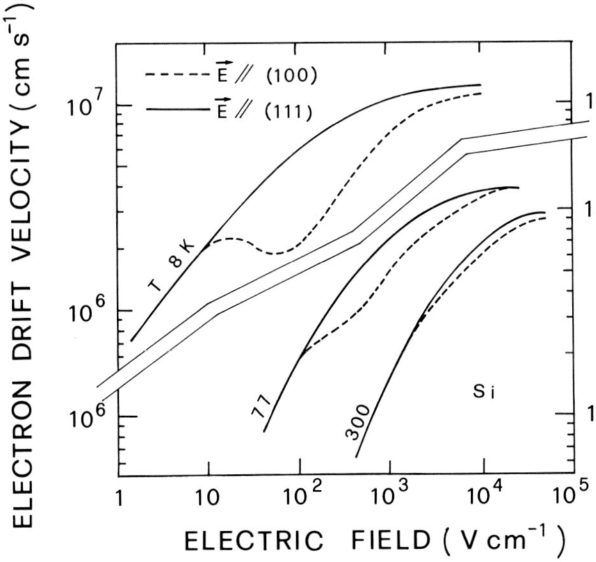

Fig. 2.30. Experimental drift velocities of electrons in Si as functions of the electric field applied parallel to ⟨111⟩ and ⟨100⟩ crystallographic directions at several temperatures [96]

separation smaller than the band gap, breakdown is prevented because electrons can populate the upper valleys in the band. At this point the new scattering mechanism of nonequivalent intervalley scattering becomes active. Since it is very effective in dissipating both energy and momentum, the energy increase with field is slowed down, and the drift velocity decreases, giving rise to the well-known phenomenon of negative differential mobility, which is the basis of Gunn effect [36, 102].

Experimental curves for n-GaAs [103–105] are presented in Fig. 2.31. A theoretical result is also shown for comparison.

Looking now at the field dependence of the diffusion coefficient, it is important to remember that Einstein relation does not hold at high fields. A rough generalization is given by the expression $D = (v_d/eE)(2/3)\langle\varepsilon\rangle$, which leads to the general behavior shown in Fig. 2.29. The field dependence of D is determined by the competitive effects of mobility reduction and energy increase. In covalent semiconductors, the former effect is slightly more efficient, causing a slow decrease of D at increasing fields. In polar semiconductors, the average energy first increases rapidly with the field, while the mobility decreases slowly. The net result is a strong initial enhancement of the diffusivity. As the field increases further, the previous situation is reversed and the mobility reduction becomes the dominant effect. In these conditions D decreases rapidly with the field.

Experimental and theoretical results are shown in Figs. 2.32 and 2.33 for n-Si [106] and n-GaAs [103, 105, 107, 108], respectively.

Almost all the hot-electron effects can be described as a refinement of the general picture presented above.

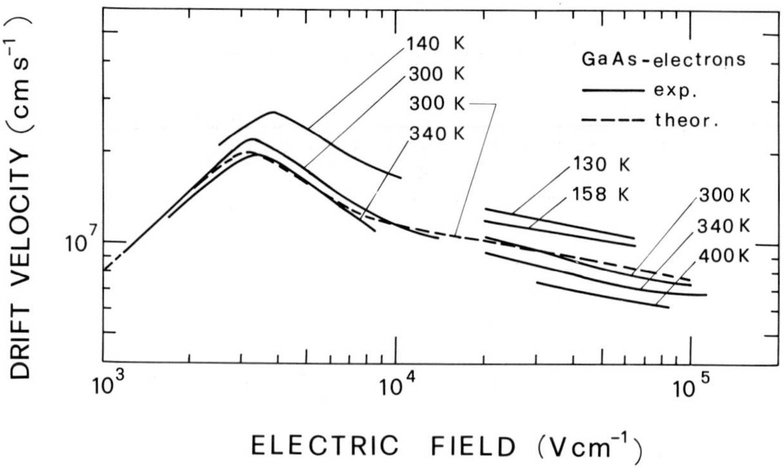

Fig. 2.31. Drift velocity as a function of field strength for electrons in GaAs at different temperatures. Continuous lines represent experimental data [103, 104]. The dashed line indicates the theoretical results of [105]

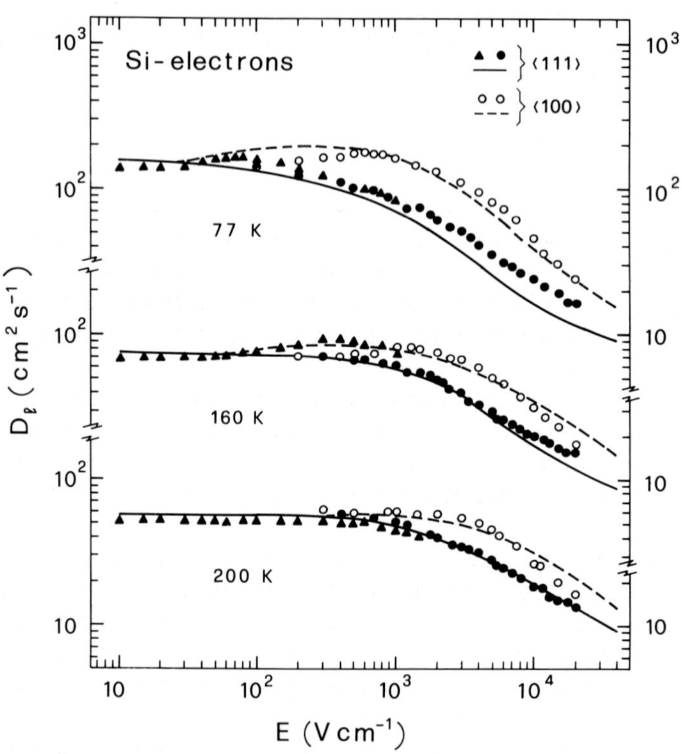

Fig. 2.32. Longitudinal diffusion coefficient of electrons in Si as a function of electric field strength at the given temperatures. Points refer to experiments and curves to theoretical Monte Carlo calculations [106]

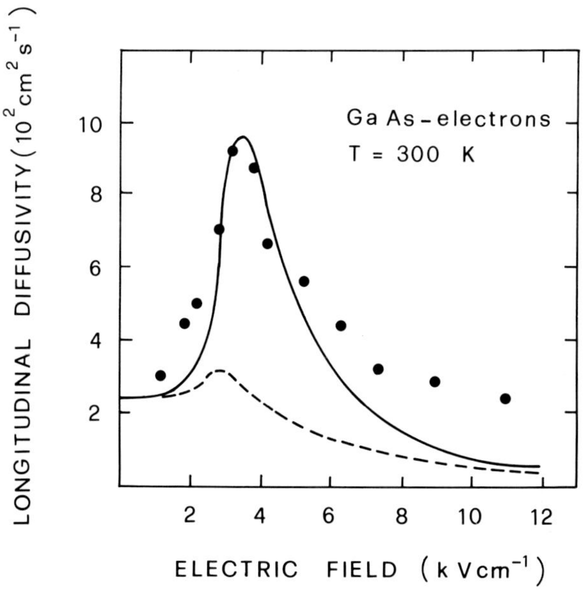

Fig. 2.33. Longitudinal diffusion coefficient of electrons in GaAs as a function of electric field strength at 300 K. Dots indicate experimental results of [103]; curves show the different theoretical results of [107] (dashed) and of [105] (continuous)

2.9 Transient Transport

The transient dynamic response of a physical system to given external forces is the response to these forces during the evolution of the system from given initial conditions to the steady state. The duration of the transient response is not known a priori and, in general, it will be at least of the order of the longest of the characteristic times of the carrier system. This time may be called the "transient transport time" and it depends upon the values of the applied field and temperature. In the case of high-field transport in semiconductors it may roughly correspond to the energy relaxation time or to the time necessary for the repopulation of the different valleys. Furthermore the transient response depends strongly upon the initial condition of the carriers.

Transient transport may occur in different physical systems. The first rather "academic" situation we may think of is that of an infinite homogeneous system with a sudden change of the applied field, for example, from zero to high value: the electron distribution requires some time to reach its steady-state nonequilibrium value. The second system of interest is a much more realistic one and can be present in ultrasmall devices: in a steady-state situation particles may enter a small region with a very high applied field and reach the end of this region so quickly that the space dependent distribution function does not reach the form equivalent to a steady-state homogeneous system.

The two situations mentioned above can be combined when the system of interest is subject to alternating fields with frequency higher than the inverse of some electronic relaxation time ($\omega\tau > 1$) and wave vector higher than the inverse of the electronic mean free path ($ql > 1$). When $\omega\tau > 1$, i.e. for fast variation of the applied field in time, the electron distribution function is still influenced by the field at previous times, while when $ql > 1$, i.e. for sharp variation of the field in space, the electron distribution function in one point is influenced by the field values at some other points so that the transport phenomenon becomes non local in time or space, respectively.

For space- or time-dependent problems the analytical solution of the Boltzmann equation is even more difficult than for homogeneous and stationary problems, while for MC programs little work needs to be added for such problems to be tackled. This confirms again the usefulness of the MC method, in particular for the analysis of small devices, where it is often necessary to consider both the transient dynamic response to voltage changes and the electronic behavior at different points of the device with different field or material properties. For such problems we cannot rely on the ergodic property of the system, and an ensemble of particles must be explicitly simulated.

The transient dynamic response which will be obtained by means of the simulation, will of course depend upon the initial conditions of the carriers, and these must be assumed according to the situation which is to be explored.

As examples of transient transport, we show in Figs. 2.34 and 2.35 the mean velocity $\langle v \rangle$ and the mean energy $\langle \varepsilon \rangle$ of an ensemble of carriers in Si and GaAs as functions of the time elapsed after the instantaneous application of a static field. In both cases electrons have initially the equilibrium Maxwellian distribution in **k**-space. For the case of Si they are randomly distributed among the six valleys, while in GaAs they are in the central valley.

As can be seen from Figs. 2.34 and 2.35, a general feature of high-field transient transport, starting from equilibrium conditions, is an overshoot of $\langle v \rangle$ and $\langle \varepsilon \rangle$ before reaching the steady-state value. The overshoot of $\langle v \rangle$ is associated with larger momentum relaxation times of "cold" electrons. The overshoot of mean energy in Si, which is found at times slightly larger than those of the overshoot of mean velocity, is associated with an initial excess of absorbed power of the electron gas, reduced later by the setup of intervalley scattering.

In the case of GaAs the transient transport is also influenced by the electron transfer to upper valleys, which further reduces the average velocity and kinetic energy. In fact, as shown in Fig. 2.35, in correspondence with the population of the upper valleys, the mean energy and velocity in the central valley become higher than the corresponding total quantities because in the upper valleys $\langle v \rangle$ and $\langle \varepsilon \rangle$ have values approximately constant and much lower than those of the central valley.

The very first part of transient transport is determined by the acceleration due to the electric force, before the effect of scattering; this regime is called

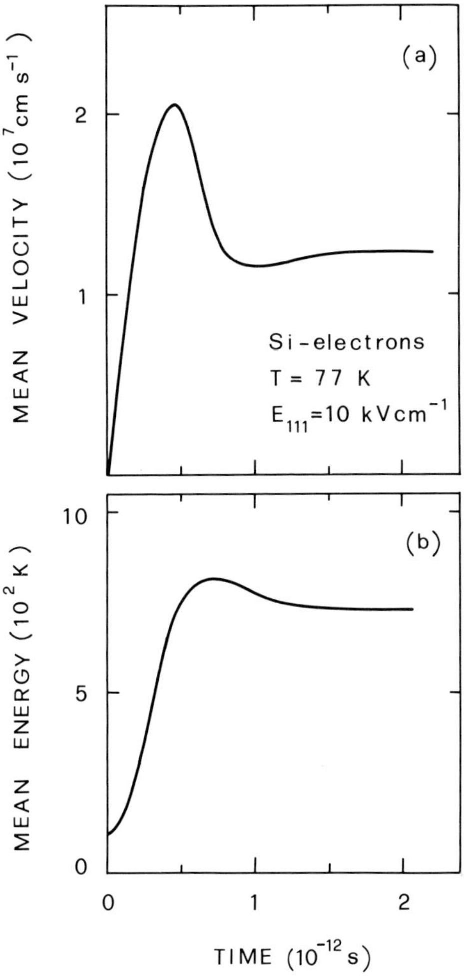

Fig. 2.34. MC results for mean velocity (*a*) and mean energy (*b*) of an ensemble of electrons in Si at 77 K as functions of the time elapsed after the application of a field $E = 10$ kV/cm along a $\langle 111 \rangle$ direction

"ballistic transport". In such conditions, if the average initial velocity of the particles is zero, then the average velocity $\langle v_z(t) \rangle$ along the field direction and the average energy $\langle \varepsilon(t) \rangle$ change with time, in a simple model, as:

$$\langle v_z(t) \rangle = at , \tag{2.9.1}$$

$$\langle \varepsilon(t) \rangle = \langle \varepsilon(0) \rangle + \tfrac{1}{2} m a^2 t^2 , \tag{2.9.2}$$

where a is the acceleration $e\mathbf{E}/m$. The linear and squared dependencies predicted by the above equations are actually present as initial behavior of the results reported in Figs. 2.34 and 2.35.

In order to show the importance of initial conditions, Fig. 2.36 reports

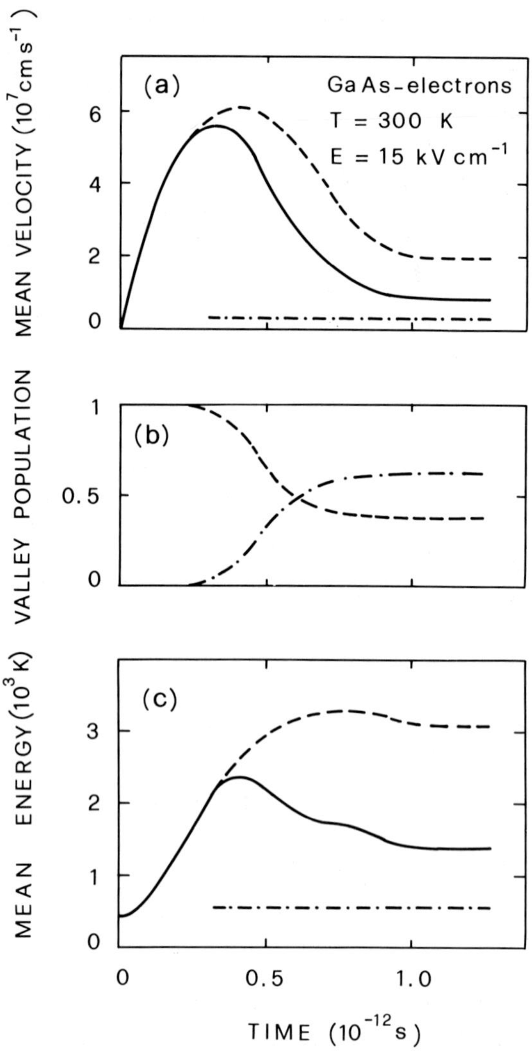

Fig. 2.35. MC results for mean velocity (*a*), valley populations (*b*), and mean energy (*c*) of an ensemble of electrons in GaAs at 300 K as functions of the time elapsed after the application of a field $E = 15$ kV/cm. Dashed and dot-dashed curves refer to quantities averaged over the central and upper valleys, respectively. Continuous curves refer to quantities averaged over all particles

ensemble MC calculations for the case of holes in Si at $T = 300$ K and $E = 50$ kV/cm. Here the transient behavior of the drift velocity and the mean energy is investigated at different initial mean energies. An overshoot in the drift velocity can be seen that is larger for the colder initial condition and disappears for the hottest case. When the electron energy is higher in the initial conditions than in the steady state, in fact, the drift velocity is at first lowered by a higher scattering and reaches its stationary value more slowly.

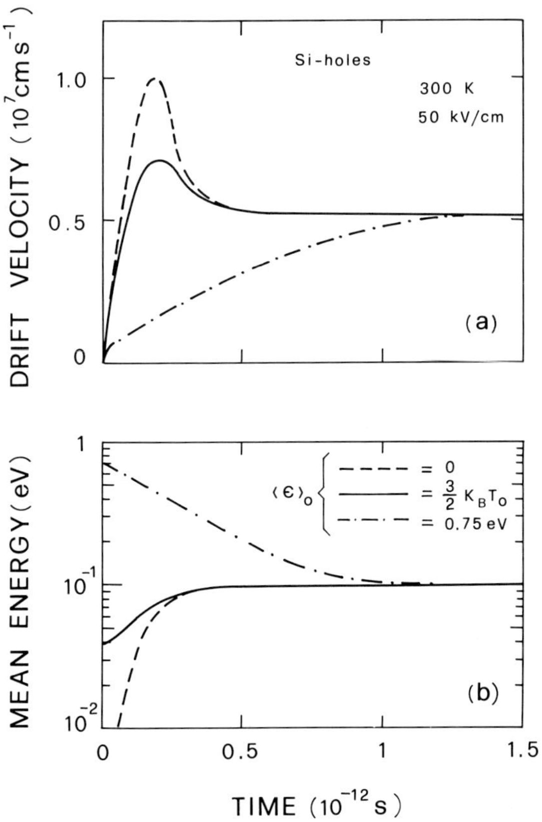

Fig. 2.36. Mean velocity (*a*) and mean energy (*b*) of an ensemble of holes in Si at 300 K as a function of the time elapsed after the application of a field $E = 50$ kV/cm. Different curves refer to the reported different values of the initial mean energy with velocity randomly distributed [109]

Space- and time-dependent phenomena may present similar characteristics and they have sometimes been confused in simulation problems. For instance the drift velocity overshoot seen above in the time-dependent homogeneous case, is often used in discussing the behavior of an electron stream coming from a low-field region and entering a region with a high applied field.

If this reasoning is correct from a qualitative point of view, in rigorous calculations the kind of average that must be done is different whether time- or space-dependent phenomena are considered. In a steady-state phenomenon with a space dependence of the applied field, it is in general necessary to consider the electron properties at given positions, so that the simulation must perform recording and averages for given points at any times, while when the object of the investigation is the evolution in time of a homogeneous system, average quantities must be evaluated at given times independent of the particle position.

2.10 The Two-dimensional Electron Gas

The theoretical treatment of electron transport presented in the previous sections referred exclusively to bulk situations, where boundaries and interfaces do not affect the electron dynamics. A radically different picture arises in the presence of heterostructures formed by different materials. Two examples of great technological interest will be presented in this section, namely the AlGaAs-GaAs and the Si-SiO_2 systems (for a complete review, see Sze [110]).

a) GaAs-AlGaAs Interface

The equilibrium energy-band diagram of three ideal heterojunctions (that is a perfect junction between two dissimilar semiconductors) are shown in Fig. 2.37 [111, 112] for *n-n*, *p-n*, and *p-p* Ge-GaAs systems, respectively. The two semiconductors (both *n*-type in the chosen case) are assumed to have different band-gaps, different permittivities and different electron affinities. Within the Shockley-Anderson model (for a general review see [113]), the interface is abrupt without interface states. The change in the energy gap at the interface leads to discontinuities (or "offsets") in the conduction and valence band edges. The magnitudes of these offsets is assumed to be independent of the doping levels. In reality, the Shockley-Anderson model is only an approximation, and the exact determination of the band offsets is still a matter of controversy. For the case of $Al_xGa_{1-x}As$-GaAs heterojunctions, it was previously stated that the band gap of the AlGaAs strongly depends on the Alluminium concentration. Thus, it is possible to control the offset of the conduction band by varying the Al content in the $Al_xGa_{1-x}As$ layer. The extraordinary advances in growth techniques favoured by the Molecular Beam Epitaxy (MBE) [114] allow the production of atomically smooth layers and a very precise control over grown layer thickness. Thus the quality of the GaAs-AlGaAs interface is much higher than the Si-SiO_2 one. Furthermore, by a proper synchronization of the source fluxes (in particular of Al and dopants), it is possible to grow heterostructures where only the $Al_xGa_{1-x}As$ layer is deliberately doped (usually with Si impurities), while the GaAs layer is grown as pure as possible (not intentionally doped). Such a principle is called "modulation doping" and is extremely important for device applications [115, 116]. Under equilibrium conditions, since the GaAs conduction band edge lies lower in energy than the AlGaAs donor states, electrons transfer from the donor into the high-purity GaAs layer. The consequent band bending leads to the formation of a two dimensional electron gas in GaAs. A similar effect will be encountered shortly for the Si-SiO_2 structure. The most important feature of the modulation doping structures is that essentially, all of the mobile carriers (electrons confined in the GaAs) and their parent donor impurities (in the AlGaAs layer) are spatially separated (see Fig. 2.38). Thus, the electron density in

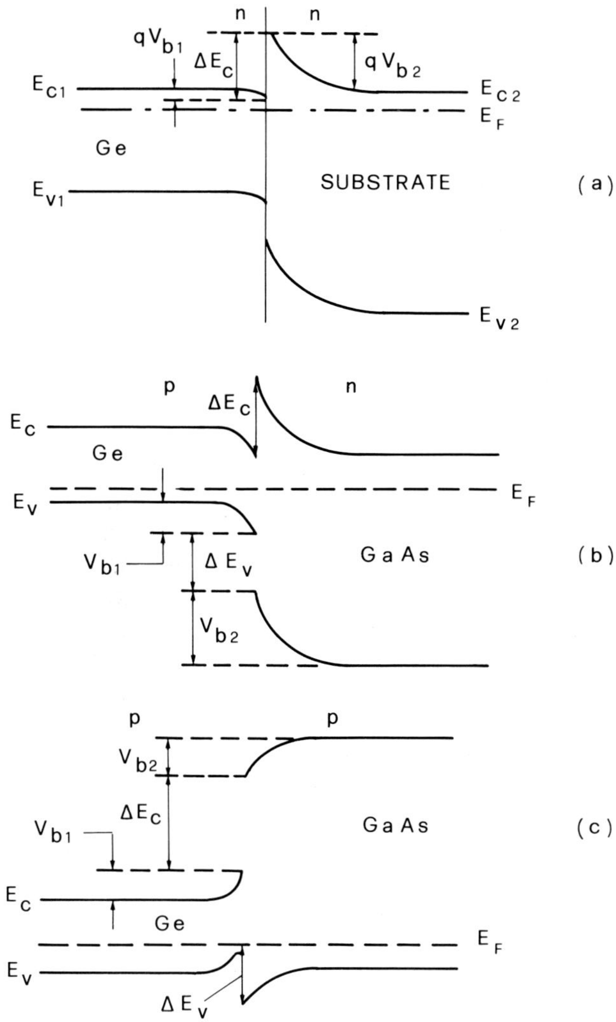

Fig. 2.37. Energy-band diagram for ideal *n-n* (*a*) [111], *p-n* (*b*) and *p-p* (*c*) [112] isotype heterojunctions

the GaAs greatly exceeds the density of unintentionally present impurity centers, leading to the extreme high mobility.

By alternating thin layers of GaAs and $Al_xGa_{1-x}As$, it is possible to grow periodic structures that are called superlattices [117]. A typical example is shown in Fig. 2.39. The periodic repetition of layers of different materials (whose thickness can reach atomic dimensions) creates an anologous periodicity in the electrical potential. Namely, each layer of the semiconductor with smaller value of the band gap constitutes a quantum well. Inside the well, the electrons in the conduction band (and the hole in the valence band) can only occupy defined energy states. Those states form "minibands" for

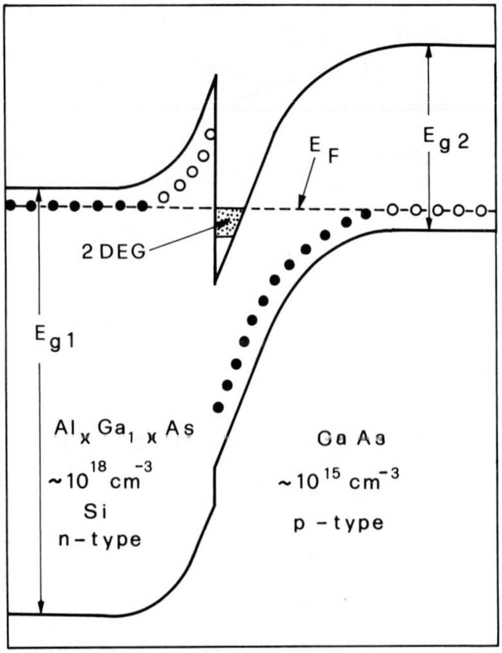

Fig. 2.38. AlGaAs-GaAs interface. Si donor levels in the alloy release electrons that enter the GaAs well at the interface, thus forming a 2D electron gas [115]. See also Fig. 2.39

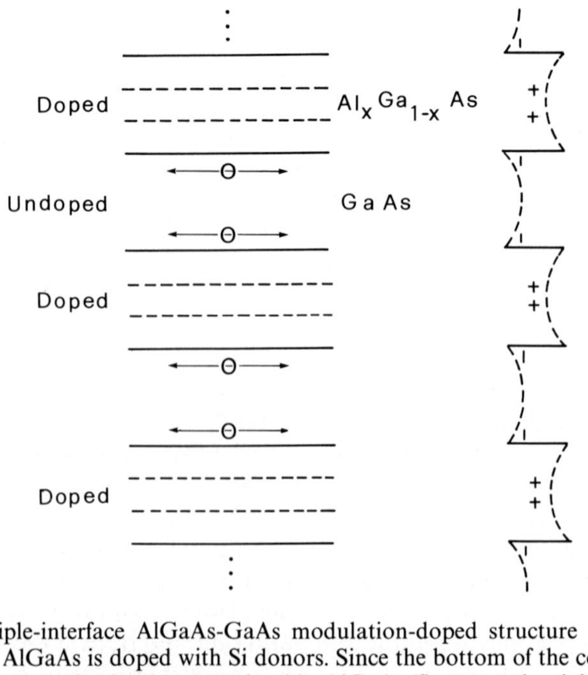

Fig. 2.39. Multiple-interface AlGaAs-GaAs modulation-doped structure where only the center region of AlGaAs is doped with Si donors. Since the bottom of the conduction band in GaAs is lower than the donor energy level in AlGaAs (figure on the right), the electrons diffuse into GaAs where they are confined by the energy barriers [116]

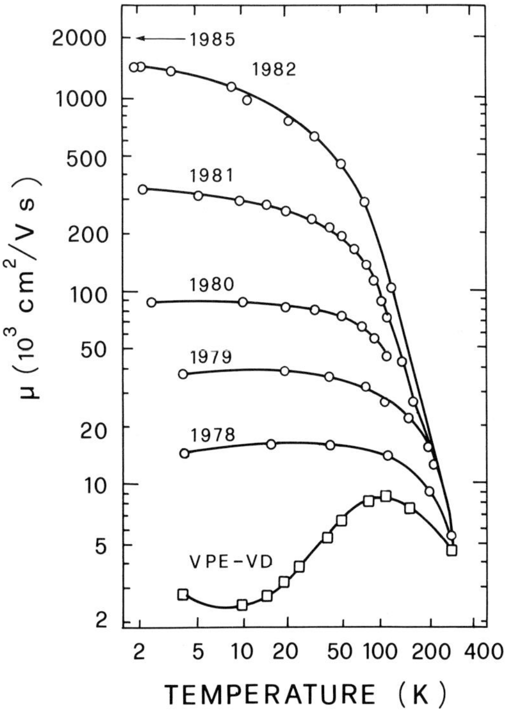

Fig. 2.40. Temperature dependence of the electron mobility in GaAs and its time evolution over the last decade [118]

infinitely extended structures, and "quasi-levels" for multiple quantum wells (MQW) with a finite number of layers.

The technological improvements in growth processes have resulted in values of the low temperature mobilities as high as 4×10^6 cm^2/V sec, that is very close to the theoretical upper limit for pure GaAs. The extent to which progress has been made over the years is clear from Fig. 2.40, which shows the temperature dependence of the mobility in its time evolution over the past two decades. The curves shown in Fig. 2.40 [118] reflect some known facts: at room temperatures, the low-field carrier transport is mainly controlled by optical phonons, while at low temperatures acoustic phonons, impurity and surface roughness are the most important scattering mechanisms. Since, in modulated structures the effect of ionized impurities is strongly reduced, and surfaces are particularly smooth, the only intrinsic limit to the electron mobility is the one set by the acoustic and the neutral-impurity interactions.

In the presence of high electric fields parallel to the interface, the electrons might gain enough energy to surmount the potential barrier that confines them. Such process, in which electrons scatter from the quantized GaAs layer into the AlGaAs region is usually called "Real Space Transfer" [119]. Since, in modulated structures, the AlGaAs is highly doped, the real space transfer moves carriers from the high mobility to the low mobility regions.

The practical implication of such process will be discussed later. As we will see in chapters 4 and 6, quantum wells and superlattices have found an extraordinary application in microelectronics, opening the way to very fast devices.

b) Si-SiO$_2$ Interface

The energy-band diagram of an ideal metal-insulator-semiconductor structure in equilibrium is shown in Fig. 2.41 for a *p*-type semiconductor [120]. At zero applied bias, the energy difference between the metal work function ϕ_m and the semiconductor work function is zero, or

$$\phi_{ms} \equiv \phi_m - \left(\chi + \frac{\varepsilon_g}{2e} + \psi_B\right) = 0 \qquad \text{for } p\text{-type semiconductors},$$

where χ is the semiconductor electron affinity, ε_g the band gap, ψ_B the potential difference between the Fermi level ε_f and the intrinsic Fermi level ε_i.

As we will see in detail in chapter 4, this structure constitutes one of the basic blocks for a wide class of semiconductor devices. The presence of the

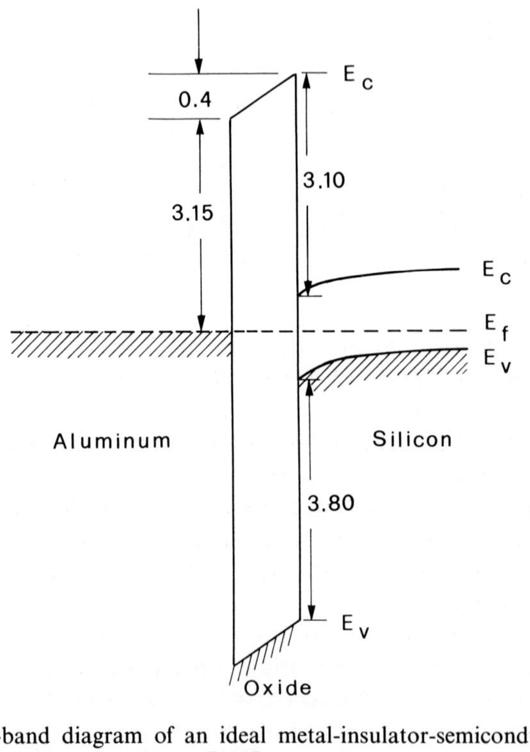

Fig. 2.41. Energy-band diagram of an ideal metal-insulator-semiconductor structure in equilibrium for a *p*-type semiconductor [120]

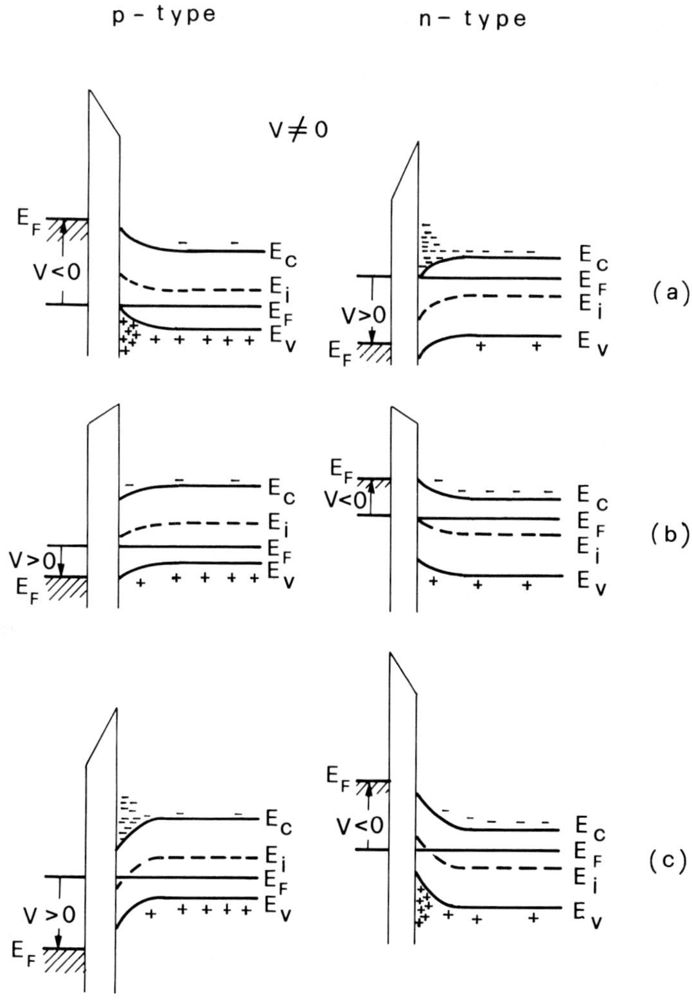

Fig. 2.42. Energy-band diagrams for ideal *p*- and *n*-type MIS diodes when $V \neq 0$, for the cases of accumulation (*a*), depletion (*b*) and inversion (*c*) [120]

metal allows the semiconductor bands to be bent with the application of a proper potential at the metal electrode.

Three important cases are shown in Fig. 2.42 for *p*-type semiconductors (the same basic features are found for the *n*-type case, also shown in the figure, with reverse polarities and carrier types). When a negative voltage ($V < 0$) is applied to the metal, the top of the valence band is tilted upwards, closer to the Fermi level which remains constant in ideal structures (no current across the insulator layer). The reduced difference between the Fermi level ε_F and the top of the valence band ε_V causes an accumulation of majority carriers (holes in this case) near the interface.

On the contrary, with the application of a positive bias ($V > 0$) on the metal gate, the energy bands bend down at the surface, and a depletion region is formed as more acceptors become exposed at the surface. As the positive bias is further increased, the conduction band edge at the surface is pushed closer to the Fermi level as a result of the enhanced band bending. Minority carriers (electrons) are induced near the surface to form what is usually referred to as "inversion layer".

Since under strong-inversion conditions the carrier wavelength is usually comparable to the thickness of the confining potential well, quantum mechanical effects are important. In particular, the carrier motion in the direction perpendicular to the surface is quantized into discrete levels (subbands). The carriers are then said to constitute a quasi two dimensional electron (or hole) gas (2DEG) confined to a region of typically few tens of angstroms at the semiconductor-insulator interface [121].

The formal general treatment of the 2DEG will be given later. Few major features will be outlined here. For the case of Si-SiO_2 interface, the quantized electron inversion layer at the surface of (100)-oriented silicon is characterized by two sets of subbands originating from the splitting of the six equivalent minima of the bulk Si conduction band. One set consists of the subbands arising from the two valleys that show the longitudinal mass in the direction perpendicular to the interface. The energy levels are labelled as E_0, E_1, E_2,... in the notation of Stern and Howard (1967) [122]. The other set of subbands originates from the four equivalent valleys that show a transverse mass in the direction normal to the surface.

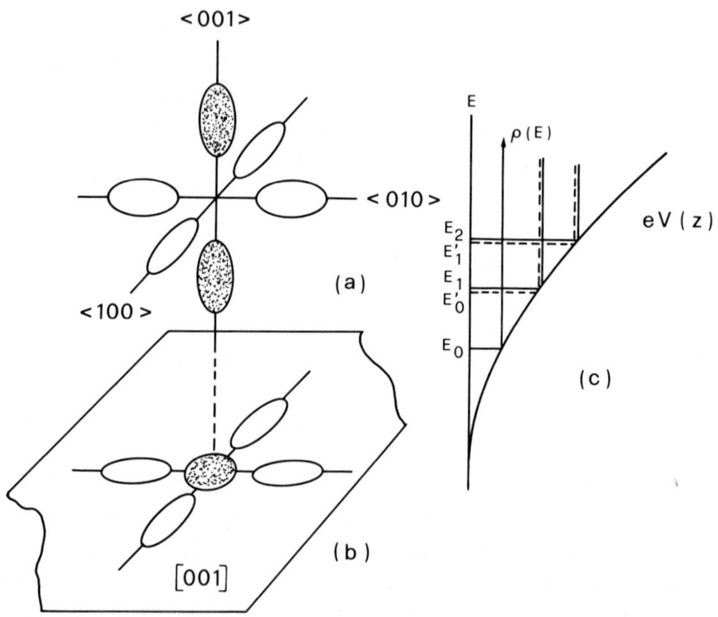

Fig. 2.43. The six equivalent ellipsoids of the Si conduction band (*a*) and their projection onto the [001] plane (*b*), which leads to two sets of subbands (*c*) [123]

These levels are labelled as $E'_0, E'_1, E'_2, \ldots$. The projection of the elipsoidal surfaces of constant energy of Si onto the [001] plane and the various subbands are shown in Fig. 2.43 [123].

The carrier transport along the interface is controlled by phonons (acoustic and optical) and by Coulomb scattering. Furthermore, the presence of the interface introduces new scattering processes, namely the carrier interaction with remote optical phonons from the strongly polar SiO_2 [124] and with the random fluctuations of the confining potential originated by fluctuations in the surface layer (surface roughness) [125].

2.10.1 Subband Levels and Wavefunctions

The two-dimensional electrons confined in a quantum well move in an potential $V(z)$ defined by

$$V(z) = -e\phi_e(z) + V_h(z) + V_{xc}(z), \tag{2.10.1}$$

where z denotes the normal direction, $\phi_e(z)$ is the electrostatic potential, $V_h(z)$ is the step function describing the interface barrier, and $V_{xc}(z)$ is the local exchange correlation potential [126]. The electrostatic potential is related to the charge distribution by Poisson's equation:

$$\frac{d^2\phi_e(z)}{dz^2} = \frac{4\pi e}{\kappa}\left[\sum_i N_i\zeta_i^2(z) + N_A(z) - N_D(z)\right], \tag{2.10.2}$$

where N_i represents the number of the electrons per unit area in subband i, and $N_A(z)$ and $N_D(z)$ are respectively the position-dependent acceptor and donor concentrations. $\zeta_i(z)$ is the envelope function normal to the layer interface. Since the potential is taken as a function of z only, the electron wavefunction can, in fact, be expressed in separable form as the product of a Bloch function parallel to the interface and an envelope function for the normal direction [127]. Within the effective mass approximation, the envelope function $\zeta_i(z)$ for the ith subband, satisfies the following Schrödinger equation

$$-\frac{\hbar^2}{2m}\frac{d^2\zeta_i(z)}{dz^2} + V(z)\zeta_i(z) = E_m\zeta_i(z). \tag{2.10.3}$$

An iterative scheme can be used to solve the above described system of Eqs. (2.10.1)–(2.10.3).

The main difference between the Si-SiO_2 and the AlGaAs-GaAs cases consists in the boundary conditions for the envelope function. In the former system, the wavefunction is assumed to vanish at the interface [121], since the potential well has a very high value. In the latter, the wavefunction penetration into the AlGaAs has to be accounted for in the calculations. The effective potential calculated from Eq. (2.10.2) by Yokoyama and Hess [128] is shown in Fig. 2.44 together with the five lowest subband energy

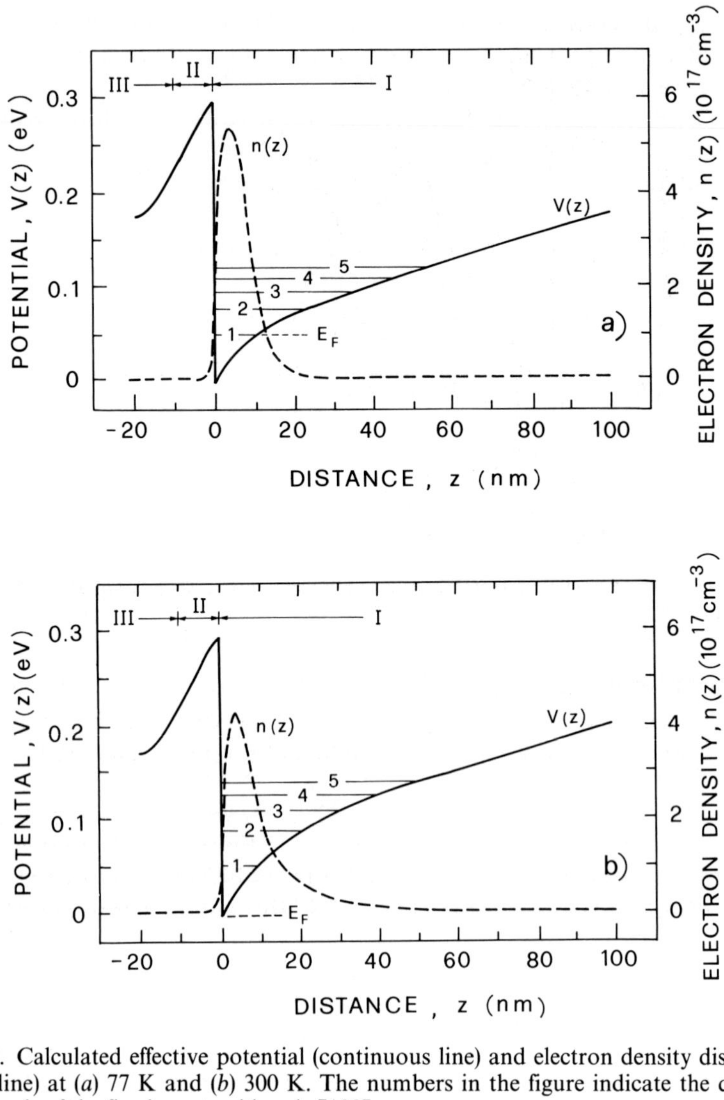

Fig. 2.44. Calculated effective potential (continuous line) and electron density distribution (dashed line) at (*a*) 77 K and (*b*) 300 K. The numbers in the figure indicate the quantized energy levels of the five lowest subbands [128]

levels at 77 and 300 K. The dashed line indicates the profile of the electron concentration. The corresponding wave functions for the first to the fifth subbands at 77 K are shown in Fig. 2.45. It can be seen from the Figs. 2.44(a) and 2.44(b) that the electron density at 300 K extends somewhat deeper into the substrate compared with the extension at 77 K, since electrons at higher temperatures tend to populate higher subbands.

Approximate expressions for triangular wells are provided by the solution of the infinite triangular barrier problem, that is the set of Hairy functions [121]. Such approximation is fairly good for the Si-SiO_2 interface, where

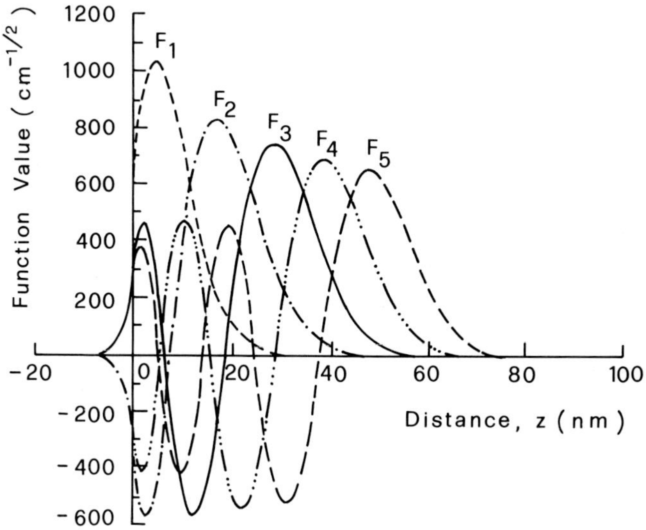

Fig. 2.45. Calculated wavefunctions for the five lowest subbands at 77 K corresponding to Fig. 2.44 a [128]

the barrier height is around 3 eV. For the AlGaAs-GaAs case, the choice of the Hairy function neglects the sizable penetration of the electron wavefunction into the AlGaAs layer. Nevertheless, the approximation can be very useful when analytical expressions are needed.

For the case of square wells, the subband levels are calculated simply from the solution of Schroedinger equation for the finite well. Bound states are characterized by wave functions which are combinations of decaying exponentials in the AlGaAs regions and oscillating functions in the confining GaAs well.

2.10.2 Scattering Rates

For two dimensional systems, the scattering rates are calculated as for the three dimensional case, starting from Fermi Golden Rule. In the following we will briefly describe the most important scattering mechanisms in quantized systems, referring to the specific references for detailed analysis. The AlGaAs-GaAs and Si-SiO_2 cases will be treated separately.

Concerning phonon scatterings, it will always be assumed that the 2D electrons interact with bulk phonon modes. Two dimensional effects associated with the phonon dispersion in AlGaAs-GaAs heterostructures have been considered to some extent by Riddoch and Ridley (1985) [129], in connection with the slab modes, and by Sawaki (1987) [130]. More work is still needed for a completely two-dimensional treatment of the electron-phonon interaction. The preliminary results indicate significant deviations in the scattering rates for well widths less than about 100 Å.

The presence of phonon perturbations in quantum wells has been considered by Lugli and Goodnick [131] and by Cai et al. [132].

a) AlGaAs-GaAs

a.1) Polar Optical Scattering

Assuming bulk screening of longitudinal modes, the matrix element for polar optical scattering is given by [133, 134]

$$|\langle \mathbf{k} \pm \mathbf{q}, j|H|\mathbf{k}, i\rangle|^2 = \frac{2\pi\hbar^2 eE_0(q^2+q_z^2)}{Vm(q^2+q_z^2+q_s^2)}\begin{bmatrix} n_{\omega_0} \\ n_{\omega_0}+1 \end{bmatrix}|G_{ij}(q_z)|^2 \tag{2.10.4}$$

where q and q_z are the parallel and normal components of the phonon wave vector, q_s is the 3D inverse screening length, m is the effective mass, i and j denote the initial and final subband indices, and the effective field eE_0 has the same expression as for the bulk case. The overlap integral $|G_{ij}(q_z)|^2$ is given by

$$|G_{ij}(q_z)|^2 = \int_{-\infty}^{\infty} dz \int_{-\infty}^{\infty} dz' \, \rho_{ij}(z)\rho_{ij}(z')e^{iq_z(z-z')}, \tag{2.10.5a}$$

$$\rho_{ij}(z) = \zeta_i^*(z)\zeta_i(z) \tag{2.10.5b}$$

between initial and final subbands. This function, which is related to the momentum uncertainty due to the spatial confinement of the electrons, is peaked for q_z corresponding to the parabolic momentum associated with the initial and final state subband energies. By converting the sum over q_z to an integral, the q_z dependence may be integrated analytically to yield a smooth function which is better suited for numerical evaluation of the scattering rate. The total scattering rate turns out to be [134]

$$\Gamma_{ij}^{op}(k) = \frac{eE_0}{2\hbar}\left\{n_{\omega_0}\int_0^{2\pi} d\theta \frac{H_{ij}(q_+)}{\sqrt{q_+^2+q_s^2}} + (n_{\omega_0}+1) \times \int_0^{2\pi} d\theta \frac{H_{ij}(q_-)}{\sqrt{q_-^2+q_s^2}}\right\}. \tag{2.10.6}$$

The function H_{ij} is defined from the integration over the envelope functions

$$H_{ij}(a) = \int_{-\infty}^{\infty} dz \int_{-\infty}^{\infty} dz' \, \rho_{ij}(z)\rho_{ij}(z')I(q, z, z') \tag{2.10.7}$$

with

$$I(q, z, z') = \frac{\pi e^{-(q^2+q_s^2)^{1/2}|z-z'|}}{\sqrt{q^2+q_s^2}}\left[1 - \frac{|z-z'|q_s^2}{2(q^2+q_s^2)^{1/2}} - \frac{q_s^2}{2(q^2+q_s^2)}\right]. \tag{2.10.8}$$

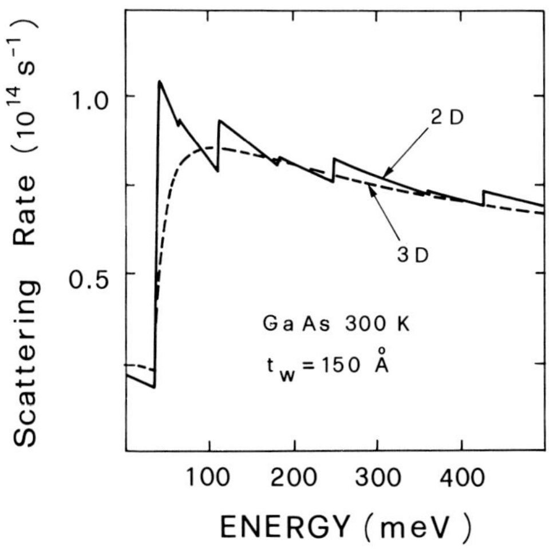

Fig. 2.46. Scattering rate of 2D electrons in GaAs interacting with polar optical phonons at 300 K, compared with the bulk case

The + and − signs on q in (2.10.6) refer to absorption and emission respectively of the scattered wave vector given by

$$q = |\mathbf{k} - \mathbf{k}'| = \left[2k^2 \pm \frac{2\omega_{ij}^* m}{\hbar} - 2k \left[k^2 \pm \frac{2\omega_{ij}^* m}{\hbar} \right]^{1/2} \cos\theta \right]^{1/2}, \tag{2.10.9a}$$

$$\hbar\omega_{ij}^* = \hbar\omega_0 \pm (E_i - E_j), \tag{2.10.9b}$$

where $\hbar\omega_{ij}^*$ serves as an effective phonon energy which depends on the initial and final subband energies E_i and E_j. The total scattering rate for the bottom subband is shown for a 150 Å single squared well in Fig. 2.46 (a value of $q_s = 0$ is used here). For comparison the 3D rate is plotted as well. As seen in the figure the 2D rate is a piecewise approximation of the 3D rate due to the discontinuities in the 2D density of states at various subband energies.

a.2) Acoustic Scattering

Deformation-potential acoustic-phonon scattering is treated similarly to the optical-phonon scattering. The problem is, however, simpler in this case, because the matrix element is independent of q. The scattering rate is given by [128].

$$\Gamma_{ij}^{ac} = \frac{m K_B T_0 C_1^2}{\hbar^3 \rho u_l^2} \int \zeta_i^2(z) \zeta_j^2(z)\, dz, \tag{2.10.10}$$

where we have used the same notations introduced for the bulk case. The acoustic-phonon scattering rates are independent of energy in the allowed energy range, and are generally much smaller than the polar optical-phonon scattering rates.

a.3) Intervalley Scattering

Assuming that the satellite valleys are quantized, the matrix element for intervalley scattering is given by [134]

$$|H_{\mathbf{kk'}}^{ij}|_i^2 = \frac{\hbar D_i^2}{2\rho A\omega_i}\begin{bmatrix} n_{\omega_i} \\ n_{\omega_i} + 1 \end{bmatrix}. \tag{2.10.11}$$

The overlap integral I_i is given by

$$I_i = \int_{-\infty}^{\infty} dz\, |\zeta_i^{v_i}(z)|^2 |\zeta_j^{v_f}(z)|^2 \tag{2.10.12}$$

where v_i and v_f refer to the initial and final valley envelope functions. The total scattering rate is thus given by

$$\Gamma_{ij}^{iv} = \sum_{v_f} m_f Z_f \frac{E_i^2 I_i}{2\rho\omega_i\hbar^2}\begin{bmatrix} n_{\omega_i} \\ n_{\omega_i} + 1 \end{bmatrix} \tag{2.10.13}$$

with m_f being the effective mass in the final state valley and the sum is over all the final state valleys.

a.4) Electron-Electron Scattering

Following the discussion of section 2.4, the e-e interaction in two dimensional systems can be studied adopting the Born approximation. Using the static screening limit in the random phase approximation, the square of the matrix element is given by [134]

$$|H^{ee}|^2 = \frac{4\pi^2 e^4\, \delta[\mathbf{k}' + \mathbf{k}_o' - \mathbf{k} - \mathbf{k}_o]}{A\kappa^2(q + q_0)^2}|F_{ijmn}(q)|^2, \tag{2.10.14}$$

which describes the scattering between an electron in the well with wave vector $\mathbf{k}$ in subband i and a second electron with wave vector $\mathbf{k}_o$ in subband j. The final states are indicated respectively by $\mathbf{k}'$ and m for the first electron and $\mathbf{k}_o'$, and n for the second electron. In Eq. (2.10.14) q_o is the two dimensional inverse screening length, A the normalization area, $q = |\mathbf{k} - \mathbf{k}'|$ and the form factor is given by

$$F_{ijmn}(q) = \int_{-\infty}^{\infty} dz \int_{-\infty}^{\infty} dz'\, \zeta_i(z)\zeta_j(z')\zeta_m^*(z)\zeta_n^*(z'). \tag{2.10.15}$$

In the Born approximation, the total scattering rate out of the first electron in an initial state $\mathbf{k}$, i to a final state in subband m is then calculated from the sum over the final states

$$\Gamma_{im}(\mathbf{k}) = \frac{2\pi e^4}{\hbar\kappa A} \cdot \sum_{\mathbf{k}_o, j, n}' f_j(\mathbf{k}_o) \int d\mathbf{k}_o' \frac{|F_{ijmn}(q)|^2}{(q+q_o)^2} \delta(E_i - E_f). \tag{2.10.16}$$

The prime summation indicates that the sum is only over electrons with antiparallel spin. Thus we neglect scattering with carriers at parallel spin which is lower in magnitude than the scattering caused by antiparallel electrons due to exchange considerations.

b) $Si\text{-}SiO_2$

b.1) Nonpolar Optical Phonons

As pointed out earlier, the many-valley Si structure is reflected directly into the subband quantization of the 2DEG. Following the work of Ferry [135] (see also [136, 137]), the interactions between the two equivalent valleys in E_0 and E_1 subbands involve g-type phonons. Interactions between the four valleys of subband E_0' involve both g- and f-type phonons, and the inter-subband scattering between the two valleys of the E_0, E_1 subbands and four valleys of the E_0' subband involve f-type phonons.
The inter-valley scattering rate from an a-valley (i-subband) to any other b-valley (j-subband) is [135]

$$\Gamma_{ia}^{op} = \frac{1}{\rho\hbar 2\omega_0} \sum_{j,b} I_{ij} D_{ij}^2 \frac{m_{1b} m_{2b}}{W_j} \begin{bmatrix} n_{\omega_0} \\ n_{\omega_0} + 1 \end{bmatrix} u(E - E_j \pm \hbar\omega_0), \tag{2.10.17}$$

where D_{ij} is the coupling constant, E_j is the energy level of the jth subband, I_{ij} is the square of the wave function overlap integral between the electron in the initial subband i and valley a and the final subband j in valley b. The $+\ -$ signs refer, respectively, to absorption and emission processes, and u is the unit step function. W_j is the level width of the jth subband for the bth set of valleys, m_{1b} and m_{2b} are the effective masses of electrons in the two directions parallel to the surface.

b.2) Acoustic Phonons

Intra-valley scattering involves long wavelength acoustic phonons. In the elastic and equipartition approximation, the scattering rate is then given by [135]

$$\Gamma_j^{ac} = \frac{(m_1 m_2)^{1/2} C_1 K_B T}{\hbar^3 \rho_m W_j u_l^2}, \tag{2.10.18}$$

b.3) Ionized Impurity Scattering

For ionized impurity and trapped interface charge scattering rates the phenomenological approach of Ref. [138] gives

$$\Gamma^{im} = \frac{GN_I e^4}{16^2 \hbar \varepsilon_m^2} \cdot \frac{1}{\varepsilon}, \tag{2.10.19}$$

where $\kappa_m = (\kappa_{\text{oxide}} + \kappa_{\text{Si}})/2$ is the effective dielectric constant at the interface, N_I is the number of oxide charges per unit area, and G is an adjusting constant which accounts for screening, position correlation between oxide charges and many other effects.

b.4) Surface Roughness

As pointed out earlier, surface-roughness scattering arises from the diffraction of the electron wavefunction off the asperities of the surface. While it is possible to give classical approaches (as we will see in chapter 5), a more rigorous treatment has to account for the quantization in the perpendicular direction. By considering the potential fluctuations induced by the surface roughness as a scattering source, the following scattering rate is found [139]:

$$\Gamma^{sr} = \frac{m\Delta^2 L^2}{2\hbar^3} \left(\frac{e^2}{\kappa_m}\right)^2 (N_s + N_{\text{depl}})^2 f(kL), \tag{2.10.20}$$

where Δ is the rms height parameter, L is the correlation length of roughness, N_s is the inversion carrier density (per unit area), N_{depl} is the substrate doping density. The function $f(kL)$ is a slowly varying function given as

$$f(kL) = \pi \exp\left(\frac{k^2 L^2}{2}\right) \left| I_0(k^2 L^2) - I_1\left(\frac{k^2 L^2}{2}\right) \right|, \tag{2.10.21}$$

where k is the two-dimensional carrier wave vector, and I_0 and I_1 are modified Bessel functions of the first kind.

References

[1] Ashcroft, N. W., Mermin, N. D.: Solid State Physics. Philadelphia: Saunders. 1976.
[2] Kohn, W., Luttinger, J. M.: Phys. Rev. *108*, 590 (1957).
[3] Kadanoff L. P., Baym, G.: Quantum Statistical Mechanics. New York: Benjamin. 1962.
[4] Barker J. R., Ferry, D. K.: Phys. Rev. Lett. *42*, 1779 (1979).
[5] Barker, J. R.: Fundamental Aspects of Quantum Transport. In: Handbook on Semiconductors (Moss, T. S., ed.), Vol. 1. North-Holland. 1982.
[6] Reggiani, L.: Physica *134B*, 123 (1985) and references therein.
[7] Mahan, G. D.: Physics Reports (Review Section Letters) *145*, 251 (1987).

[8] Jacoboni, C., Ferry, D. K. (eds.): Quantum Transport in Semiconductors. Plenum (to be published).
[9] Kane, E. O.: J. Phys. Chem. Solids *1*, 82 (1956).
[10] Ottaviani, G., Reggiani, L., Canali, C., Nava, F., Alberigi-Quaranta, A.: Phys. Rev. *B12*, 3318 (1975).
[11] Dresselhaus, G., Kipp, A. F., Kittel, C.: Phys. Rev. *98*, 368 (1955).
[12] Costato, M., Gagliani, G., Jacoboni, C., Reggiani, L.: J. Phys. Chem. Solids *35*, 1605 (1974).
[13] Bosi, S., Jacoboni, C., Reggiani, L.: J. Phys. *C12*, 1525 (1979).
[14] Smith, R. A.: Semiconductors. Cambridge: Cambridge University Press 1964.
[15] Conwell, E. M., Vassel, M. O.: Phys. Rev. *166*, 797 (1968).
[16] Fawcett, W., Boardman, D. A., Swain, S.: J. Phys. Chem. Solids *31*, 1963 (1970).
[17] Paige, E. G. S.: The Electrical Conductivity of Germanium (Gibson, A. F., Burgess, R. E., eds.). London: Heywood. 1964.
[18] Jacoboni, C., Reggiani, L.: Rev. Mod. Phys. *55*, 645 (1983).
[19] Gagliani, G., Reggiani, L.: Nuovo Cimento *B30*, 207 (1975).
[20] Jacoboni, C., Nava, F., Canali, C., Ottaviani, G.: Phys. Rev. *B24*, 1014 (1981).
[21] Herring, C., Vogt, E.: Phys. Rev. *101*, 944 (1956).
[22] Jacoboni, C., Reggiani, L.: Advances in Physics *28*, 493 (1979).
[23] Chelikowsky, J. R., Cohen, M. L.: Phys. Rev *B14*, 556 (1976).
[24] Ivey, J. L., Mieher, R. L.: Phys. Rev. Letters *29*, 176 (1972).
[25] Jacoboni, C., Minder, R., Majni, G.: J. Phys. Chem. Solids *36*, 1129 (1975).
[26] Littlejohn, M. A., Houser, J. R., Glisson, T. H.: J. Appl. Phys. *48*, 4587 (1977).
[27] Eastman, L. F.: Physics Today, October 1986, p. 77.
[28] Milnes, A. G.: Solid-State Electr. *29*, 99 (1986).
[29] Hartstein, A., Ning, T. H., Fowler, A. B.: Surf. Sci. *58*, 178 (1976).
[30] Hauser, J. R., Littlejohn, M. A., Glisson, T. H.: Appl. Phys. Lett. *28*, 458 (1976).
[31] Bir, G. L., Pikus, G. E.: Symmetry ad Strain-Induced Effects in Semiconductors. (Translated by Shelnitz, P., translation edited by Louvish, D.) New York: Wiley. 1974.
[32] Froehlich, H.: Proc. R. Soc. (London) *A188*, 521 and 532 (1947).
[33] Stratton, R.: Proc. R. Soc. (London) *A246*, 406 (1958).
[34] Eaves, L., Hoult, R. A., Stradling, R. A., Tidey, R. J., Portal, J. C., Askenazy, S.: J. Phys. *C8*, 1034 (1975).
[35] Fawcett, W., Paige, E. G. S.: J. Phys. *C4*, 1801 (1971).
[36] Alberigi-Quaranta, A., Jacoboni, C., Ottaviani, G.: Riv. Nuovo Cimento *1*, 445 (1971).
[37] Norton, P., Levinstein, H.: Phys. Rev. *B6*, 470 (1972).
[38] Asche, M., Boichenko, B. L., Bondar, V. M., Sarbei, O. G.: Phys. Stat. Sol. (b) *44*, 173 (1971).
[39] Dienys, V., Kancleris, Z.: Phys. Stat. Sol. (b) *67*, 317 (1975).
[40] Jacoboni, C.: Proc., 13th Intern. Conf. Phys. Semic. (Fumi, G., ed.), p. 1195. Roma: Marves. 1976.
[41] Nash, J. C., Holm-Kennedy, J. W.: Appl. Phys. Lett. *24*, 139 (1974).
[42] Lugli, P., Ferry, D. K.: Physica *B129*, 532 (1985); IEEE El. Dev. Lett. *EDL6*, 25 (1985).
[43] Ziman, J. M.: Principles of the Theory of Solids. Cambridge: Cambridge University Press. 1972.
[44] Reggiani, L., Calandra, C.: Phys. Lett. *A43*, 339 (1973).
[45] Wiley, J. D.: Phys. Rev. *B4*, 2485 (1971).
[46] Kittel, C.: Quantum Theory of Solids. New York: Wiley. 1963.
[47] Harrison, W. A.: Phys. Rev. *104*, 1281 (1956).
[48] Bir, G. L., Pikus, G. E.: Fiz. Tverd. Tela (Leningrad) *2*, 2287 (1961) [Sov. Phys. Solid State *2*, 2039 (1961)].
[49] Ziman, J. M.: Electrons and Phonons. Oxford: Oxford University Press. 1960.
[50] Lawaetz, P.: Phys. Rev. *183*, 730 (1969).
[51] Vogl, P.: Phys. Rev. *13*, 694 (1976).

[52] Poetz, W., Vogl, P.: Phys. Rev. *B24*, 2025 (1981).
[53] Poetz, W., Kocevar, P.: Phys. Rev. *B28*, 7040 (1983).
[54] Bordone, P., Jacoboni, C., Lugli, P., Reggiani, L., Kocevar, P.: J. Appl. Phys. *61*, 1460 (1987).
[55] Lugli, P., Jacoboni, C., Reggiani, L., Kocevar, P.: Appl. Phys. Lett. *50*, 1251 (1987).
[56] Costato, M., Reggiani, L.: Nuovo Cimento Lett. *4*, 1179 (1970).
[57] Conwell, E. M.: High Field Transport in Semiconductors. (Solid State Phys., Suppl. 9.) New York: Academic Press. 1967.
[58] Tiersten, M.: IBM J. Res. Dev. *5*, 122 (1961).
[59] Tiersten, M.: J. Phys. Chem. Solids *25*, 1151 (1964).
[60] Reggiani, L., Canali, C., Nava, F., Ottaviani, G.: Phys. Rev. *B16*, 2781 (1977).
[61] Lawaetz, P.: Doctoral Dissertation, University of Lyngby, 1967.
[62] Costato, M., Reggiani, L.: Phys. Status Solidi *B58*, 47 and 471 (1973).
[63] Streitwolf, H. W.: Phys. Status Solidi *B37*, K47 (1970).
[64] Lax, M., Birman, J. L.: Phys. Status Solidi *B49*, K153 (1972).
[65] Asche, A., Sarbei, O. G.: Phys. Status Solidi *B103*, 11 (1981).
[66] Paige, E. G. S.: The Electrical Conductivity of Germanium (Gibson, A. F., Burgess, R. E., eds.). London: Heywood. 1964.
[67] Birman, J., Lax, M., Loudom, R.: Phys. Rev. *145*, 620 (1966).
[68] Eaves, L., Hoult, R. A., Stradling, R. A., Tidey, R. J., Portal, J. C., Askenazy, S.: J. Phys. *C8*, 1034 (1975).
[69] Ferry, D. K.: Phys. Rev. *B14*, 1605 (1976).
[70] Ridley, B. K.: Quantum Processes in Semiconductors. Oxford: Clarendon. 1982.
[71] Froehlich, H.: Proc. R. Soc. (London) *A188*, 521 (1947).
[72] Conwell, E. M., Weisskopf, V. F.: Phys. Rev. *77*, 388 (1950).
[73] Brooks, H., Herring, C.: Phys. Rev. *83*, 879 (1951).
[74] Chattopadhyay, D., Queisser, H. J.: Rev. Mod. Phys. *53*, 745 (1981).
[75] Reggiani, L.: Phys. Rev. *B17*, 2800 (1978).
[76] Bohm, D., Pines, D.: Phys. Rev. *92*, 609 (1953).
[77] Madelung, O.: Introduction to Solid State Physics. Berlin-Heidelberg-New York: Springer. 1978.
[78] Lugli, P.: Ph. D. Dissertation, Colorado State University, 1985.
[79] Kumenov, S. E., Perel, V. I.: Sov. Phys. Semicon. *16*, 1982 (1982).
[80] Davidov, B. I.: In: Plasma Physics and the Problem of Controlled Thermonuclear Reactions, Vol. 1. AN SSR. 1958.
[81] Ehrenreich, H.: J. Phys. Chem. Solids *8*, 1996 (1965).
[82] Varga, B. B.: Phys. Rev. *137*, 1896 (1965).
[83] Mooradian, A., Wright, G. B.: Phys. Rev. Lett. *16*, 999 (1966).
[84] Kim, M. E., Das, A., Centuria, S. D.: Phys. Rev. *B18* 6890 (1978).
[85] Wolf, P. A.: In: Proc., Int. Conf. Light Scatt. Spectra Solids. Berlin-Heidelberg-New York: Springer. 1969.
[86] Asche, M., Sarbei, O. G.: Phys. Status Solidi *B126*, 607 (1984).
[87] Osman, M. A.: Ph. D. Dissertation, Arizona State University, 1986.
[88] Reif, F.: Fundamentals of Statistical and Thermal Physics. New York: McGraw-Hill. 1965.
[89] Budd, H.: In: Proc., Int. Conf. Semiconductors, Kyoto. J. Phys. Soc. Japan Suppl. *21*, 420 (1966).
[90] Price, P. J.: In: Proc., 9th Int. Conf. Phys. Semiconductors (Ryvkin, S. M., ed.), p. 753. Leningrad: Nauka. 1968.
[91] Rees, H. D.: J. Phys. Chem. Solids *30*, 643 (1969).
[92] Kurosawa, T.: In: Proc., Int. Conf. Phys. Semiconductors, Kyoto. J. Phys. Soc. Jpn. Suppl. *21*, 424 (1966).
[93] Price, P. J.: Semiconductors and Semimetals, p. 249. New York: Academic Press. 1979.
[94] Ehrenreich, H.: J. Phys. Chem. Solids *2*, 131 (1957) and *12*, 97 (1959); Phys. Rev. *120*, 1951 (1960).

[95] Long, D., Meyers, J.: Phys. Rev. *115*, 1107 (1959).
[96] Canali, C., Jacoboni, C., Nava, F., Ottaviani, G., Alberigi-Quaranta, A.: Phys. Rev. *B12*, 2265 (1975).
[97] Fletcher, K., Butcher, P. N.: J. Phys. *C5*, 212 (1972).
[98] Costato, M., Reggiani, L.: Nuovo Cimento *68B*, 64 (1970).
[99] Costato, M., Jacoboni, C., Reggiani, L.: Phys. Stat. Sol. (b) *52*, 461 (1972).
[100] Jacoboni, C., Gagliani, G., Reggiani, L., Turci, O.: Solid State Electron. *21*, 315 (1978).
[101] Kittel, C.: Elementary Statistical Physics. New York: Wiley. 1958.
[102] Gunn, J. B.: Int. J. Sci. Technol. *46*, 43 (1965).
[103] Ruch, J. G., Kino, G. S.: Phys. Rev. *174*, 921 (1968).
[104] Houston, P. A., Evans, A. G. R.: Solid State Electron. *20*, 197 (1977).
[105] Pozhela, J., Reklaitis, A.: Solid State Commun. *27*, 1073 (1978).
[106] Brunetti, R., Jacoboni, C., Nava, F., Reggiani, L., Bosman, G., Zijlstra, R. J. J.: J. Appl. Phys. *52*, 6713 (1981).
[107] Fawcett, W., Rees, H. D.: Phys. Lett. *11*, 731 (1969).
[108] Ohmi, T., Hasuo, S.: Proc., 10th Int. Conf. Phys. Semiconductors, Cambridge, Massachusetts, p. 60, 1970.
[109] Reggiani, L., Vaissière, J. C., Nougier, J. P., Gasquet, D.: In: Proc., 3rd Int. Conf. Hot Carriers in Semiconductors. J. Phys. (Paris) Colloq. *42*, C7-357 (1981c).
[110] Sze, S. M.: Physics of Semiconductor Devices, 2nd edn. New York: Wiley. 1981.
[111] Chang, L. L.: Solid State Electron. *8*, 721 (1965)
[112] Anderson, R. L.: Solid State Electron. *5*, 341 (1962).
[113] Milnes, A. G., Feucht, D. L.: Heterojunctions and Metal-Semiconductor Junctions. New York: Academic Press. 1972.
[114] Gossard, A. C.: IEEE J. Quantum Electron. *QE22*, 1649 (1986).
[115] Störmer, H. L.: Surf. Sci. *132*, 519 (1983).
[116] Solomon, P. M., Morkoc, H.: IEEE Trans. El. Dev. *ED31*, 1015 (1984).
[117] Dingle, R., Störmer, H. L., Gossard, A. C., Wiegmann, W.: Appl. Phys. Lett. *33*, 655 (1978).
[118] Dingle, R.: IEEE Trans. El. Dev. *ED31*, 1662 (1984).
[119] Hess, K.: Solid State Electron. *31*, 319 (1988); and references therein.
[120] Muller, P. S., Kamins, T. I.: Device Electronics for Integrated Circuits, 2nd edn. New York: Wiley. 1986.
[121] Ando, T., Fowler, A. B., Stern, F.: Rev. Mod. Phys. *54*, 437 (1982).
[122] Stern, F., Howard, W. E.: Phys. Rev. *163*, 816 (1967).
[123] Ferry, D. K., Hess, K., Vogl, P.: In: VLSI Electronics: Microstructure Science, Vol. 2, p. 67. New York: Academic Press. 1981.
[124] Hess, K., Vogl, P.: Solid State Comm. *30*, 807 (1979).
[125] Harstein A., Ning, T. H., Fowler, A. B.: Surf. Sci. *58*, 178 (1976).
[126] Stern, F., Das Sarma, S.: Phys. Rev. *B30*, 840 (1984).
[127] Ando, T.: J. Phys. Soc. Jap. *51*, 3893 (1982).
[128] Yokoyama, K., Hess, K.: Phys. Rev. *B33*, 5595 (1986).
[129] Riddoch, F. A., Ridley, B. K.: Physica *134B*, 342 (1985).
[130] Sawaki, N.: J. Phys. *C19*, 4965 (1986).
[131] Lugli, P., Goodnick, S. M.: Phys. Rev. Lett. *59*, 716 (1987).
[132] Cai, W., Marchetti, C. M., Lax, M.: Phys. Rev. *B35*, 1369 (1987).
[133] Price, P. J.: Ann. Phys. *133*, 217 (1981).
[134] Goodnick, S. M., Lugli, P.: Phys. Rev. *37*, 2578 (1988).
[135] Ferry, D. K.: Phys. Rev. *B14*, 5364 (1976).
[136] Hao, C., Zimmermann, J., Charef, M., Fauquembergue, F., Constant, E.: Solid State Electr. *28*, 733 (1985).
[137] Terashima, K., Hamaguchi, C., Taniguchi, K.: Superl. and Microstr. *1*, 15 (1985).
[138] Ning, T. H., Sah, C. T.: Phys. Rev. *B6*, 4605 (1972).
[139] Ferry, D. K.: Thin Sol. Films *56*, 243 (1979).

3 The Monte Carlo Simulation

3.1 Fundamentals

The Monte Carlo method, as applied to charge transport in semiconductors, consists of a simulation of the motion of one or more electrons inside the crystal, subject to the action of external forces due to applied electric and magnetic fields and of given scattering mechanisms [1–4]. The durations of the carrier free flights between two successive collisions and the scattering events involved in the simulation are selected stochastically in accordance with some given probabilities describing the microscopic processes. As a consequence, any Monte Carlo method relies on the generation of a sequence of random numbers with given distribution probabilities. Such a technique takes advantage of the fact that nowadays any computer generates sequences of random numbers evenly distributed between 0 and 1 at a sufficiently fast rate. The generation of random variables with any distribution, starting from random numbers evely distributed between 0 and 1, is discussed in Appendix B.

When the purpose of the analysis is the investigation of a steady-state, homogeneous phenomenon, it is sufficient in general to simulate the motion of one single electron; from ergodicity we may assume that a sufficiently long path of this sample electron will give information on the behaviour of the entire electron gas. When, on the contrary, the transport process under investigation is not homogeneous or is not stationary, then it is necessary to simulate a large number of electrons and follow them in their dynamic histories in order to obtain the desired information on the process of interest.

Let us summarize here the structure of a typical Monte Carlo program suited to the simulation of a stationary, homogeneous, transport process. The details of each step of the procedure will be given in the following sections. For the sake of simplicity we shall refer to the case of electrons in a simple semiconductor subject to an external electric field **E**. The simulation starts with one electron in given initial conditions with wavevector $\mathbf{k}_0$; then the duration of the first free flight is chosen with a probability distribution determined by the scattering probabilities. During the free flight the external forces are made to act according to the relation

$$\hbar\dot{\mathbf{k}} = -e\mathbf{E}. \tag{3.1.1}$$

In this part of the simulation all quantities of interest, velocity, energy, etc., are recorded. Then a scattering mechanism is chosen as responsible for the end of the free flight, according to the relative probabilities of all possible scattering mechanisms. From the differential cross section of this mechanism a new **k** state after scattering is randomly chosen as initial state of the new free flight, and the entire process is iteratively repeated. The results of the calculation become more and more precise as the simulation goes on, and the simulation ends when the quantities of interest are known with the desired precision. A simple way to determine this precision is based on the separation of the entire history into a number of successive subhistories and is described in section 3.7.3.
Figure 3.1. shows a flowchart of a simple Monte Carlo program suited for the simulation of a stationary, homogeneous transport process.
Figure 3.2 illustrates the principles of the method by showing the simulation

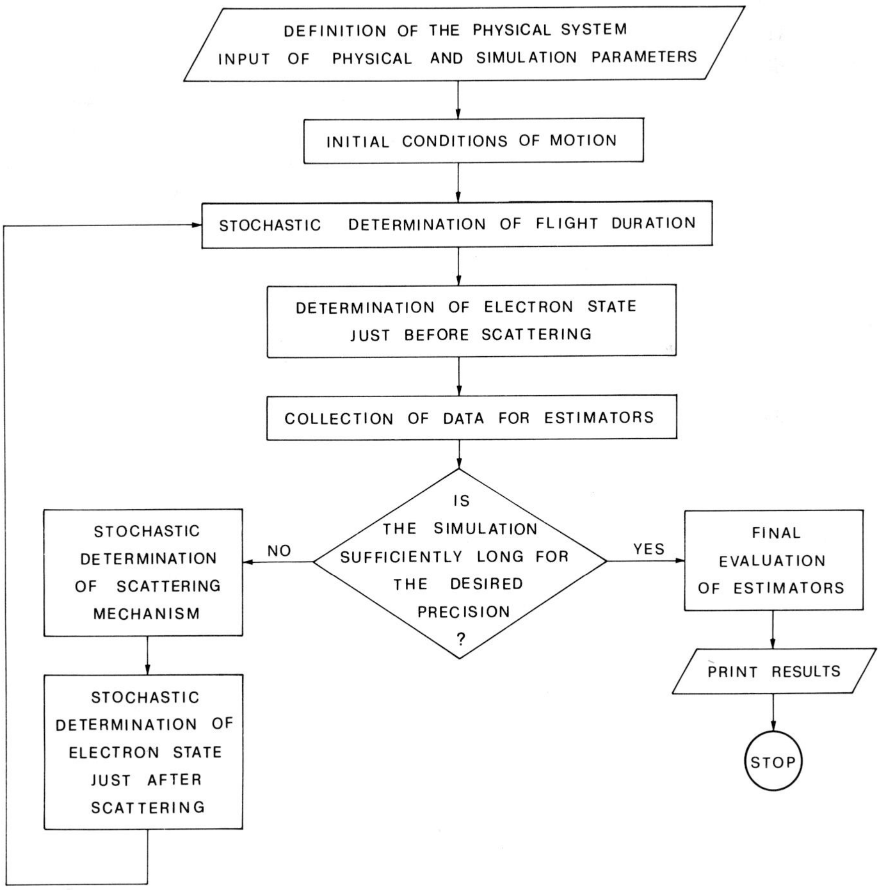

Fig. 3.1. Flowchart of a typical Monte Carlo program [4]

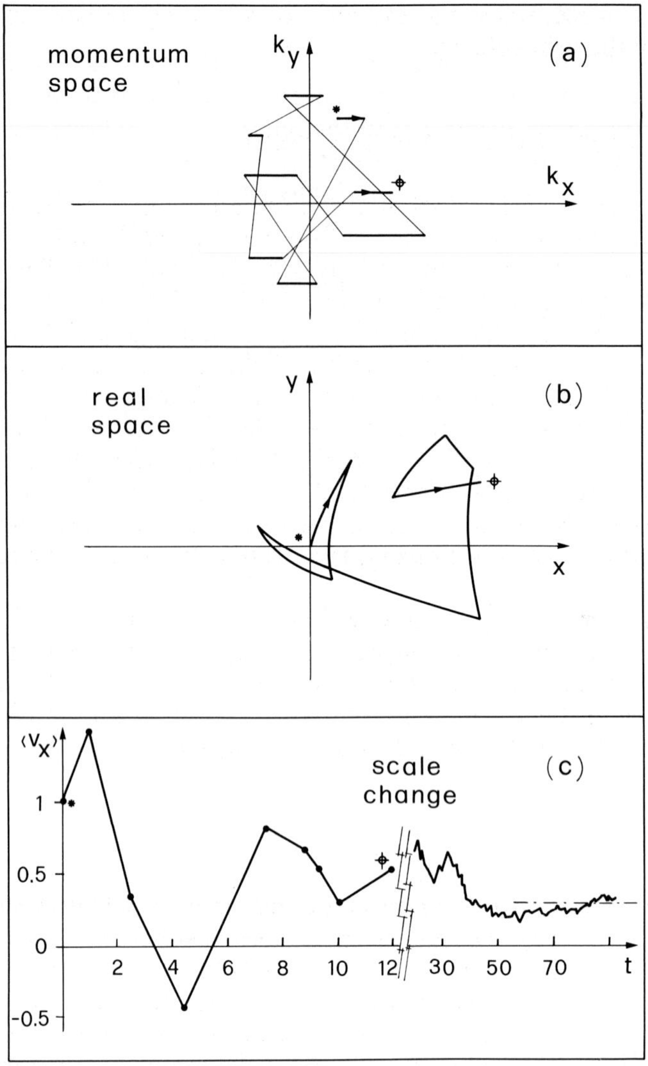

Fig. 3.2. The principles of the Monte Carlo method. *a* Simulation in *k*-space of the particle, subject to an accelerating force along the positive *x* direction; the heavy segments represent the change of *k* due to the force during the free flights, separated by discontinuous variations due to scattering processes. *b* The corresponding path of the particle in real space. *c* Average velocity of the particle as a function of simulation time. The left section of the curve corresponds to the simulation illustrated in the upper parts of the figure. The horizontal dot-dashed line represents the limiting result of a very long simulation. Special symbols indicate corresponding points in the three parts of the figure. Arbitrary units are used

in **k** space and real space and the effect of collecting statistics in the determination of the drift velocity.

3.2 Definition of the Physical System

The starting point of the program is the definition of the physical system of interest, including the parameters of the material and the values of physical quantities, such as lattice temperature T and electric field. The dispersion relation $\varepsilon(\mathbf{k})$ is usually given in the form of Eqs. (2.2.3), (2.2.4) or (2.2.5), in conjunction with the non-parabolicity correction of Eq. (2.2.18). It is also possible to use the full band structure $\varepsilon(\mathbf{k})$, calculated for example by pseudopotential methods and tabulated in a 3-dimensional grid [5, 6]. It is worth noting that, among the parameters that characterize the material, the least known are the coupling strengths describing the interactions of the electron with the lattice and/or extrinsic defects inside the crystal.

At this level we also define the parameters that control the simulation, such as the duration of each subhistory, the desired precision of the results, and so on.

The next step in the program is a preliminary calculation of each scattering rate as a function of electron energy. This step will provide information on the maximum value of these functions, which will be useful for optimizing the efficiency of the simulation. Finally, all cumulative quantities must be put at zero in this preliminary part of the program.

3.3 Initial Conditions

In the case under consideration at the moment, in which a steady-state situation is simulated, the time of simulation must be long enough that the initial conditions of the electron motion do not influence the final results. The choice of a "good" time of simulation is a compromise between the need for ergodicity ($t \to \infty$) and the request to save computer time. When a highly improbable initial value for the electron wave vector **k** is chosen, the first part of the simulation can be strongly influenced by this inappropriate choice.

If a very high electric field is applied, and an energy of the order of $K_B T$ is initially given to the electron, this energy will be much lower than the average energy in steady-state conditions, and during the transient it will increase towards its steady-state value. As a consequence, the electron response to the field, in terms of mobility, may be initially much higher than that of stationary conditions. This phenomenon corresponds to the transient overshoot discussed in section 2.9 and it is reflected in real space by initial free flights which are much longer than in stationary conditions because of an abnormally large momentum relaxation time (see Fig. 3.3).

The longer the simulation time, the less influence the initial conditions will

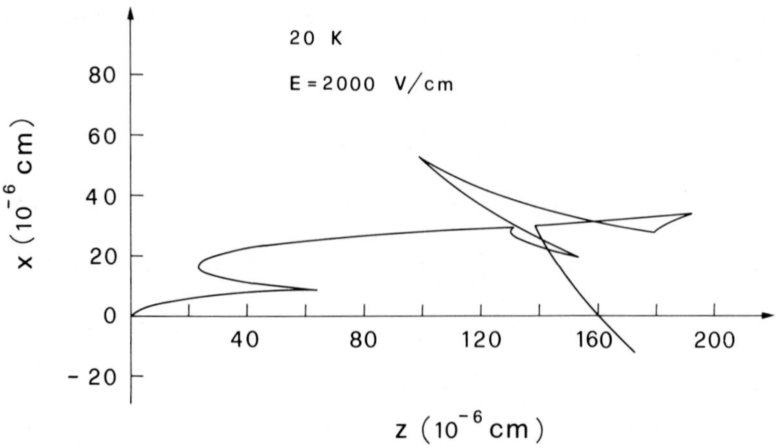

Fig. 3.3. Projection in the xz plane of real space of the trajectory of a carrier starting with thermal energy, under the influence of a high electric field. The well oriented initial free trajectories reflect values of the carrier scattering times longer than those under steady-state conditions, as a consequence of the lower energy

have on the average results; however, in order to avoid the undesirable effects of an inappropriate initial choice and to obtain a better convergence, the elimination of the first part of the simulation from the statistics may be advantageous. When the simulation is divided into many subhistories, better convergence to steady state can be achieved by taking the initial state of each new subhistory as equal to the final state of the previous one. In this way, only the initial condition of the first subhistory will influence the final results in a biased way.

On the other hand, when a simulation is made to study a transient phenomenon and/or a transport process in a nonhomogeneous system (as in the case of device simulation), then it is necessary to simulate many electrons separately; in this case the distribution of the initial electron states for the particular physical situation under investigation must be taken into account and the initial transient becomes an essential part of the problem to be solved.

3.4 The Free Flight, Self Scattering

The electron wave vector $\mathbf{k}$ changes continuously during free flight because of the applied field, according to Eq. (3.1.1). Thus if $P[\mathbf{k}(t)]\,dt$ is the probability that an electron in the state $\mathbf{k}$ suffers a collision during the time interval dt, the probability that an electron which had a collision at time $t = 0$ has not yet undergone another collision after a time t is

$$\exp\left[-\int_0^t P[\mathbf{k}(t')]\,dt'\right]. \tag{3.4.1}$$

Consequently, the probability $\mathscr{P}(t)$ that the electron will suffer its next collision during dt around t is given by

$$\mathscr{P}(t)\,dt = P[\mathbf{k}(t)]\exp\left[-\int_0^t P[\mathbf{k}(t')]\,dt'\right]dt. \tag{3.4.2}$$

Because of the complexity of the integral at the exponent, it is impractical to generate stochastic free flights with the distribution of Eq. (3.4.2), starting from evenly distributed random numbers r and applying the techniques discussed in Appendix B: with this approach an integral equation would need to be solved for each scattering event [1]. Rees [7, 8] has devised a very simple method of overcoming this difficulty. If $\Gamma \equiv 1/\tau_0$ is the maximum value of $P(\mathbf{k})$ in the region of $\mathbf{k}$ space of interest, a new fictitious "self-scattering" is introduced such that the total scattering probability, including this self-scattering, is constant and equal to Γ. If the carrier undergoes such a self-scattering, its state $\mathbf{k}'$ after the collision is taken to be equal to its state $\mathbf{k}$ before the collision, so that in practice the electron path continues unperturbed as if no scattering at all had occurred. Generally it is sufficient that Γ be not less than the maximum value of $P(\mathbf{k})$; furthermore, as we shall see below, Γ can be extended to a convenient function of energy.
Now, with a constant $P(\mathbf{k}) = \tau_0^{-1}$, Eq. (3.4.2) reduces to

$$\mathscr{P}(t) = \frac{1}{\tau_0}\exp(-t/\tau_0), \tag{3.4.3}$$

and random numbers r can be used very simply to generate stochastic free flights t_r with the direct technique described in Appendix B. They will be given by

$$t_r = -\tau_0 \ln(1-r). \tag{3.4.4}$$

However, since r is evenly distributed between 0 and 1, so also is $(1-r)$, and in practice in place of Eq. (3.4.4) we generally use

$$t_r = -\tau_0 \ln(r). \tag{3.4.5}$$

The computer time "wasted" in taking care of self-scattering events is, in general, more than compensated for by the simplification in calculating the free-flight duration.
As regards the choice of the constant Γ, we note that in general $P(\mathbf{k})$ is a function of the electron energy ε; a suitable choice for Γ is then the maximum value of $P(\varepsilon)$ in the region of energies which are expected to be sampled during the simulation. When $P(\varepsilon)$ is not a monotonic function of ε, its maximum value must be estimated in some way, for example, with a tabulation at the beginning of the computer program. When $P(\varepsilon)$ is an increasing function of ε, as is often the case, one can take $\Gamma = P(\varepsilon_M)$, where ε_M is a maximum electron energy with negligible probability of being achieved by the carrier during the simulation. It must be observed, however, that the range of energy "visited" by the electron during the simulation is not known

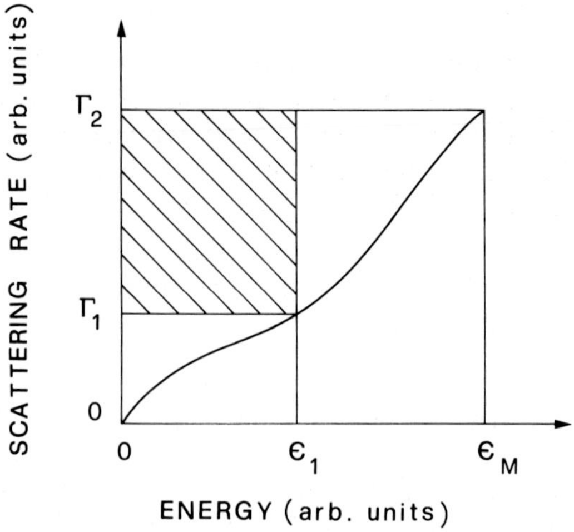

Fig. 3.4. Sketch of a two-level step-shaped total scattering rate, including self-scattering, appropriate for reducing the number of self-scattering events with respect to a single-level choice (Γ_2). Γ_1 is used for energies between 0 and ε_1. The shaded region illustrates the increased efficiency resulting from the two-level choice

at the beginning, when Γ is to be chosen. Therefore, an estimate must be made for ε_M, keeping in mind that ε_M cannot be taken too large, if one is to prevent an unnecessarily large value of Γ that would result in a waste of computer time for self-scattering events. Appropriate action must be taken if the energy during the simulation becomes larger than ε_M [4].

For a semiconductor model that contains several valleys, a different value of ε_M may be appropriately taken in each type of valley.

Sometimes the total scattering probability $P(\varepsilon)$ has a large variation around some threshold value due to a strong scattering mechanism with a given activation energy (a typical case is intervalley scattering from central to upper valleys in polar semiconductors). In this case, a single value of Γ may result in a very large number of self-scattering events at low electron energies (see Fig. 3.4). It is then possible to introduce a step-shaped scattering rate $\Gamma(\varepsilon)$ given by

$$\Gamma(\varepsilon) = \begin{cases} \Gamma_1 = \tau_1^{-1}, & \varepsilon \leqslant \varepsilon_1, \\ \Gamma_2 = \tau_2^{-1}, & \varepsilon > \varepsilon_1, \end{cases} \tag{3.4.6}$$

where ε_1 is a suitable threshold energy, and Γ_1 and Γ_2 are the maximum values of $P(\varepsilon)$ in the two corresponding energy ranges.

In using $\Gamma(\varepsilon)$ as given by Eq. (3.4.6), was should remember that during a free flight, the electron energy may exceed the value ε_1. The following two cases may occur according to the initial state of the free flight.

(i) An electron begins a free flight with an energy below ε_1. A random number r is then used to generate with Γ_1 a free flight with duration t_r given by

$$t_r = -\tau_1 \ln(r). \tag{3.4.7}$$

If at the end of this free flight, the electron energy is still less than ε_1, t_r is retained, and the simulation proceeds as usual. If, on the contrary, at the end of this free flight, the electron energy is above ε_1, t_r as given by Eq. (3.4.7) is not correct. The same random number r is then used to generate the correct duration of the free flight as follows. From Eq. (3.4.6) we have, instead of Eq. (3.4.3),

$$\mathscr{P}(t) = \begin{cases} \dfrac{1}{\tau_1} \exp(-t/\tau_1), & t < \tilde{t}, \\ \dfrac{1}{\tau_2} \exp\{-[\tilde{t}/\tau_1 + (t - \tilde{t})/\tau_2]\}, & t \geqslant \tilde{t}, \end{cases} \tag{3.4.8}$$

where $\tilde{t}$ is the time necessary for the electron to reach the energy ε_1 and can be easily evaluated from the electron dynamics. By application of the direct technique (see Appendix B), t_r is then given by

$$t_r = -\tau_2 \ln(r) + \tilde{t}(1 - \tau_2/\tau_1). \tag{3.4.9}$$

(ii) An electron begins a free flight with an energy above ε_1. The duration of this free flight can always be determined by Eq. (3.4.6) with $\tau = r_2$. In fact, even if the final energy is below ε_1, one is justified in considering the upper value Γ_2 of $\Gamma(\varepsilon)$ extended to lower energies, as long as this is done independently of the particular value of r. Of course, Γ_2 must then be consistently considered in the determination of the scattering mechanism (in particular for self-scattering) which caused the end of the free flight (see next section).
The only situation not accounted for in the above discussion is when, during a single free flight, the electron energy starting from values below ε_1 first increases to above ε_1 and then decreases to below ε_1; but this situation cannot occur with static fields and normal band structures.
The above idea can be extended to piecewise functions $\Gamma(\varepsilon)$ with more than one step. In this case, the straightforward extension of Eq. (3.4.9) is given by [9]

$$t_r = t_{i-1} + \tau_i \left[-\ln(r) - \sum_{j=1}^{i-1} (t_j - t_{j-1})/\tau_j \right], \tag{3.4.10}$$

where the subindices refer to the discretization of the energy domains. The effectiveness of the procedure depends critically on the additional computation time required by the calculation of the free flight compared to the time saved through the reduction of self-scattering events.
An attractive iterative-gamma method has been proposed [10], where the lowest possible tabulated value of $\Gamma(\varepsilon_i)$ is chosen for the flight generation, with an automatic self-adjusting procedure. A careful analysis has showed [11] that such a method might under certain circumstances produce wrong distributions.
In general, the most appropriate method strongly depends on the specific

simulation to be performed and on the type of computer available (in terms of speed and memory). However, any adopted procedure should be compared, from the point of view of saving computer time, with the so-called fast self-scattering technique (see next section).

3.5 The Scattering Process

During a free flight, the electron dynamics is governed by Eq. (3.1.1) so that at its end the electron wave vector and energy are known, and all scattering probabilities $P_i(\varepsilon)$ can be evaluated, where i indicates the ith scattering mechanism. The probability of self-scattering will be the complement to Γ of the sum of the P_i's. A mechanism must then be chosen among all those possible: given a random number r, the product $r\Gamma$ is compared with the successive sums of the P_i's, and a mechanism is selected as described in Appendix B.

If all scattering mechanisms have been tried and none of them has been selected, it means that $r\Gamma > P(\varepsilon)$, and a self-scattering occurs. A self-scattering event is therefore most time-consuming, since all P_i's must be explicity calculated. However, the process may be shortened by use of an expedient that might be called fast self-scattering. It consists of setting up a mesh of the energy range under consideration at the beginning of the simulation and then recording in a vector the maximum total scattering probability $P^{(l)}$ in each energy interval $\Delta\varepsilon^{(l)}$ (energy intervals equally distributed on a logarithmic scale may be useful). At the end of the flight, if the electron energy falls in the nth interval, before trying all P_i's separately, one compares $r\Gamma$ with $P^{(n)}$. At this stage if $r\Gamma > P^{(n)}$ then a self-scattering certainly occurs; otherwise all P_i's will be successively evaluated. Thus only when $P(\varepsilon) < r\Gamma < P^{(n)}$ does a self-scattering occur which requires the evaluation of all P_i's.

In modern computers it is in general more convenient to tabulate all scattering probabilities in a mesh of the energy axis and to use look-up tables during the simulation. This is particularly true when very sophisticated scattering probabilities are used which require a large amount of computation, as, for example, when the full three-dimensional band structure is taken into account [5].

As a final consideration we may note that some scattering mechanisms may depend on the distribution function f itself, which is the final result of the simulation, and therefore they are not known at the beginning, when the program is set up. This is the case, for example, of intercarrier scattering and of phonon or impurity scattering when screening is to be considered, since the screening depends on the carrier distribution. In these cases it is necessary to establish some self-consistent procedure, as can be done with the Ensemble Monte Carlo technique, presented in section 3.8 (see, in particular, section 3.9).

The problem is even more difficult when not only a dependence of the scattering mechanisms upon f is present, but f itself depends on the position in real space, as happens in devices. In these cases a very careful choice of Γ is to be made (see section 5.5.4).

3.6 The Choice of the State After Scattering

Once the scattering mechanism that caused the end of the electron free flight has been determined, the new electron state after scattering, $\mathbf{k}_a$, must be chosen as the final state of the scattering event. If the free flight ended with a self-scattering, $\mathbf{k}_a$ must be taken as equal to $\mathbf{k}_b$, the state before scattering. When, in contrast, a true scattering occurred, then $\mathbf{k}_a$ must be generated stochastically, according to the differential cross-section of that particular mechanism. The techniques used to select the state after each particular type of scattering will now be discussed.
The same classification and structure of Sect. 2.4 will be kept.

3.6.1 Phonon Scattering, Deformation-Potential Interaction

a) Electron Intravalley Scattering. Acoustic Phonons

a.1) Elastic, Equipartition Approximation. Spherical, Parabolic Bands

Owing to the isotropic character of the scattering, any state $\mathbf{k}'$ belonging to the energy-conserving sphere has the same probability of occurrence as the final state of the scattering, independently of the angle formed with the initial state $\mathbf{k}$. The probability that the polar angles of $\mathbf{k}'$ with respect to any convenient directions will be contained in the intervals $d\theta$ and $d\varphi$ is proportional to $\sin\theta\, d\theta\, d\varphi$. By application of the direct technique (see Appendix B), θ and φ can be chosen with two random numbers r_1 and r_2 as

$$\cos\theta = 1 - 2r_1\,, \tag{3.6.1}$$

$$\varphi = 2\pi r_2\,. \tag{3.6.2}$$

When this simple model is used, the polar axis is usually taken along the direction of the applied field.

a.2) Elastic, Equipartition Approximation. Ellipsoidal, Nonparabolic Bands

The scattering is still isotropic in the space of $\mathbf{k}^*$ transformed with Herring-Vogt transformation. Thus Eqs. (3.6.1) and (3.6.2) above can still be used for the determination of $\mathbf{k}^{*\prime}$ after scattering.

a.3) Inelastic Acoustic Scattering. Spherical, Parabolic Bands

In order to determine the final states of the scattering, the following procedure can be used: first the value q involved in the transition is chosen with the aid of the rejection technique (see Appendix B), applied to the probability in Eq. (2.4.23). Then the magnitude of the electron wave vector $\mathbf{k}'$ after scattering is obtained by energy conservation, while the angle β between $\mathbf{k}'$ and $\mathbf{k}$ is obtained by momentum conservation. The angle φ of rotation around the direction of $\mathbf{k}$ is chosen at random according to Eq. (3.6.2).
If the components of $\mathbf{k}'$ are to be determined with respect to fixed directions $\hat{x}$, $\hat{y}$ and $\hat{z}$, as is necessary when full three-dimensional simulation is performed, the following formulas can be used (see Fig. 3.5):

$$k'_z = k'(\cos\beta\cos\gamma + \sin\beta\cos\varphi\sin\gamma), \tag{3.6.3}$$

$$k'_y = k'(\cos\beta\cos\gamma' - \sin\beta\cos\varphi\cos\gamma\cos\eta - \sin\beta\sin\varphi\sin\eta), \tag{3.6.4}$$

$$k'_x = k'(\cos\beta\cos\gamma'' - \sin\beta\cos\varphi\sin\eta\cos\gamma + \sin\beta\sin\varphi\cos\eta), \tag{3.6.5}$$

where $\gamma = \widehat{kz}$, $\gamma' = \widehat{ky}$, $\gamma'' = \widehat{kx}$, η is the angle between y and the component

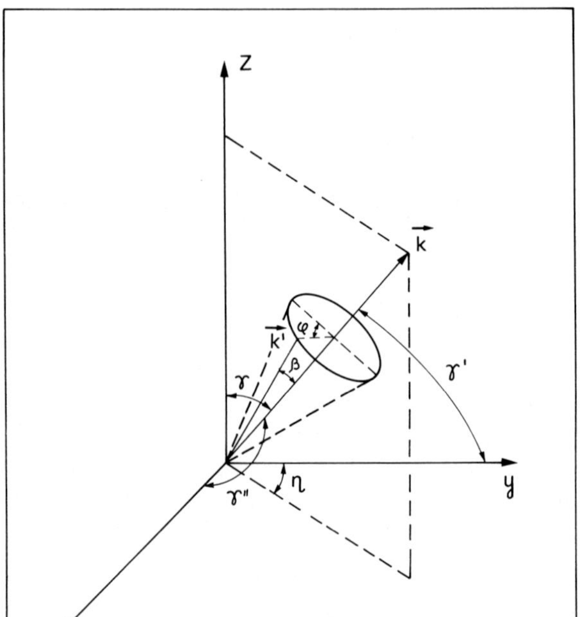

Fig. 3.5. Geometry of the determination of the state after scattering when full three-dimensional simulation is performed (see text)

of **k** in the (x, y) plane; the origin of the angle φ is taken along the plane formed by **k** and $\hat{z}$.

a.4) Inelastic Acoustic Scattering. Ellipsoidal, Nonparabolic Bands

A procedure similar to that described for the previous case can be followed in order to obtain the state of the electron $\mathbf{k}^{*\prime}$ after scattering, transformed with the Herring-Vogt transformation.

b) Electron Intravalley Scattering. Optical Phonons

b.1) Spherical, Parabolic Bands

Due to the isotropic character of the scattering, the method given by Eqs. (3.6.1) and (3.6.2) can be applied for the choice of the final state.

b.2) Ellipsoidal, Nonparabolic Bands

The final state in the Herring-Vogt transformed space is still random, with constant probability over the energy-conserving sphere of $\mathbf{k}^{*\prime}$. Thus the standard method of Eqs. (3.6.1) and (3.6.2) can be used.

c) Electron Intervalley Scattering

In this case, the final valley in the transition is different than that of the initial state. If the bottoms of the two valleys are different, the new kinetic energy is to be evaluated with respect to the bottom of the new valley.
All states in the energy-conserving sphere are equally probable, and the standard method of Eqs. (3.6.1) and (3.6.2) can be used for the determination of the final state $\mathbf{k}'$. Furthermore, when several equivalent valleys are possible as final valleys of the transition, one of them must be selected at random. Sometimes it may be convenient not to distinguish between final valleys which are totally equivalent in the simulation, i.e., equivalent from the crystallographic point of view, and equally oriented with respect to the applied field. In this case appropriate weights must be given to the possible "representative" final valleys for each type of intervalley scattering.

d) Hole Intraband Scattering. Acoustic Phonons

For the determination of the final state after scattering, Eq. (2.4.39) can be used with the following procedure. First the values of the final angles ϑ' and

ψ' are generated according to Eqs. (3.6.1) and (3.6.2) and the phonon wave vector q is taken as $2^{1/2}k(1 - \cos\theta)^{1/2}$, k being the absolute value of the wave vector before scattering. Then the above final angles are accepted or rejected according to the rejection technique (Appendix B) applied to the probability in Eq. (2.4.39). Finally the absolute value of the wave vector k' after scattering is obtained by energy conservation.

e) Hole Intraband Scattering. Optical Phonons

Similarly to the previous case, for the determination of the final state after scattering, final angles are first generated according to Eqs. (3.6.1) and (3.6.2), and then accepted or rejected according to the probability in Eq. (2.4.46).

3.6.2 Phonon Scattering, Electrostatic Interaction

a) Acoustic, Piezoelectric Phonons

For the generation of the state after scattering the following procedure can be used [12] in the elastic approximation. From Eq. (2.4.53) the probability that the electron is scattered through θ is found to be

$$P(\theta) \propto \frac{(1 - \cos\theta)\sin\theta}{[\beta^2 + 2k^2(1 - \cos\theta)]^2}; \qquad (3.6.6)$$

the application of the direct technique leads to the following relation between the random number r and the corresponding θ_r,

$$r\left[\ln\left(1 + \frac{4k^2}{\beta^2}\right) - \frac{1}{1 + \beta^2/4k^2}\right] = \ln\left(1 + \frac{2k^2}{\beta^2}(1 - \cos\theta_r)\right) - \frac{1 - \cos\theta_r}{1 - \cos\theta_r + \dfrac{\beta^2}{2k^2}}, \qquad (3.6.7)$$

which can be solved numerically for θ_r.

b) Polar Optical Phonons

For parabolic bands with $\alpha = 0$, Eq. (2.4.58) reduces to:

$$P(\theta)\, d\theta \sim \frac{\sin\theta\, d\theta}{(\varepsilon + \varepsilon' - 2\sqrt{\varepsilon\varepsilon'}\cos\theta)}. \qquad (3.6.8)$$

Angles θ can be generated according to this distribution by means of the direct technique, with the result

$$\cos\theta_r = [(1+f) - (1+2f)^r]/f, \tag{3.6.9}$$

where

$$f = 2\sqrt{\varepsilon\varepsilon'}/(\sqrt{\varepsilon} - \sqrt{\varepsilon'})^2 . \tag{3.6.10}$$

For the non parabolic case the rejection technique can be used to generate angles θ according to Eq. (2.4.58). When θ has been determined, a further random number determines the azimuthal angle between 0 and 2π as described by Eq. (3.6.2), and then $\mathbf{k}'$ is completely specified.

3.6.3 *Ionized Impurity Scattering*

a) Electrons

a.1) Spherical, Parabolic Bands

For the determination of the state after scattering, the angle θ between $\mathbf{k}$ and $\mathbf{k}'$ can be determined by application of the direct technique to the distribution given by the integrand in Eq. (2.4.63). Straightforward calculations lead to

$$\cos\theta_r = 1 - \frac{(1+4\varepsilon/\varepsilon_\beta)\sin^2(\theta_m/2) + r\cos^2(\theta_m/2)}{1+4\frac{\varepsilon}{\varepsilon\beta}[1 - r\cos^2(\theta_m/2)]} . \tag{3.6.11}$$

This equation becomes

$$\cos\theta_r^{CW} = \frac{(\varepsilon/\varepsilon_b)^2 r - 1}{(\varepsilon/\varepsilon_b)^2 r + 1} , \tag{3.6.12}$$

in CW approach, and

$$\cos\theta_r^{BH} = 1 - \frac{2(1-r)}{1+4\dfrac{\varepsilon}{\varepsilon\beta} r} \tag{3.6.13}$$

in BH approach.
The azimuthal angle φ can be determined at random according to Eq. (3.6.2).

a.2) Ellipsoidal, Nonparabolic Bands

The determination of the final state after scattering can be performed in a straightforward way, once the Herring-Vogt transformation has been applied. The results are

$$\cos \theta_r^{CW} = \frac{\left[\dfrac{\gamma(\varepsilon)}{\varepsilon_b(1 + 2\alpha\varepsilon)}\right]^2 r - 1}{\left[\dfrac{\gamma(\varepsilon)}{\varepsilon_b(1 + 2\alpha\varepsilon)}\right]^2 r + 1}, \tag{3.6.14}$$

$$\cos \theta_r^{BH} = 1 - \frac{2(1 - r)}{1 + 4r\dfrac{\gamma(\varepsilon)}{\varepsilon'_\beta}}. \tag{3.6.15}$$

b) Holes

b.1) Spherical, Parabolic Bands

With respect to the electron case, the overlap factor is now to be accounted for. For the determination of the state after scattering, the following combined procedure (Appendix B) can be used. The angle θ between $\mathbf{k}$ and $\mathbf{k}'$ is first generated according to Eq. (3.6.13). Then the overlap factor $\frac{1}{4}(1 + 3\cos^2\theta)$ is used in the rejection technique to accept or reject this angle.
The azimuthal angle φ is determined as usual with Eq. (3.6.2).

b.2) Warped Bands

A similar procedure as for the previous case can be applied to the angular dependence of the integrand function in Eq. (2.4.79).

3.6.4 Carrier-Carrier Scattering

a) Screened Electron-Electron Interaction

Following the calculations of section 2.4.4, the two-body screened electron-electron interaction changes the initial states $\mathbf{k}$ and $\mathbf{k}_0$ into the final states $\mathbf{k}'$ and $\mathbf{k}'_0$. The angular distribution of the final relative wave vector $\mathbf{g}' = \mathbf{k}' - \mathbf{k}'_0$ with respect to $\mathbf{g} = \mathbf{k} - \mathbf{k}_0$ is given by Eq. (2.4.90). The angle θ between $\mathbf{g}$ and $\mathbf{g}'$ can be generated from [13]

$$\cos \theta_r = 1 - \frac{2r}{1 + g^2(1 - r)/\beta^2}, \tag{3.6.16}$$

where $\mathbf{r}$ is a random number between 0 and 1. It can be noticed that, in the relative frame of the colliding electrons, the angular distribution is the same as the one obtained previously for ionized impurity scattering.
The azimuthal angle φ of $\mathbf{g}'$ around $\mathbf{g}$ is taken at random between 0 and 2π.

The relative wave vector $\mathbf{g}'$ after collision is now completely determined since $|\mathbf{g}| = |\mathbf{g}'|$. The final state $\mathbf{k}_0'$ and $\mathbf{k}'$ are given by

$$\begin{aligned} \mathbf{k}_0' &= \mathbf{k}_0 - \tfrac{1}{2}(\mathbf{g}' - \mathbf{g}), \\ \mathbf{k}' &= \mathbf{k} + \tfrac{1}{2}(\mathbf{g}' - \mathbf{g}). \end{aligned} \tag{3.6.17}$$

b) Electron-Plasmon Interaction

As shown in section 2.4.4, the angular dependence of the electron-plasmon scattering is the same as the one of the LO phonon scattering. Therefore, the cosine of the angle θ between $\mathbf{k}$ and $\mathbf{k}'$ is found from the relation (3.6.9) with $f = 2\sqrt{\varepsilon\varepsilon'}/(\sqrt{\varepsilon} - \sqrt{\varepsilon'})^2$. In this case, the upper limit on the wave vector set by q_c imposes an additional condition on θ [14]. This additional limit is found by isolating the q-integration in the calculation of the total scattering rate in Eq. (2.4.93), to obtain

$$\begin{aligned} P(\theta) \propto \int_0^{q_c} & dq[N_q \delta(\varepsilon_{k+q} - \varepsilon_k - \hbar\omega_p) + (N_q + 1) \\ & \times \delta(\varepsilon_{k-q} - \varepsilon_k + \hbar\omega_p)]. \end{aligned} \tag{3.6.18}$$

From the requirement that the argument of the δ-function has at least one zero within the range of integration, the following conditions on θ are found: for emission

$$\cos\theta > \frac{2m^*\omega_p/\hbar - q_c^2}{2kq_c}, \tag{3.6.19}$$

and, for absorption,

$$\cos\theta < \frac{q_c^2 + 2m^*\omega_p/\hbar}{2kq_c}. \tag{3.6.20}$$

c) Screened Electron-Hole Interaction

The final state for electron-hole interaction within the screened potential model is formally the same as in Eq. (3.6.16) for electron-electron collision. The only difference is in the definition of the relative wave vector, which is now given by Eq. (2.4.103).

3.7 Collection of Results for Steady-State Phenomena

The data collected at each free flight will form the base for the determination of the quantity of interest.

3.7.1 Time Averages

Generally speaking, we may obtain the average value of a quantity $A(\mathbf{k})$ (e.g., the drift velocity, the mean energy, etc.) during a single history of duration T as

$$\begin{aligned}\langle A\rangle_T &= \frac{1}{T}\int_0^T A[\mathbf{k}(t)]\,dt \\ &= \frac{1}{T}\sum_i \int_0^{t_i} A[\mathbf{k}(t')]\,dt'\,,\end{aligned} \tag{3.7.1}$$

where the integral over the whole simulation time T has been separated into

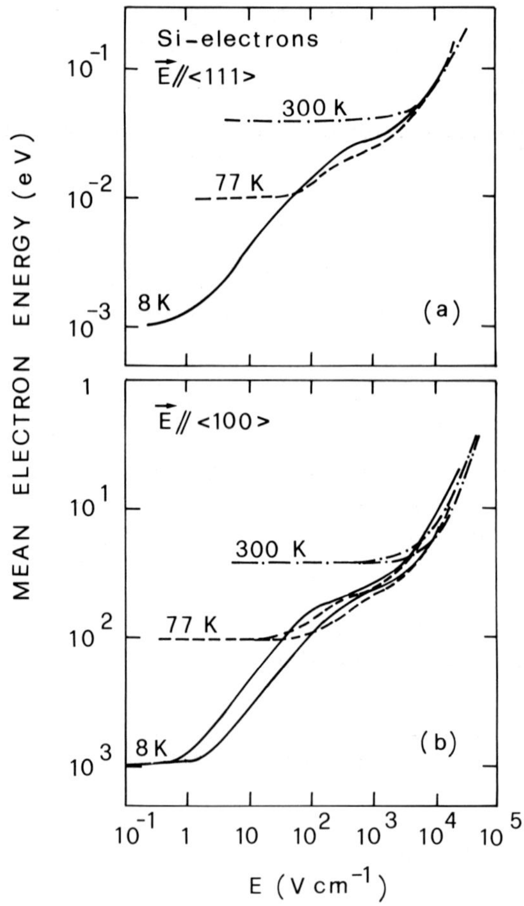

Fig. 3.6. Mean electron energy of electrons in Si as a function of electric field at the temperatures indicated, obtained with MC calculations. *a* Results obtained with E parallel to a $\langle 111\rangle$ direction; *b* results obtained for hot and cold valleys with E parallel to a $\langle 100\rangle$ direction [15]

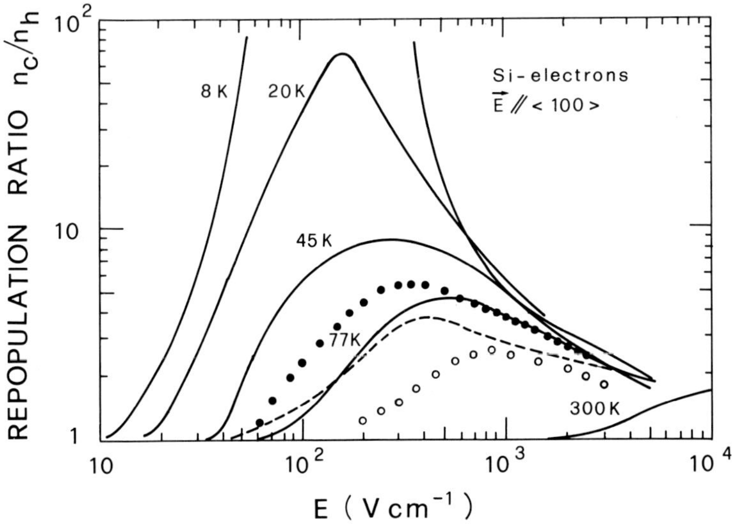

Fig. 3.7. Repopulation ratio of cold-to-hot valleys of electrons in silicon as a function of field applied along a $\langle 100 \rangle$ direction at various temperatures [15]

the sum of integrals over all free flights of duration t_i. When a steady state is investigated, T should be taken sufficiently long for $\langle A \rangle_T$ in Eq. (3.7.1) to represent an unbiased estimation of the average of the quantity A over the electron gas.

We may also obtain the electron distribution function for a homogeneous system in steady state: a mesh in **k** space (or in energy) is set up at the beginning of the computer run; during the simulation the time spent by the sample electron in each cell of the mesh is recorded and, for large T, this time conveniently normalized will represent the electron distribution function, that is, the solution of the Boltzmann equation [2]. This evaluation of the distribution function can be considered a special case of Eq. (3.7.1) in which we choose for A the functions $n_j(\mathbf{k})$ with value 1 if $\mathbf{k}$ lies inside the jth cell of the mesh and zero otherwise.

This method is the one most generally used to obtain transport quantities from Monte Carlo simulations.

Figures 2.30, 2.32, 2.33 of chapter 2 and Figs. 3.6 to 3.14 show some typical examples of results obtained with steady-state MC calculations [15, 16].

3.7.2 *Synchronous Ensemble*

The synchronous-ensemble method of obtaining an average quantity $\langle A \rangle$ from the simulation of a steady-state phenomenon has been introduced by Price [17, 18]. The average value of a quantity $\langle A \rangle$ is defined as the ensemble average at time t over the N electrons of the system,

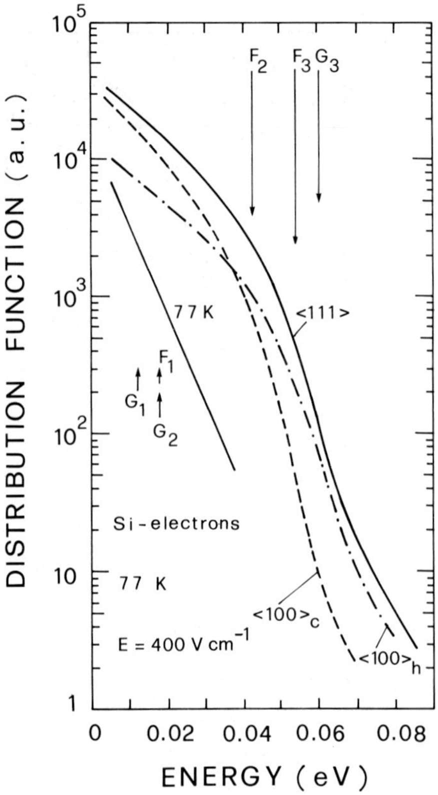

Fig. 3.8. Energy distribution function of electrons in Si obtained at 77 K from MC calculations with an electric field $E = 400$ V/cm. The continuous line refers to all electrons for the field parallel to a $\langle 111 \rangle$ direction. The dashed and dot-dashed lines refer to electrons in cold and hot valleys, respectively, for the field parallel to a $\langle 100 \rangle$ direction. The slope of the straight line indicates the lattice temperature. The arrows indicate the energy of the intervalley phonons used in the model, and their lengths are proportional to the corresponding coupling constants [15]

$$\langle A \rangle = \frac{1}{N} \sum_i A_i(t). \tag{3.7.2}$$

The steady-state distribution function is proportional to the number of electrons $n(\mathbf{k})\,\mathbf{\Delta k}$ that at time t are found to be in a cell of fixed volume $\mathbf{\Delta k}$ around $\mathbf{k}$. Then Eq. (3.7.2) can also be evaluated as

$$\langle A \rangle = C \sum_k n(\mathbf{k}) A(\mathbf{k}), \tag{3.7.3}$$

where C is an appropriate normalization constant, and $n(\mathbf{k})$ can be considered as proportional to the probability of finding any given electron in state $\mathbf{k}$.

If we now consider the before-scattering distribution function $n_b(\mathbf{k})$, propor-

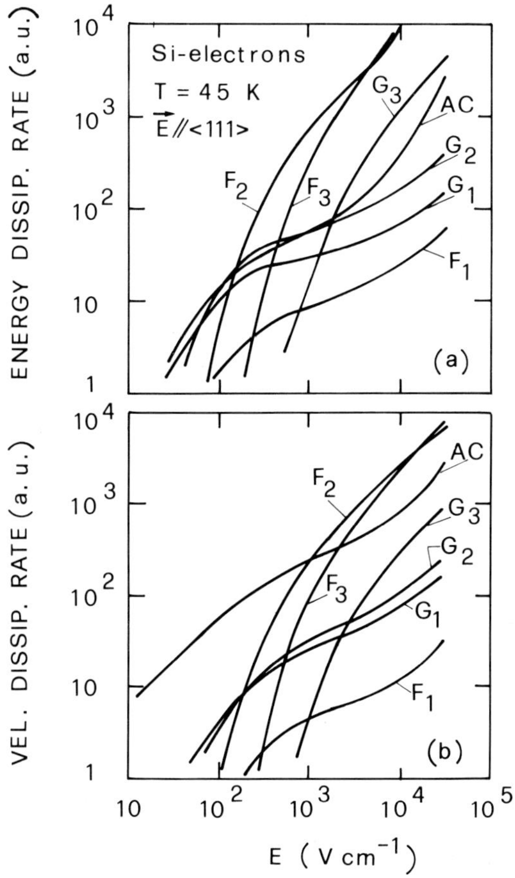

Fig. 3.9. MC results for the energy (a) and velocity (b) dissipation rates due to the various scattering mechanisms of electrons in Si, as functions of field, applied along a $\langle 111 \rangle$ direction at 45 K [15]

tional to the probability that an electron is found in $\mathbf{k}$ immediately before a scattering event, it is easy to convince ourselves that $n_b(\mathbf{k})$ is proportional to the number of electrons with wave vector $\mathbf{k}$ times the scattering rate $1/\tau(\mathbf{k})$ for such electrons. Thus the steady-state distribution function is given by

$$n(\mathbf{k}) = \frac{1}{\tau_0} n_b(\mathbf{k}) \tau(\mathbf{k}), \tag{3.7.4}$$

where τ_0 is an appropriate normalization constant. If, by including self-scattering, we use a constant $\tau(\mathbf{k})$, then the steady-state distribution function becomes proportional to the before-scattering distribution

$$n(\mathbf{k}) \propto n_b(\mathbf{k}), \tag{3.7.5}$$

and Eq. (3.7.3) can be used in a MC simulation in the following form:

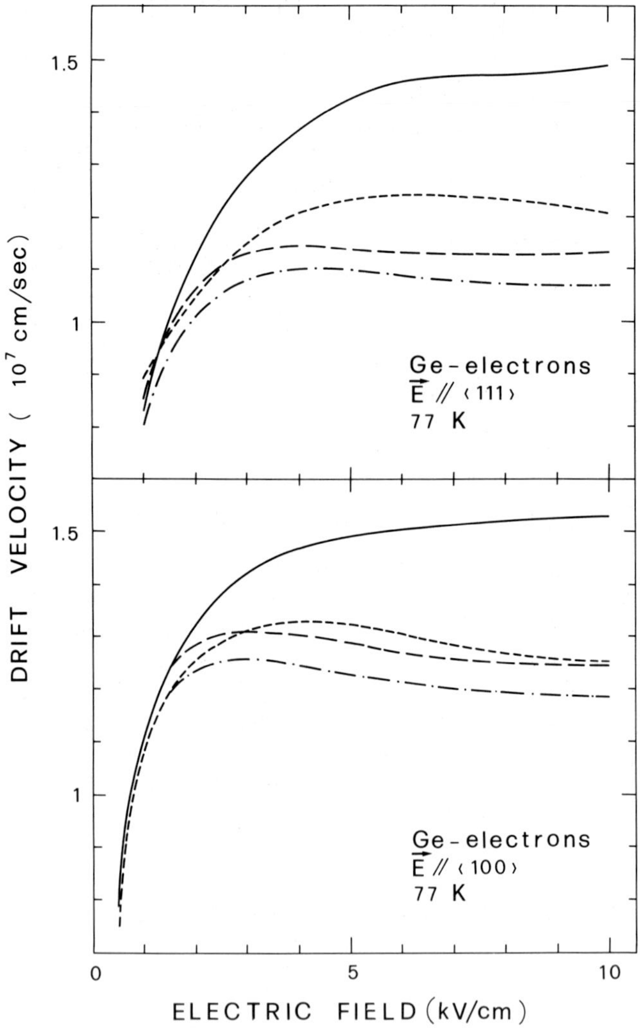

Fig. 3.10. MC results for electron drift velocity in Ge at 77 K as a function of field strength, with the field applied along a $\langle 111 \rangle$ and a $\langle 100 \rangle$ direction. Different lines refer to different models used in the calculations [16]

$$\langle A \rangle = \frac{1}{N} \sum_i A_{bi}, \tag{3.7.6}$$

where the sum covers all N electron free flights, and A_{bi} indicates the value of the quantity A evaluated at the end of the free flight immediately before the ith scattering event.

If self-scattering with a step-shaped total scattering probability, as described in section 3.4, is used, then the various terms in the sum of Eq. (3.7.6) must be weighted with a factor $\Gamma^{-1}(\mathbf{k})$; but in this case particular care must be

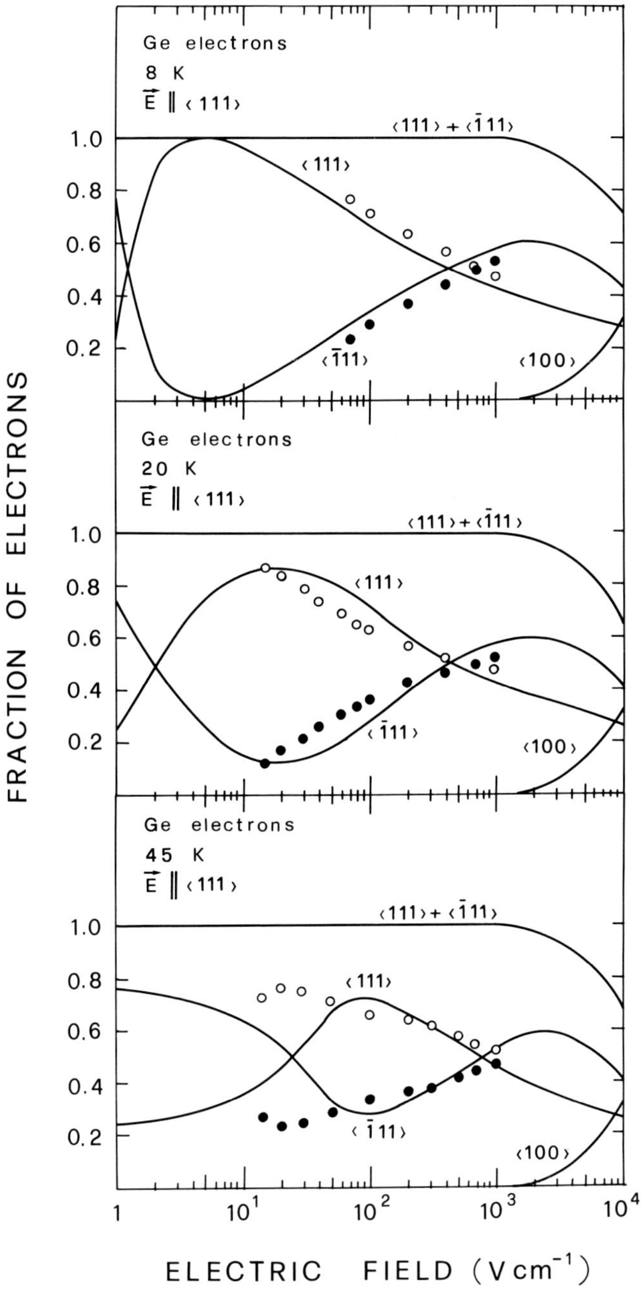

Fig. 3.11. Population of the different valleys as a function of electric field strength in Ge at the given temperatures. Points indicate experimental results, curves theoretical MC calculations [16]

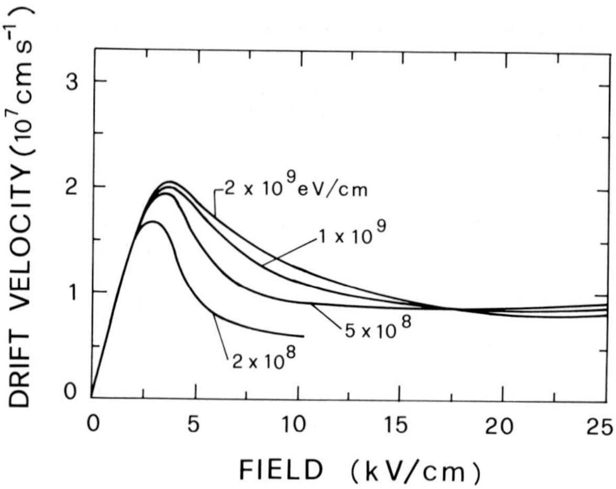

Fig. 3.12. Drift velocity of electrons in GaAs at 300 K as a function of field strength, obtained with MC calculations with the indicated values for the coupling constant of the intervalley deformation-potential scattering [2]

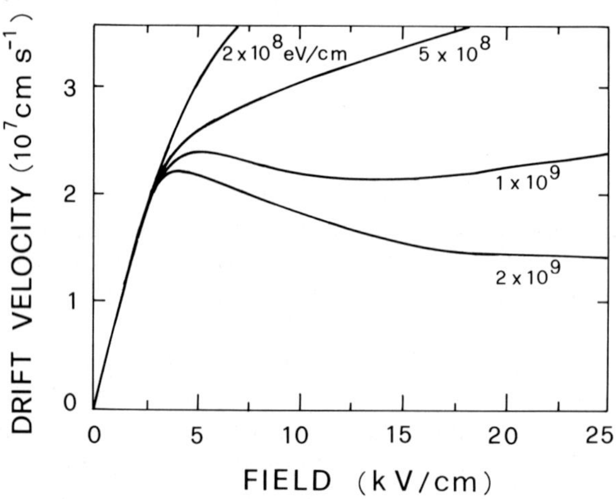

Fig. 3.13. Drift velocity of electrons in the Γ valley of GaAs at 300 K as a function of field strength, obtained with MC calculations with the indicated values for the coupling constant of the intervalley deformation-potential scattering. With high coupling, a negative differential mobility is present for electrons in the central valley, due to the increasing efficiency of intervalley scattering at increasing field strengths [2]

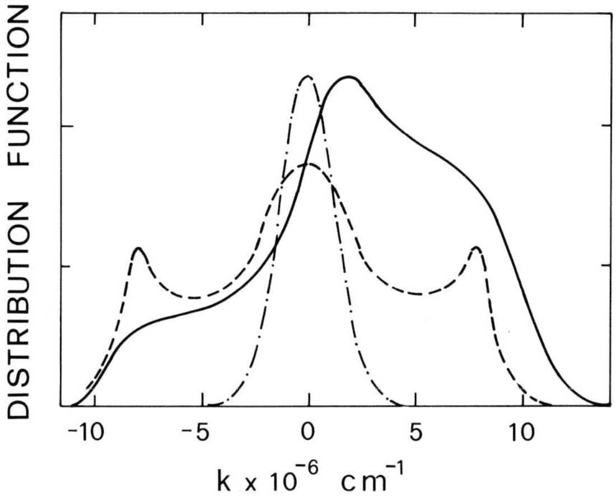

Fig. 3.14. Electron distribution function in the Γ valley of GaAs at 300 K as a function of wave vector obtained with MC calculations along lines through $k = 0$ parallel (continuous line) and perpendicular (dashed line) to a field of 15 kV/cm. The dot-dashed line shows, for comparison, the zero-field distribution function [2]

taken when, during a flight, $\mathbf{k}(t)$ crosses the border from one value of Γ_i, to another value, since in Eq. (3.7.4) $\tau(\mathbf{k})$ must be a unique, well-defined function of $\mathbf{k}$.

3.7.3 *Statistical Uncertainty*

In order to estimate the statistical uncertainty of a time-average result $\langle A \rangle_T$ due to the finite value of the simulation time T, we can divide the whole history in N sub-histories of duration T/N; for each of them a value $\langle A \rangle_l$ is obtained. We may then take the average of the $\langle A \rangle_l$'s (which will be equal to $\langle A \rangle_T$) as the most probable value of $\langle A \rangle$ and its standard deviation as the statistical uncertainty on $\langle A \rangle_T$. The sub-histories must be sufficiently long to be considered as independent of each other, but sufficiently short to allow us to perform a large number N of them. In this case the uncertainty decreases as $1/\sqrt{N}$. Furthermore, it is possible to carry out, on the series of partial results $\langle A \rangle_l$, those tests [19] which verify the statistical nature of their fluctuations.
The same separation of the sample in sub-ensembles can be performed when the synchronous-ensemble method is used.

3.8 The Ensemble Monte Carlo (EMC)

As we anticipated in section 2.9, little work needs to be added to MC programs in order to deal with time and/or space-dependent problems.

Since for such problems we cannot rely on the ergodicity of the system, an ensemble of particles must be explicitly simulated [20, 21].
One exception to this rule will be discussed in 3.11.1 dealing with phenomena which are periodic in time. In this case the different situations experienced by the electron at equivalent times will take into account the many possibilities of various particle, and again the simulation of a single electron may yield the necessary information about the entire electron gas.

a) Transients

We shall consider here the case of a homogeneous electron gas with time-dependent behavior. In particular, it is of interest to study the transient dynamic response to a sudden change in the value of an applied field. In this situation many particles must be independently simulated with appropriate distributions of initial conditions. Provided the number of simulated particles is sufficiently large, histograms obtained by reading the electron wave vectors or energies at regular intervals of time provide the distribution function $f(\mathbf{k}, t)$ or $f(\varepsilon, t)$, respectively. If a single quantity A is of interest, its average value can be obtained on this sample ensemble as a function of time. It must be stressed that the estimations given by Eqs. (3.7.1) and (3.7.6) are based on the hypothesis of steady-state conditions and cannot be used when a time-dependent phenomenon is analyzed: the ensemble average of a quantity A must actually be estimated according to its definition, given by Eq. (3.7.2).
The duration of the transient response is not known a priori and will be of the order of the largest of the characteristic times of the electron system. This time may be called the "transient-transport time" and in general depends upon the values of the applied field and temperature; in the case of high-field transport in semiconductors, it may roughly correspond to the energy relaxation time or to the time for repopulation of different valleys.
To determine the precision of the results obtained, one separates the entire ensemble into a certain number of subensembles and estimates for each of them the quantity of interest A. Then their average value and standard deviation can be taken, respectively, as the most probable value and the statistical uncertainty of A.
The transient dynamic response obtained by means of the simulation will, of course, depend upon the initial conditions of the carriers, and these must be assumed according to the situation to be explored. Figures 2.34 to 2.35 in chapter 2 are examples of results obtained with EMC.

b) Space-dependent Phenomena

The simulation of a steady-state phenomenon in a physical system where electron transport depends upon the position in space is of particular interest for the analysis and modeling of devices. For this case, too, an ensemble of independent particles must be used, and averages must be taken over particles at given positions. The statistical uncertainties of the results can be obtained with subensembles, as indicated in the preceeding section.

In a steady-state situation, particles enter the region of interest continuously, and in the simulation an initial wave vector must be assumed when a new particle is considered, according to the momentum distribution of the particles in the physical system under investigation. For example, a cold hemi-Maxwellian distribution, that is, one with only positive components along a given direction, may be convenient in simulating the metallic contact of a device. If the simulated electron entering the device goes back to the contact, a new electron must be generated, but the "lost" electron must be accounted for in the final results in a way that depends on the particular analysis being carried out. The problem of MC simulation of particles in a space-dependent system is the essence of the simulation of electron devices and will be discussed in full detail in the following chapters.

3.9 Many Particle Effects

3.9.1 Carrier-Carrier Scattering

We have seen in section 2.4.4 that the definition of an interaction hamiltonian for carrier-carrier scattering allows the calculation of total scattering rates for various types of Coulomb interactions, within the random phase approximation. The separation into short and long-range contributions of the Coulomb Hamiltonian leads to the definition of a screened two-body interaction on one side and of the collective behavior (described in terms of plasmons) on the other side, which leads in turn to the carrier-plasmon scattering. Such scattering is treated in the MC as the other scatterings (phonons and impurities). More difficult is the inclusion of the screened Coulomb interaction into the standard algorithm. The total scattering rates obtained in section 2.4.4 depend on the carrier-distribution function. In the one particle MC, the distribution function is one of the outputs of the calculation. Thus, some sort of self-consistent calculation must be performed in which an assumed $f(\mathbf{k})$ is used to evaluate scattering probabilities and the same $f(\mathbf{k})$ results as solution.

Bacchelli and Jacoboni [22] took an iterative approach consisting of the following procedure: in the first iteration, $f_1(\mathbf{k})$ is obtained from a MC simulation without *e-e* interaction. Then in the successive iterations the MC simulation is repeated, introducing *e-e* interaction with a constant scattering probability in which the final state of the collision is chosen randomly according to a drifted Maxwellian distribution, with the drift velocity and the mean energy obtained in the previous iterative step. The procedure ends when convergence to a stable result ensures self-consistency of the solution. A somewhat different approach was proposed by Matulionis et al. [23], in which two electrons are simulated, each representing the whole system. For each of them, a record is kept of the most recent distribution function obtained from the simulation over a specified period. In the simulation of one electron, among the scattering mechanisms, and *e-e* interaction with a screened Coulomb potential is introduced, and evaluated by means of the memory of the other electron. At each collision, the memory of the simulated particle is updated by the new state, while the memory of the other electron is not changed. When the other electron is considered as a MC electron, the memory of the former particle now represents the ambient electron gas. The procedure is recurrent, so that a sufficiently long simulation provides convergence of the final result.

The use of the Ensemble Monte Carlo technique greatly simplifies the treatment of carrier-carrier scattering. Since the distribution function is built in the simulated ensemble of carriers, the sum over f_n in Eqs. (2.4.89) and (2.4.105) can be properly taken into account at any point of the simulation, even under transient conditions.

To achieve this aim two alternative methods have been proposed.

The first one [24, 25] starts from the realization that a sum over the distribution function is simply an ensemble average of a given quantity. Thus the scattering rate for a carrier-carrier scattering can be expressed as

$$\Gamma_{c\text{-}c}(\mathbf{k}_0) = \frac{2\mu e^2}{\hbar^3 \kappa^2 N} \sum_{\text{all carriers}} \frac{g}{\beta^2[g^2 + \beta^2]}, \tag{3.9.1}$$

where μ, g and β are given in general by Eqs. (2.4.102)–(2.4.104), which include the case of one-carrier plasma as well as that of a many-component plasma. The sum is taken over all wave vectors of the sub-ensemble formed by the carriers of a given type. It is obvious that the scattering rate defined from the ensemble average does not require any assumption on the form of the distribution function. Furthermore, it is not limited to steady-state situations, but applies as well to the study of transient phenomena. The main limitation comes from the computational requirement of Eq. (3.9.1), which involves a tridimensional sum over the ensemble of carriers and depends on $\mathbf{k}$ rather than on its amplitude.

In the implementation in Refs. [24] and [25], the scattering rates are tabulated over a two-dimensional grid at fixed times during the simulation (a three-dimensional grid could also be used). The time step for the tabulation depends critically on the specific situation of interest. If the distribution function changes rapidly over short period of times (such as for examples following laser photoexcitation), the scattering rate should be calculated in such a way as to reproduce the changes in the distribution. Typically, time-steps of 100 femtoseconds are used. On the contrary, if the time scale of the relevant physical processes is of the order of picoseconds or longer, such a fine discretization in time is not necessary. In the limit of steady-state situations, no update is required any longer in the tabulation of the scattering rate. The carrier-carrier scattering falls into the category of interactions whose rate changes in time. It is therefore important to let the maximum scattering rate adjust itself during the simulation (that is to have a self-scattering variable in time), to avoid the unphysical occurrence of negative rates for real processes.

A second method of dealing with the carrier-carrier scattering is based on a rejection algorithm [26]. A self-scattering internal to the interparticle scattering is introduced by substituting the term $g/(g^2 + \beta^2)$ in Eq. (2.4.89) with its maximum value $1/2\beta$. This enhanced carrier-carrier scattering rate is introduced among the other mechanisms. When an intercarrier collision is selected during the simulation, the counterpart carrier is chosen at random from the ensemble. The internal rejection is performed by comparing a random number r between 0 and 1 with $2\beta g/(g^2 + \beta^2)$. If the collision is accepted, the final state is calculated following the procedure outlined in section 3.6.4.

With respect to the previous method, the self-scattering algorithm for the carrier-carrier scattering does not require the tabulation of the scattering rates, which can be a strong limitation in terms of computer time and memory requirements. However, a large number of rejections occurs.

When an interparticle scattering is selected during the ensemble simulation, the **k**-vector of the second carrier involved in the binary collision can be taken at random from the ensemble. The selection of the final state $\mathbf{k}'_0$ of the simulated carrier is performed according to the prescription given earlier.

Two possibilities exist concerning the partner carrier involved in the collision. Since it has been selected from the ensemble, its state can be also changed to become $\mathbf{k}' = \mathbf{g}' + \mathbf{k}'_0$. In this way, the total energy would be conserved in the process. On the other hand, this would force the second carrier to end its free flight at a time different from the one determined from its own scattering probability. To circumvent such a problem, the second electron state could be left unchanged. This procedure creates however another fault, as it introduces an artificial energy dissipation in the collision process.

The algorithms just presented have been applied to a variety of situations, in the presence of external electric fields [24–26] or of optical excitations. Some results for the latter case will be shown in section 3.11.6.

Electron-electron and electron-plasmon scatterings are particularly important when fast electrons are injected into heavily doped regions, as for example in some semiconductor devices. The injected electrons lose energy and momentum through the interaction with the cold degenerate electron gas. In order to simulate the injection over a barrier (which corresponds to the interface between different regions of a device), the EMC electrons can be started as a monoenergetic beam at energy ε_0 and momentum $\mathbf{k}_0$ [27]. Figure 3.15 shows the average drift velocity as a function of the distance travelled by the injected packet of electrons, for two different values of initial energy ε_0. In Fig. 3.16, the energy (curve *a*) and parallel momentum (curve *b*) distribution functions are plotted, evaluated when the center of the packet is 200 nm from the point of injection, for the highest injection level ($\varepsilon_0 =$ 0.24 eV). The dotted curves show the distribution functions in the absence

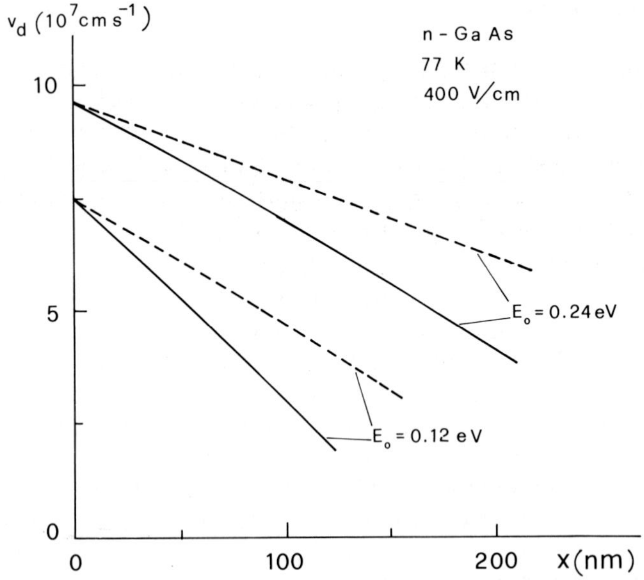

Fig. 3.15. Average drift velocity as a function of the distance travelled by an injected packet of the electrons in GaAs at 77 K, obtained with MC calculation with (solid curve) and without (dashed curve) electron-electron interaction, with an applied field of 400 V/cm. The numbers on the curves indicate the initial electron energies [27]

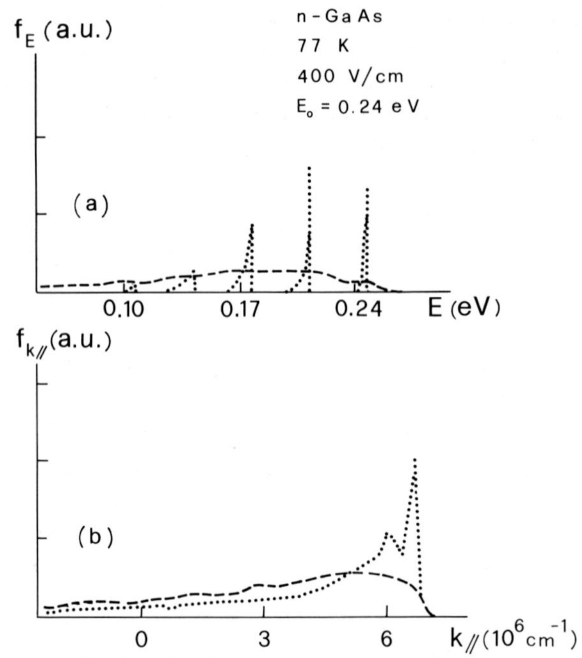

Fig. 3.16. Energy (curve *a*) and parallel momentum (curve *b*) distribution functions of a packet of electrons injected in GaAs at 77 K with an initial energy of 0.24 eV, with an applied field of 400 V/cm, evaluated with the MC simulation when the center of the packet is 200 nm from the point of injection with (dashed curve) and without (dotted curve) electron-electron interaction [27]

of the complete electron-electron interaction; the distinct peaks in the energy d.f. are related to discrete energy losses by the emission of polar-optical phonons. The contribution of the *e-e* and *e-pl* scattering is evident from the dashed curves where, as a result of a much faster randomization of the fast electrons, the d.f.'s are completely smeared out. The importance of carrier-carrier scattering on the performance of semiconductor devices will be discussed in chapter 6.

3.9.2 Molecular Dynamics and Monte Carlo Method

An alternative method for the treatment of the electron-electron interaction has been proposed by Jacoboni [28]. It is based on a molecular dynamics simulation for the real-space trajectories of the ensemble of electrons interacting via a bare Coulomb potential, coupled to the usual EMC algorithm for the **k**-space dynamics.

The strength of the method lies in the fact that it does not require any assumptions on the screening between carriers. Furthermore, it naturally accounts for a fundamental component of a classical electron gas, namely the carrier density fluctuations [29]. The main extension with respect to the traditional EMC routine is found in the splitting of the free flights between successive scattering-events. The trajectories of the carriers moving in an external electric field are now subdivided into shorter paths, in order to properly account for the changes in the interparticle Coulomb forces as the electrons move in the system. The classical equations of motion are solved within a finite difference scheme:

$$\mathbf{r}_i(t + \Delta t) = \mathbf{r}_i(t) + \mathbf{v}_i \, \Delta t + \frac{1}{2} \frac{\mathbf{F}_i(t)}{m} \Delta t^2 , \tag{3.9.2}$$

$$\mathbf{v}_i(t + \Delta t) = \mathbf{v}_i(t) + \frac{\mathbf{F}_i(t)}{m} \Delta t , \tag{3.9.3}$$

where $\mathbf{r}_i$ and $\mathbf{v}_i$ are respectively the position and the velocity of the i-th carrier. The time step Δt has to be taken small enough so that the interparticle forces can be considered constant during that time interval. On the other hand, exceedingly small Δt's lead to unacceptable long computation times. The solution has been found in terms of variable time steps.

The total force $\mathbf{F}_i(t)$ in Eq. (3.9.3) is given by the sum of interparticle Coulomb interaction and the applied electric field. A positive neutralizing background has also been used [14].

The real-space part of the simulation is identical to the one found in the study of classical plasmas [30]. The techniques adopted to deal with the long-range of the Coulomb interaction are also derived from plasma simulations. The simulated electrons are confined to a box with periodic boundary conditions. The calculation of the interparticle forces is the limiting factor, since it requires the calculation of the relative distance of each pair of

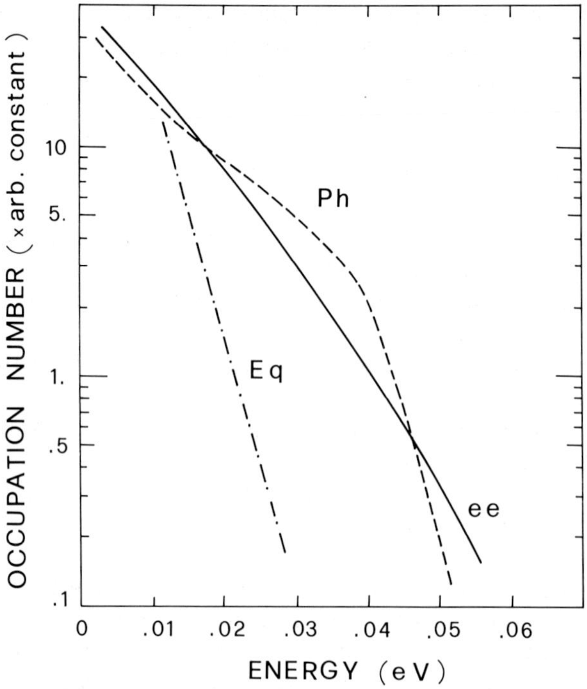

Fig. 3.17. Occupation number of electrons in a simple semiconductor modelled on Si obtained with MC calculations at $T = 45$ K and $E = 300$ V/cm. The dashed curve (Ph) has been obtained by considering only phonon scattering; the solid curve (*e-e*) includes electron-electron interaction with 10^{17} cm^{-3} electrons; the dot-dashed curve (Eq) indicates the equilibrium distribution [28]

electrons. Therefore, the number of simulated particles has to be kept low (typically, a few hundred).

The molecular dynamics MC technique has been applied to various transport problems in semiconductors.

Figure 3.17 shows the effect of the electron-electron scattering on the energy distribution function using a simplified model of *n*-Si [28]. The dashed curve (obtained by considering only phonon scatterings) shows a depletion of the energy states above the threshold for optical phonon emission. The inter-carrier scattering pushes the distribution towards a Maxwellian shape (continuous curve) at a hotter temperature than the lattice. The equilibrium temperature is also shown in the figure.

The tendency of the electron-electron scattering to randomize the distribution function, leading it to a hot Maxwellian, will also be encountered later in the description of ultrafast phenomena in the presence of optical excitation.

The electron-electron interaction creates correlations among the carriers. The real space simulation has allowed the study of such cross-correlations on the noise properties of semiconductors in the presence of external fields

[31]. Furthermore, the two particle correlation function has been calculated [32], indicating that the simple Debye screening models might not be accurate under hot electron conditions.

3.9.3 Degeneracy in Monte Carlo Calculations

As shown in the previous sections, the Ensemble Monte Carlo method is a semiclassical technique, in that it simulates electrons as classical particles undergoing scattering events that are calculated according to the quantum mechanical transition probabilities. Classical statistics are used, and the Pauli exclusion principle is neglected. On the other hand, many of the situations of interest in semiconductor devices involve high concentrations of electrons, where degeneracy is expected to play an important role. In GaAs at 300 K, this occurs for $n > 5 \times 10^{17}\ \mathrm{cm}^{-3}$.

A powerful algorithm to include the Pauli exclusion principle in the MC technique has been proposed by Bosi and Jacoboni [33] and extended to EMC by Lugli and Ferry [34].

The Pauli exclusion principle can be thought of as a many-body effect that influences the transport properties of degenerate semiconductors by limiting the phase space available for electronic transitions. The probability of an electronic transition from a state $\mathbf{k}$ into a state $\mathbf{k}'$ is in general proportional to the probability $f(\mathbf{k})$ that the initial state is occupied and the probability $[1 - f(\mathbf{k}')]$ that the final state is unoccupied. Thus, the transition rate from $\mathbf{k}$ to $\mathbf{k}'$ is $S(\mathbf{k}, \mathbf{k}') = P(\mathbf{k}, \mathbf{k}')f(\mathbf{k})[1 - f(\mathbf{k}')]$, where $P(\mathbf{k}, \mathbf{k}')$ is the scattering probability per unit time usually calculated from the Fermi golden rule. The standard MC procedure works within the approximation $f(\mathbf{k}') = 0$. That is, it considers all final states as being available. The latter method is only applicable to non-degenerate conditions, where the effect of the Pauli principle is negligible. The problem of working with the factor $[1 - f(\mathbf{k}')]$ in the Monte Carlo algorithm lies in the fact that $f(\mathbf{k}')$ is not known a priori. However, with the EMC technique, the distribution function evolves with the ensemble and is known at any time. In principle, one possibility that would allow us to overcome the problem is to up-date at fixed time intervals during the simulation the scattering probabilities evaluated numerically, taking into account the current knowledge of $f(\mathbf{k})$ in the factor $(1 - f)$.

The alternative method of Bosi and Jacoboni, suggested for a one particle simulation, is based on the self-consistent iterative algorithm that resorts to the rejection technique to account for the occupation probability of the final state at each scattering event. At the time of the selection of the particular scattering process, the final state of the transition is still not known. The exclusion principle can at first be neglected for the determination of the length of the free flight, and for the choice of the scattering mechanism and of the final state. However, once the final state $\mathbf{k}'$ is selected, $f(\mathbf{k}')$ is, at least in principle, known from the simulation, and a random number between 0 and 1 can be used to accept or reject the transition. If $r > f(\mathbf{k}')$ the transition is accepted; if instead $r < f(\mathbf{k}')$, the scattering is treated as a self-scattering. This corresponds to the introduction of an auxiliary self-scattering mechanism with probability proportional to $f(\mathbf{k}')$. This rejection procedure must be included since the factor $[1 - f(\mathbf{k}')]$ was ignored in earlier steps. As the procedure is iterated, a steady state will eventually be reached, with the probability of transitions into the state $\mathbf{k}'$ proportional to the occupancy of that state. In applying this procedure the normalization of the distribution function must be carefully accounted for in such a way that the evaluated $f(\mathbf{k})$ represents the occupation number of the state $\mathbf{k}$. Such a normalization is to be performed by taking into account the number of states in the considered cell of $\mathbf{k}$-space and the density of carriers in the physical system to be simulated.

The extension of the procedure to ensemble simulation is straightforward. In this case the distribution function is evaluated over the grid in $\mathbf{k}$-space by counting the number of electrons in each cell. The distribution function so defined is then properly normalized as indicated above, taking also into account the number of particles in the ensemble.

When a scattering event occurs in the EMC simulation, the cell c containing the final state for the selected transition, is found. The normalized distribution f_c, corresponding to that cell, is compared with a number r randomly chosen between 0 and 1, and the scattering event is accepted or rejected as indicated above.

As it is always the case for an EMC simulation, the algorithm just described allows the analysis of the transient evolution of any given initial distribution to its steady-state.

Figure 3.18 shows the effect of the exclusion principle on the electron dynamics. The continuous curves refer to the classical case, that is they are obtained with the standard EMC simulation. The starred curves indicate the results of the inclusion of degeneracy in the manner described in this section.

The energy distribution function $f(\varepsilon)$ in Fig. 3.18a is the number of electrons in an energy interval $\Delta\varepsilon$ around ε. The occupation number $n(\varepsilon)$ in Fig. 3.18b is obtained by normalizing $f(\varepsilon)$ with respect to the number of allowed states in the corresponding energy cell. The momentum distribution function $f(k_z)$, for the component of the electron momentum parallel to the field, is the projection of $f(\mathbf{k})$ onto the z axis (Fig. 3.18c).

As a consequence of the reduced maximum occupancy of each allowed state, electrons are pushed into the tail of the distribution. This shift corresponds to an increase of the electron temperature of about 50 K (from 140 to 210 K), when the Pauli principle is included. The increase in the electron temperature is also reflected in the broadening of the momentum distribution $f(k_z)$ in Fig. 3.18c.

At lower fields (90 V/cm), the effect of degeneracy is even more pronounced, as shown in Fig. 3.19. The electrons have a very small drift velocity and tend to remain in the lowest portion of the conduction band, with an average temperature nearly equal to the lattice temperature (continuous curves).

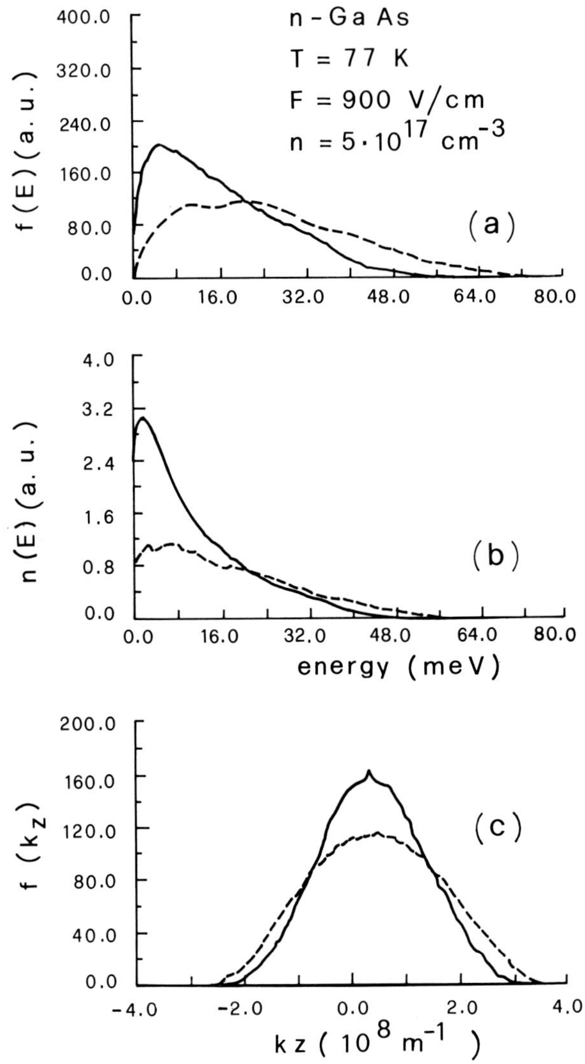

Fig. 3.18. Effect of exclusion principle in electron dynamics, as obtained with MC calculations Parts *a* and *b* of the figure show the electron distribution function and occupation number, respectively, as functions of energy in GaAs at 77 K with an applied field of 900 V/cm. *c* shows the electron distribution function as a function of the electron momentum parallel to the field. Continuous curves refer to classical statistics; dashed curves are obtained taking into account Pauli exclusion principle. The electron (and impurity) concentration is 5×10^{17} cm^{-3} [34]

Since it is impossible to allocate more than two electrons per state, a migration of the electrons towards higher energy states (starred curves) occurs. In the degenerate EMC calculation, a very high number of transitions, which would be allowed when classical statistics are used, are instead forbidden and thus rejected due to the occupation of the final state. In the

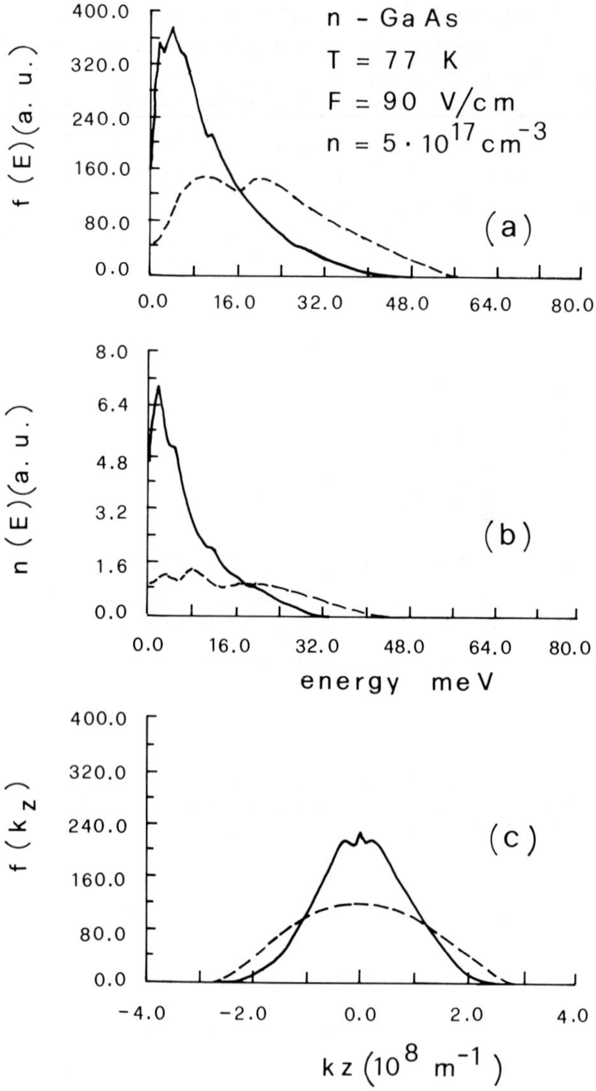

Fig. 3.19. The same as in Fig. 3.18 with an applied field of 90 V/cm [34]

simulations illustrated in Fig. 3.19, the Pauli exclusion principle prevented more than 80% of the attempted transitions.

3.10 Monte Carlo Simulation of the 2DEG

The structure of a MC program for the two dimensional electron gas follows the same lines presented in the previous sections.
The quantization in the direction perpendicular to the interface restricts the carrier motion to two dimensions. Thus, within each subband, electrons are

characterized by the two-dimensional wave vector $\mathbf{k} \equiv (k_z, k_y)$, which changes as a result of external fields and scattering events in the same fashion described earlier. As pointed out in section 2.10, phonons (or other scatterers) can introduce transition between different subbands. In this case, the energy difference between the initial and final levels is added or subtracted to the electron kinetic energy. The wavefunction character of the 2DEG (induced by the confinements into the quantum well) is contained in the scattering rates, which are calculated as illustrated in section 2.10. The most

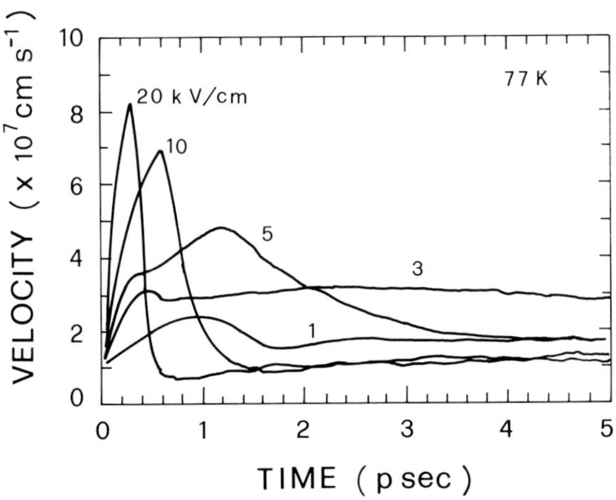

Fig. 3.20. Time evolution of the electron drift velocity at 77 K in a single AlGaAs-GaAs quantum well for different applied field strengths, applied at $t = 0$. [After Yokoyama, K., Hess, K.: Phys. Rev *B33*, 5595 (1986)]

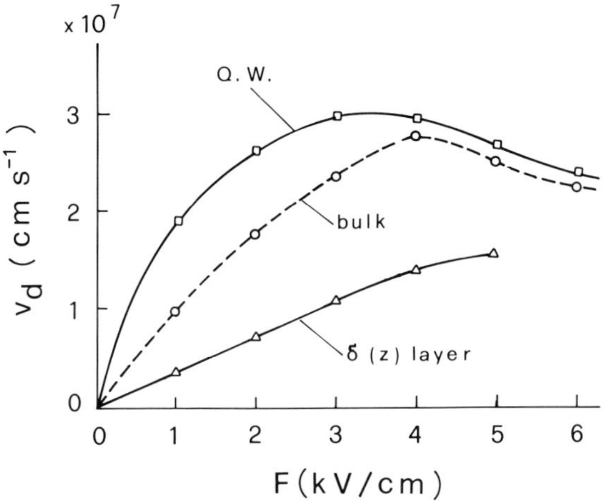

Fig. 3.21. Velocity-field curve for a 150 Å AlGaAs-GaAs-AlGaAs single quantum well (squares) and for a $\delta(z)$ doping layer in GaAs (triangles) at 300 K. The dashed curve reports the result for bulk GaAs with an impurity concentration of 10^{17} cm^{-3}. [After Lugli, P., et al.: Superlattices and Microstructures *2*, 335 (1986)]

general approach should be based on a self-consistent solution of Schroedinger and Poisson equations coupled to the MC simulation. In this way, the potential profile of the confining well and the subband separation would be consistent with the electron distribution in each subband. In practice, a fully self-consistent simulation requires enormous resources, and the MC simulation is usually performed on a given fixed potential well [36].

The selection of the final state follows the following rule: for completely randomizing scatterings (acoustic in the elastic approximation, intervalley, non-polar optical), the two-dimensional final state is generated with a uniform angular dispersion. For polar-optical and electron-electron a rejection technique is applied to the distribution in the integrands in Eq. (2.10.6) and (2.10.16), respectively. The Pauli exclusion principle can be included through a generalization of the rejection method presented earlier. For two-dimensional systems, a distribution function $f(k_z, k_y)$ has to be used for each subband.

Figure 3.20 to 3.23 show some results of MC simulations of the 2DEG.

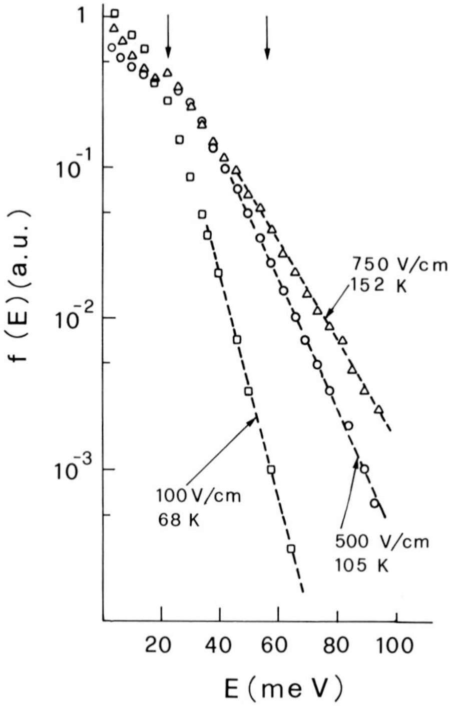

Fig. 3.22. Steady-state electron distribution function for a 258 Å AlGaAs-GaAs-AlGaAs single quantum well at three different field strengths. The lattice temperature is 5 K. The vertical arrows indicate the energies of the first and second excited subbands. The electron temperatures calculated from the slope of the high-energy tails are also indicated. [After Goodnick, S. M., Lugli, P.: In: High-Speed Electronics (Källbäck B., Beneking H., eds.), p. 117. Berlin-Heidelberg-New York: Springer. 1986.

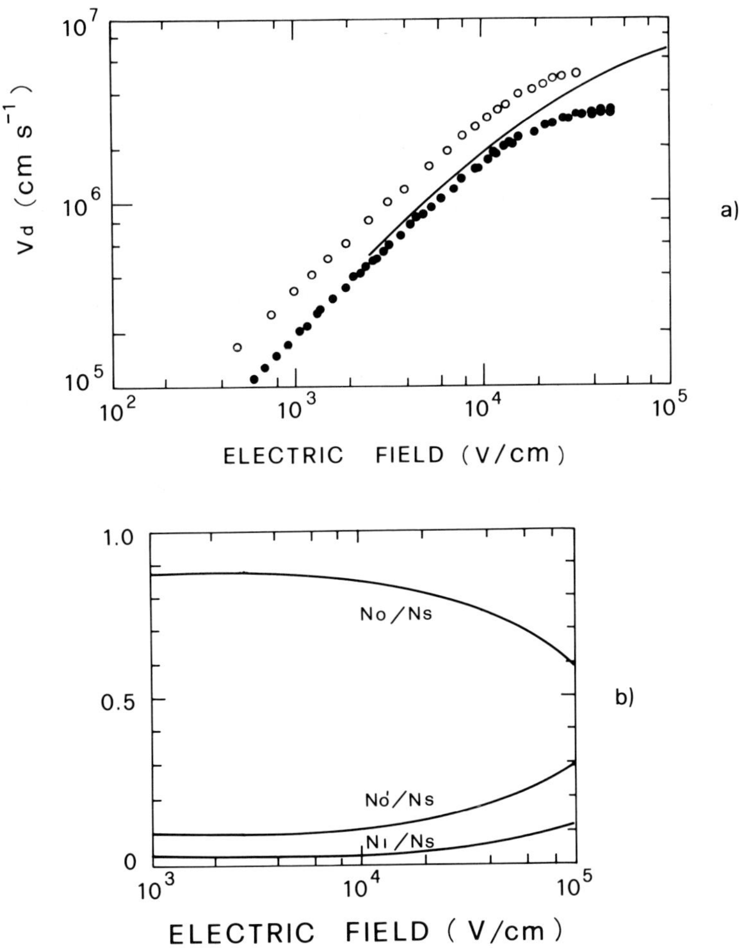

Fig. 3.23. (*a*) Drift velocity of hot electrons in Si-MOS inversion layers at 300 K as a function of electric field. Points indicate experimental results and the line MC results for 2D electrons. (*b*) Relative populations in the first three subbands as a function of electric field obtained from a MC simulation of 2D electrons. [After Hamaguchi, C: Physica 134b, 87 (1985)]

3.11 Special Topics

3.11.1 Periodic Fields

Monte Carlo simulation can be extended to permit calculation of the response of charge-carriers to periodic external fields of any strength [36, 37]. If a field

$$\mathbf{E} = \mathbf{E}_0 + \mathbf{E}_1 \sin(\omega t) \tag{3.11.1}$$

is applied, and the ac term is small enough to be in the linear-response regime, the average electron velocity will be of the form

$$\langle \mathbf{v}(t) \rangle = \mathbf{v}_0 + \mathbf{v}_1 \sin(\omega t) + \mathbf{v}_2 \cos(\omega t). \tag{3.11.2}$$

The coefficients $\mathbf{v}_1$ and $\mathbf{v}_2$ of the fundamental response in Eq. (3.11.2) can be obtained as sine and cosine Fourier transforms, respectively, of the velocity of the simulated electron over its history. Since the equation of motion of the particle subject to a field given by Eq. (3.11.1) is known in explicit terms, the Fourier coefficients $\mathbf{v}_1$ and $\mathbf{v}_2$ can easily be obtained by the simulation [36].

For large periodic fields, the periodic part of the current will contain higher harmonics, besides the fundamental frequency. These components can also be obtained by Fourier analysis of the simulated velocity, but in this case statistical noise problems become severe.

It is also possible to obtain the total response of the electron gas without Fourier analysis, by sampling the electron velocity at fixed times, corresponding to definite phases in the period of the external force. To be more explicit, as illustrated in Fig. 3.24, let us suppose that we "read" from the simulation the electron velocity at times given by 0, Δt, 2 Δt, ..., l Δt, ...,

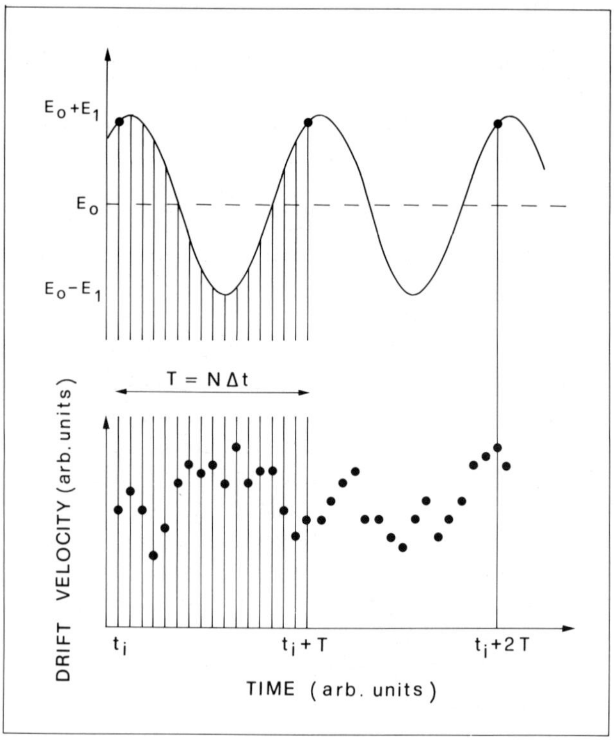

Fig. 3.24. Simulation of transport with periodic fields. A random variable is recorded in response to the periodic input signal $E(t)$ [37]. See text

where Δt is a fraction $1/N$ of the period $2\pi/\omega$ of the ac field:

$$\Delta t = \frac{1}{N} \frac{2\pi}{\omega} . \tag{3.11.3}$$

Then we average the values of $\mathbf{v}$ obtained at times $l\,\Delta t$, $(l + N)\,\Delta t$, $(l + 2N)\,\Delta t, \ldots$. The result is an estimation of the average electron velocity $\langle \mathbf{v} \rangle$, which is a periodic function of t with the same period $2\pi/\omega$, at the times given above.
The synchronous ensemble method described in section 3.7.2 for static fields has been extended by Lebwohl [36] to include the case of periodic fields. The result is a simple parametric dependence of Eq. (3.7.4) upon time.

3.11.2 Diffusion, Autocorrelation Function, and Noise

Diffusion may be considered a special, important case of space dependent phenomenon which is in general also time dependent, and a great deal of work has been done in this field with the MC technique (see for example, Refs. [38] to [42]).
In the linear-response regime, diffusion D and mobility μ are related by the Einstein relation (Eq. 2.7.8). Thus an independent determination of the diffusion coefficient does not add any particular information about the transport properties of a material that one could not obtain by determining the Ohmic mobility.
At high fields, however, the Einstein relation fails and the study of hot-electron diffusion becomes useful for two reasons: (i) from a basic point of view it provides an independent check, in addition to the drift velocity, of theoretical models of the material under investigation; (ii) from a more applied point of view, it provides useful information for the analysis of most solid-state devices.
It must be added that deviations of the diffusion coefficient from its equilibrium value can also occur at fields lower than those at which hot-electron phenomena are usually present, because of the possibility of intervalley diffusion, which occurs in many-valley semiconductors [40].
Equation (2.7.11) can be used to determine the diffusion coefficient by means of a MC simulation: a number of particles are independently simulated and their positions are recorded at fixed times. For large enough simulation times, the second central moment shows the linearity predicted by Eq. (2.7.11), and from its slope D_l is obtained. For short simulation times Eq. (2.7.11) does not hold, since the space-velocity correlations discussed in Sec. 2.7.1 are not yet established [42].
Other components of D can be obtained in a similar way with the Monte Carlo method, since, in general,

$$D_{ij} = \tfrac{1}{2}(d/dt)\langle (x_i - \langle x_i \rangle)(x_j - \langle x_j \rangle) \rangle . \tag{3.11.4}$$

A Monte Carlo determination of D can also be performed by using Eq.

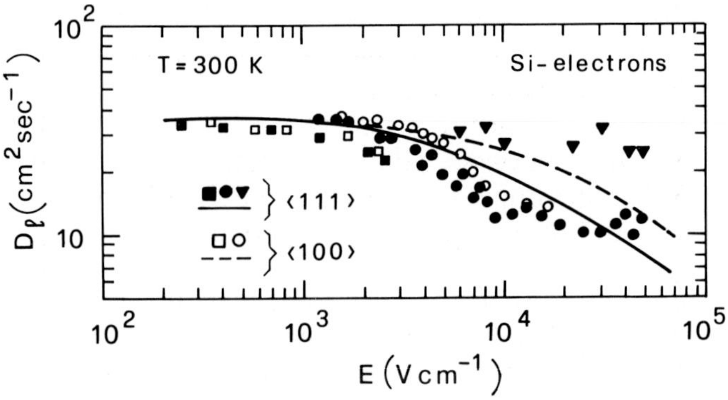

Fig. 3.25. Longitudinal diffusion coefficient of electrons in Si as a function of field at 300 K. Points refer to experimental data and lines to MC calculations [4]

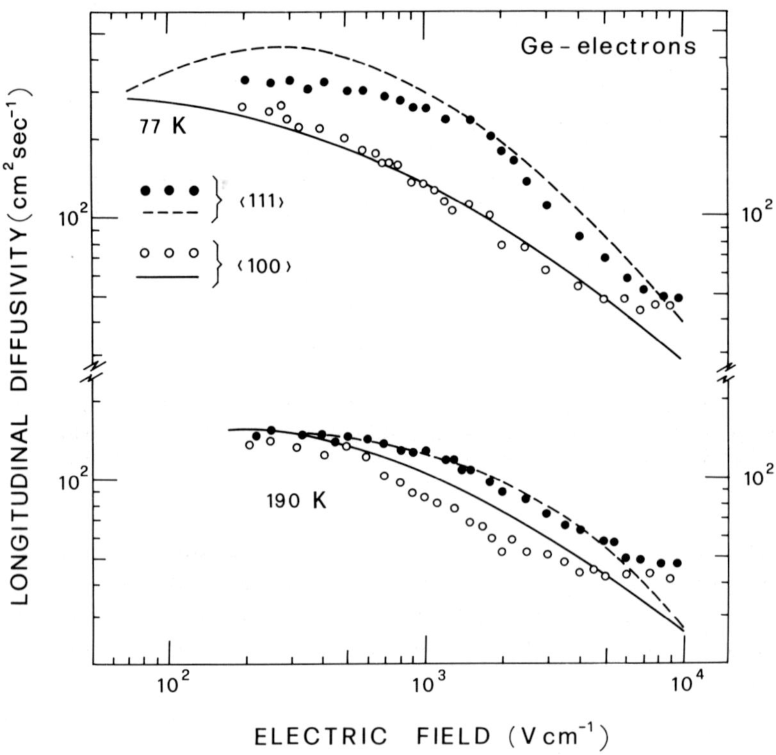

Fig. 3.26. Longitudinal diffusion coefficient of electrons in Ge as a function of field at different temperatures. Circles indicate experimental data; lines indicate MC calculations

(2.7.12) with a many-particle simulation. Figure 2.33 of chapter 2 and Figs. 3.25 to 3.27 show some examples of diffusion coefficients evaluated with MC techniques.

The evaluation of the autocorrelation function of δv as given by Eq. (2.7.13) in steady-state conditions can be easily performed from a MC simulation

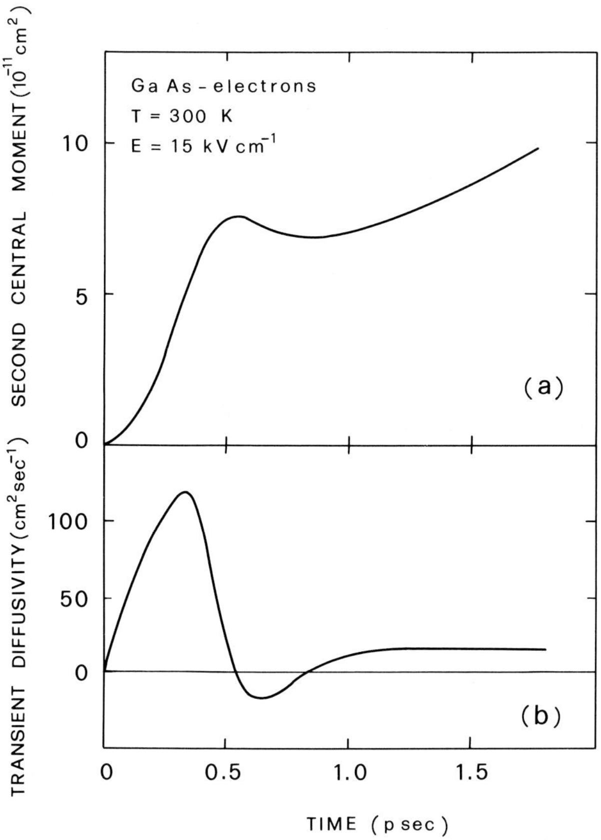

Fig. 3.27. (*a*) Second central moment and (*b*) longitudinal diffusion coefficient as functions of time for electrons in GaAs, as obtained from a MC simulation, at 300 K with an applied field of 15 kV/cm. The final slope of the curve in (*a*) corresponds to the steady state value of the diffusion coefficient shown in (*b*) at sufficiently long times [42]

as follows: let T be the time interval in which the autocorrelation function is to be sampled. In general T must be taken as much longer than the autocorrelation time, that is, such that $C(t) \sim 0$ for $t \geqslant T$. Then T is divided into a number M of intervals of duration $\Delta T = T/M$ to determine $C(t)$ at the times

$$0, \Delta T, 2\,\Delta T, \ldots, M\,\Delta T = T. \tag{3.11.5}$$

During the simulation, the velocity of the sampled particle is recorded at the time values $i\,\Delta T$, $i = 0, 1, \ldots$. When i becomes greater than M, the products

$$v(i\,\Delta T)v((i-j)\,\Delta T), \qquad j = 0, 1, \ldots, M, \tag{3.11.6}$$

are evaluated for each i and j. Products corresponding to the same value of j are averaged over the simulation, thus obtaining

$$\overline{v(t)v(t+j\,\Delta T)} = C(j\,\Delta T) + v_d^2, \tag{3.11.7}$$

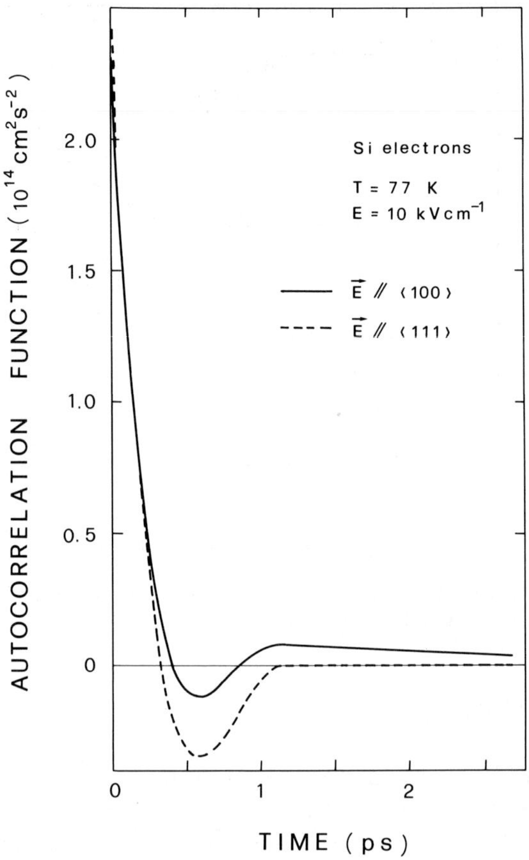

Fig. 3.28. Velocity autocorrelation function for electrons in Si, as obtained from a MC simulation, at 77 K for an applied field of 10 kV/cm. The long-time tail for the field parallel to a ⟨100⟩ direction is due to intervalley transfer [42]

since in a steady-state situation the time average (indicated by the overbar) is equivalent to the ensemble average.

Figure 3.28 shows an example of velocity autocorrelation functions determined by the technique described above.

As shown in chapter 2 the velocity autocorrelation function can be used to determine the diffusion coefficient [see Eq. (2.7.14)] or the noise spectrum [see Eq. (2.7.16)]. Therefore the noise spectrum can be obtained from a MC procedure either as the Fourier transform of $C(t)$ or by taking the average of the absolute squared Fourier transform of the electron velocity over a particle ensemble [43, 44]. Figure 3.29 shows an example of such an application.

The MC technique has also been applied to the analysis of the different physical sources of noise [45, 46].

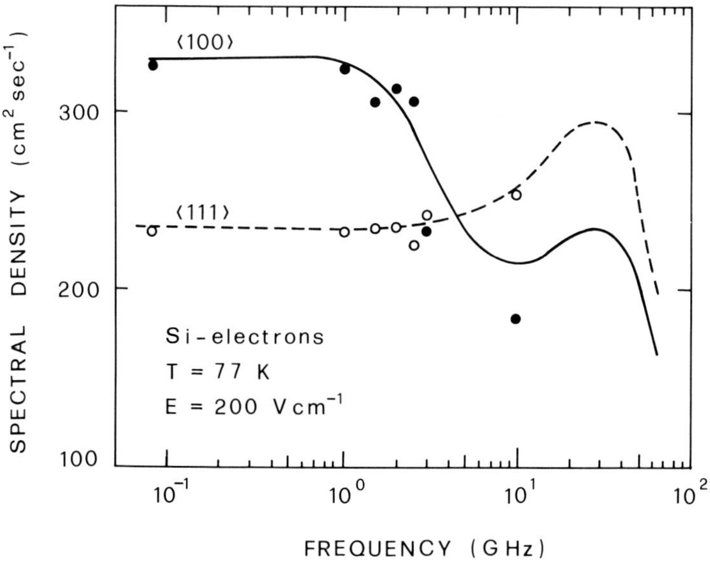

Fig. 3.29. Spectral density of velocity fluctuations for electrons in Si at 77 K with an applied field of 200 V/cm. Points refer to experimental data and lines to MC calculations [46]

3.11.3 Ohmic Mobility

When the Monte Carlo method is used to obtain the drift velocity of charge carriers at low applied fields, the statistical uncertainty originating from thermal motion may become particularly large. Of course, in principle, apart from the round-off errors of the computer, any desired precision can be obtained, but the computer time necessary for a high degree of precision often makes such a method impractical and expensive at very low fields.

The uncertainty due to thermal motion is particularly bothersome when the Ohmic mobility is sought. In this case, however, it is possible to evaluate the diffusion coefficient at zero field by one of the methods discussed in the previous section and then to obtain the Ohmic mobility by means of the Einstein relation. It is worth noting that when no external field is applied, the energy, and therefore the scattering probability, of the particle is constant during a free flight, so that self-scattering is not necessary. The Ohmic mobilities shown in Figs. 2.23 and 2.25 are evaluated with this method.

At low fields, another difficulty may arise when we deal with many-valley semiconductors. Intervalley scattering is due to phonons with a minimum equivalent temperature of the order of 10^2 K. If the lattice temperature is well below such a value, intervalley phonon absorption is almost absent because very few phonons are present in the crystal with that energy and, at low fields, the electron energy very rarely reaches values such as to make intervalley phonon emission possible. As a consequence, the MC simulation may proceed for very long times without any valley change of the sampling electron. In this situation it is very difficult to obtain a correct estimate of valley populations. When Ohmic conditions are analyzed and no intervalley transition is present at all, this difficulty may be overcome by simulating one electron in each valley, since we know a priori that the populations of all valleys must be the same. On the other hand, when we enter the field region in which a repopulation occurs and intervalley transitions are very rare, the MC method becomes inefficient. We shall return to this point in section 3.12 when we consider several variance-reducing techniques.

3.11.4 Impact Ionization

The impact ionization rate α_I can be obtained from the MC method by introducing the probability of impact ionization as an independent scattering mechanism which is added to phonon and impurity scatterings [20, 47–49].

In an EMC method, the impact ionization process has been examined by the following procedure [20]. The energy of the minority carrier (of the created electron-hole pair) relative to the band edge is neglected and the carrier itself disregarded. Therefore the sum of the two majority-carrier final energies $(\varepsilon_{f1} + \varepsilon_{f2})$ equals $(\varepsilon - \varepsilon_g)$, where ε is the initial energy and ε_g the energy gap. After each ionization process, one of the resulting $(N + 1)$ particles, chosen at random, is eliminated to maintain a fixed sample size. The pair generation rate per particle per unit time, g_I, is obtained by counting ionization events; it is converted to impact ionization rate by

$$\alpha_I = g_I/v_d\,. \tag{3.11.8}$$

In other papers [47–49], a single-particle MC simulation is used, and when impact ionization occurs the created electron-hole pair is disregarded. The impact ionization rate has been obtained by averaging the distance to impact ionization over a sufficient number of ionizations, as in Eq. (3.11.8) [50, 51].

3.11.5 Magnetic Fields

The scope of the MC method can be extended to include the case where both magnetic and electric fields are present. In this case the equation of motion of a carrier during a free flight will be

$$\hbar\dot{\mathbf{k}} = -e\left[\mathbf{E} + \frac{\mathbf{v}}{c} \times \mathbf{B}\right], \tag{3.11.9}$$

where **B** is the magnetic field, c the light velocity in vacuum, and **v** the group velocity of the particle.

Apart from this change in electron dynamics, the whole MC simulation

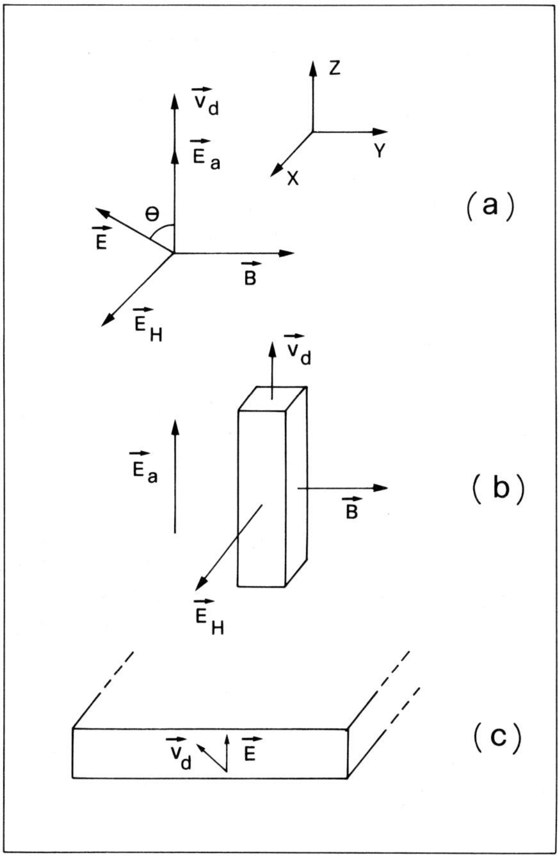

Fig. 3.30. Geometry for B, E, and v_d (see text)

proceeds as usual [52, 53] but the drift velocity will not, in general, be parallel to the applied electric field. Loss of cylindrical symmetry requires in this case a full three-dimensional simulation, even with very simple models. With a spherical band and a magnetic field **B** orthogonal to **E**, the drift velocity $\mathbf{v}_d$ is also in the plane perpendicular to **B** as shown in Fig. 3.30. A Monte Carlo simulation performed with such a geometry of fields for an infinite homogeneous system yields the drift velocity $\mathbf{v}_d$, and in particular the two components v_l and v_t parallel and perpendicular to **E**, and the angle θ formed by $\mathbf{v}_d$ and **E**.

At this point we may interpret the results of the simulation in two different physical ways [52], according to the particular system under consideration. One interpretation corresponds to the case of a long sample. The electric field **E** is decomposed into two components $\mathbf{E}_a$, parallel to $\mathbf{v}_d$, which must be considered as the applied field, and $\mathbf{E}_H$, perpendicular to both **B** and $\mathbf{v}_d$, which must be considered as the Hall field [see parts (a) and (b) of Fig. 3.30]. In this case the drift mobility μ_d is determined by

$$\mu_d = \frac{v_d}{E_a} = \frac{v_d}{E \cos \theta}, \tag{3.11.10}$$

while the Hall mobility is

$$\mu_H = \frac{E_H}{BE_a}. \tag{3.11.11}$$

Therefore, from the MC simulation it is possible to obtain a direct determination of the Hall scattering factor r_H given by [54]

$$r_H = \frac{\mu_H}{\mu_d} = \frac{E_H}{Bv_d}. \tag{3.11.12}$$

This number is of interest, since it depends upon the particular type of scattering mechanism that controls the transport process.
The second interpretation corresponds to the case of very short flat samples in which the electrons can move from one contact to the other at the Hall angle θ with respect to the applied field, thus producing a transverse current and no Hall field [see part (c) of Fig. 3.30]. The longitudinal mobility in this case is obtained as

$$\mu_E = \frac{v_l}{E} = \frac{v_d}{E} \cos \theta = \mu_d \cos^2 \theta. \tag{3.11.13}$$

For complicated systems, such as nonspherical bands, multivalley bands, and bipolar conduction, the usual modifications of the Hall-effect analysis [52] must be applied to the interpretation of MC results.

3.11.6 *Optical Excitation*

Optical measurement has become in recent years one of the most important and effective tools in the investigation of carrier dynamics in semiconductors. As the resolution of optical techniques has been pushed below the picosecond limit (laser pulses with a time spread of a few hundred femtoseconds can nowadays be obtained by various methods [55, 56]), it has become possible to perform time-resolved measurements that give information on the evolution of photoexcited carrier distributions. The Ensemble Monte Carlo provides an excellent theoretical probe of the ultrafast phenomena that are active on subpicosecond and picosecond time scales. With minor modifications with respect to the general algorithm described in the present chapter, it is possible to have a one-to-one correspondence between the simulation and the experiment. Such modifications involve mainly the initial condition of the MC simulation. In the experiments, carriers (electrons and holes) are created by the incoming laser pulse, then they approach equilibrium through an internal thermalization (via carrier-carrier scattering) and an equilibration with the lattice (via carrier-phonon scattering). The number of simulated carriers varies with the law [57].

$$n(t) = n_0 + 2C \operatorname{arctg}(e^{\omega t}), \tag{3.11.14}$$

which describes the excitation by a laser pulse intensity

$$I(t) = I_0 \cosh^{-1}(\omega t), \tag{3.11.15}$$

where ω and C are parameters related to the width and power of the laser pulse.

Density-dependent quantities (such as the screening parameter and the carrier-carrier scattering rate) have to be adjusted during the simulation to follow the change in carrier concentration due to the laser pulse.

The **k**-space distribution is determined by the allowed transitions activated by light absorption. In polar materials, vertical transitions from the valence to the conduction band lead to *e-h* pair creations. In general, all three hole bands are involved, and the initial carrier distribution can be very complicated.

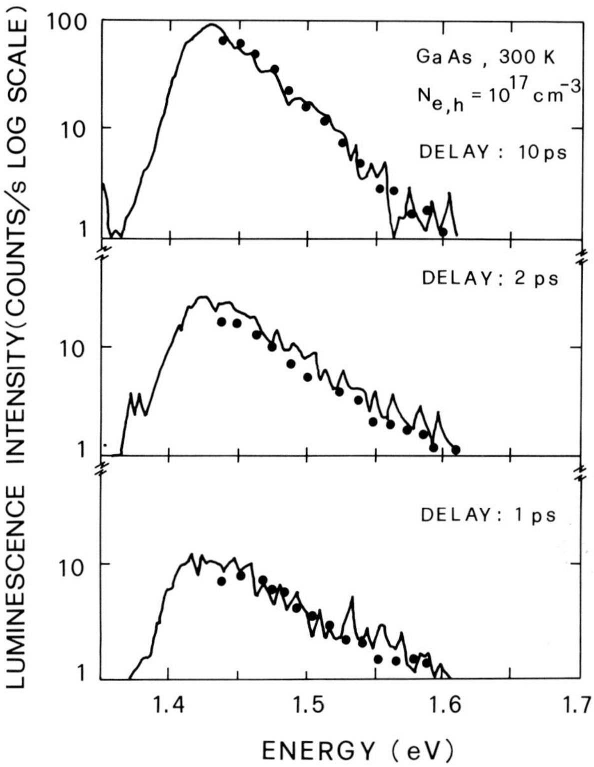

Fig. 3.31. Luminescence spectra of GaAs at 300 K excited by 400 fs pulses at 2.04 eV for three different delays after excitation. Zero delay corresponds to the center of the laser pulse. Continuous lines refer to the experimental spectra, while full circles are calculated with an ensemble MC simulation [58]

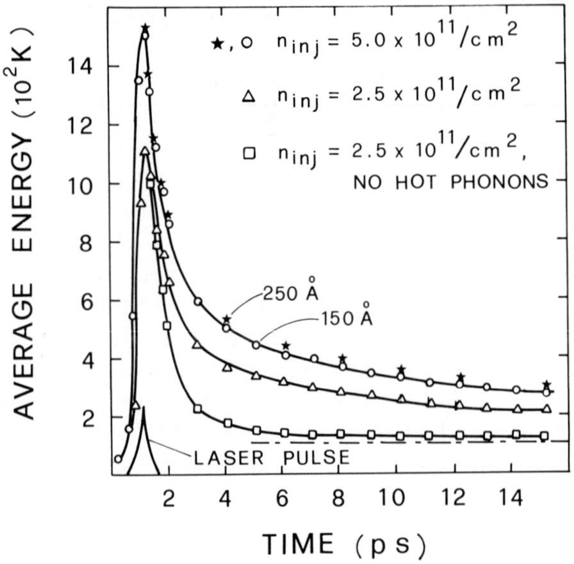

Fig. 3.32. Average electron energy vs time for two different injection densities in a 150 Å AlGaAs-GaAs-AlGaAs single quantum well at 5 K, obtained from an ensemble MC calculation that includes the effect of non-equilibrium LO phonons. For the highest injection case, results are shown for well widths of 150 Å (circles) and 250 Å (asterisks). The dashed curve represents the equilibrium energy for a total concentration of 5×10^{11} cm^{-2} [59]

An example of MC results for optical excitation in bulk semiconductors and quantum well is presented in Figs. 3.31 and 3.32 [58, 59].

3.11.7 Quantum Mechanical Corrections

The Monte Carlo procedure presented in the present section is a semi-classical technique, in that it combines classical features (free flights calculated from Newton's law) with quantum-mechanical ones (scattering rates taken from the first-order perturbation theory). Collisions are treated as local events in space and time. The possibility of going beyond such an approximation by extending the standard Monte Carlo procedure has been investigated by several authors. Two main streams can be identified. In the first type, spectral density functions are used to describe the scattering event. The energy-conserving δ-function typical of Fermi golden rule is substituted by the joint spectral-density function [60]. The one-to-one correspondence between ε and $\mathbf{k}$ given by the dispersion relation $\varepsilon(\mathbf{k})$ is relaxed, and a many-body energy is introduced, which corresponds to a broadening of the one-electron energy due to the effect of the interaction. Figures 3.33 and 3.34 give a pictorial description of the scattering process in the standard Monte Carlo procedure and in the modified one, respectively. The justification of the use of spectral densities comes from a fully quantum mechanical ap-

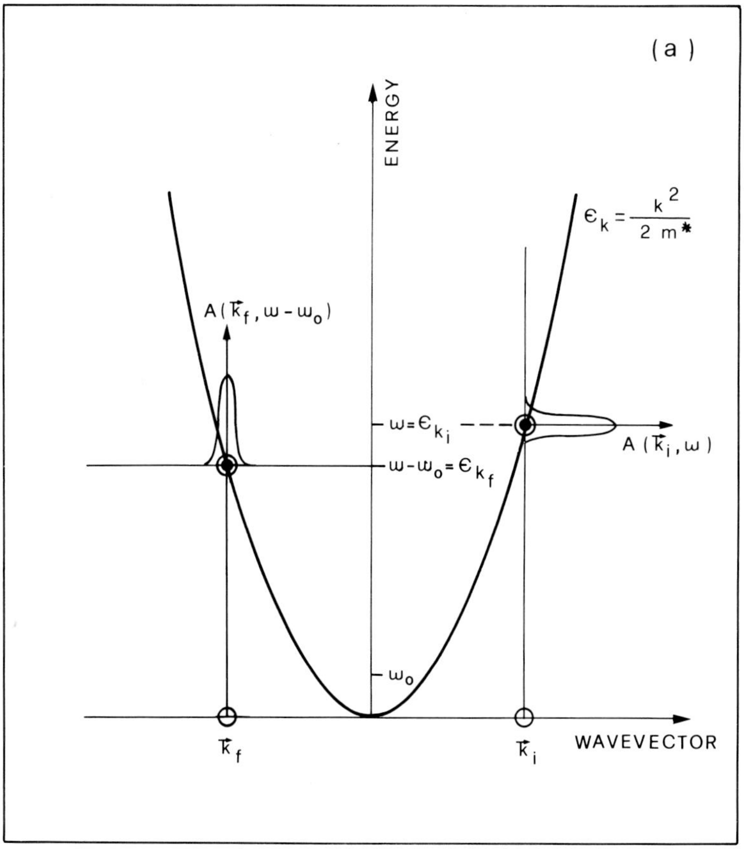

Fig. 3.33. Schematic picture of a transition due to optical phonon emission as described in terms of spectral density. In the non-interacting model displayed here, the spectral densities are δ functions of $\varepsilon_{kf} - (\varepsilon_{ki} - \hbar\omega_0)$; therefore kinetic energy conservation is fulfilled [60]

proach based on the generalized Kadanoff-Baym kinetic equation [60]. The accelerative part of the MC algorithm is left unchanged, while the selection of the final state of the scattering process is performed using Lorenzian functions fitted to the exact joint spectral density for the given transition. The effect of collisional broadening is an enhancement of the low- and high-energy tails of the distribution function, especially at very high electric fields. Such an effect can be of great importance for phenomena with a definite threshold at high-energy, such as electron injection from Si into SiO_2 [5] or impact ionization [61].

An alternative approach is based on the numerical solution of Lionville's equation for the density matrix in an interacting electron-phonon system [62, 63]. There, a Monte Carlo sampling of various terms of the perturbative series expansion of Lionville's equation is used. Thus, the approach does not represent an extension of the traditional MC procedure, but rather a

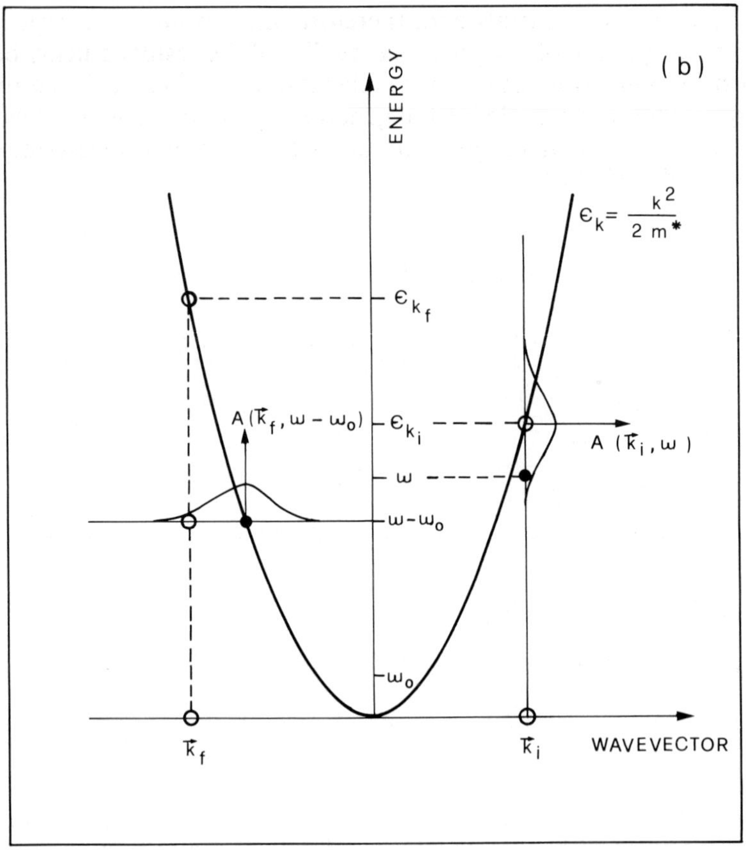

Fig. 3.34. Same as Fig. 3.33 for the interacting model. Now the spectral densities are broadened and $\varepsilon_{kf} \neq (\varepsilon_{ki} - \hbar\omega_0)$ so that the kinetic energy is not conserved

novel quantum Monte Carlo approach along the same lines as those developed to solve Dyson's equation. [64]. The semiclassical limit of this procedure yields a new "backward" MC technique for the solution of the Boltzmann equation, in which the state **k** where the distribution function is to be evaluated, is chosen arbitrarily at the beginning of the procedure [65].

3.12 Variance-reducing Techniques

The Monte Carlo method is a statistical procedure for the solution of mathematical problems. Therefore, the results obtained by means of such a method are always affected by some statistical uncertainty. In secion 3.7.3 we discussed the standard algorithms used to evaluate this uncertainty for the solution of transport problems.

As a general rule, the statistical precision of the results increases as the square root of the number of trials and, therefore, the amount of computer time necessary to appreciably improve the quality of the results quickly becomes very considerable. It is thus of particular interest to find variance-reducing techniques, which can be defined in general as procedures which "change or at least distort the original problem in such a way that uncertainty in the answer is reduced" [19].

3.12.1 Variance Due to Thermal Fluctuations

In MC simulation the drift velocity v_d of the sample electron is obtained as an average of its velocity during the simulation. If v_d is a small fraction of the thermal velocity v_{th}, an estimate of v_d is, in general, substantially influenced by the variance due to v_{th}. In many cases this contribution to the uncertainty on v_d can be considerably reduced by a modification of the MC procedure [3, 66], as described below.

When the transition probability per unit time is velocity randomizing, i.e., is such that

$$P(\mathbf{k}, \mathbf{k}') = P(\mathbf{k}, -\mathbf{k}'), \tag{3.12.1}$$

then a value $\mathbf{k}'$ for the state after scattering is as probable as its opposite $-\mathbf{k}'$. The random number used to generate $\mathbf{k}_a$, by selecting one of them, introduces a variance that, in turn, builds up the thermal fluctuation of the electron velocity.

It is possible to eliminate this fluctuation by splitting the simulation into two different histories, corresponding to the two equally probable initial conditions $\mathbf{k}_a$ and $-\mathbf{k}_a$ for the new free flight. If a constant Γ is used and the same random number is taken to generate the two flights, the latter have equal weights in the simulation. Thus both can be considered by taking into account the averages of the quantities evaluated in the two flights, thereby reducing the variance by eliminating thermal contribution. One of the flights in finally chosen at random to continue the simulation.

This procedure can be extended to more than two states after scattering and, with appropriate weights, to the case of non-velocity-randomizing processes.

3.12.2 Variance Due to Valley Repopulation

As is well known [67], in a many-valley semiconductor when the valleys are differently oriented with respect to the applied field, they contribute with different valley drift velocities $\mathbf{v}_{dv}$ to the total average electron drift velocity $\mathbf{v}_d$:

$$\mathbf{v}_d = \sum_v n_v \mathbf{v}_{dv}, \tag{3.12.2}$$

where n_v is the relative concentration of carriers in the vth valley.

Under hot-electron conditions, this situation corresponds to different carrier heating, and a redistribution of the electron among the different valleys occurs, called "valley repopulation".

At low fields and temperatures, intervalley transitions are very rare, and the variance of the concentrations n_v and, in turn, of the drift velocity $\mathbf{v}_d$ may be very large. The situation may be improved by using the distribution function, as obtained in each type of valley with the MC calculation, and the transition probabilities from the valleys of different types, to evaluate the valley repopulation: [15, 66]. From a balance of the transition rates from hot to cold valleys, one obtains the repopulation ratio n_c/n_h. Such an estimate exploits all the information obtained during the simulation (i.e., information acquired when the electron "visits" the region of energy where intervalley scattering is possible), not only the information obtained from transitions that have actually happened. In this way the variance associated with the repopulation ratio is reduced. However, it must be noted that, in order to get correct results with the above procedure, one must know the tails of the distribution function in hot and cold valleys with sufficient accuracy at energies above threshold for intervalley phonon emission. This again can become a severe problem, which can be approached with the variance-reducing technique described next.

3.12.3 Variance Related to Improbable Electron States

When a physical phenomenon of interest is due to the occurrence of improbable electron states, the standard MC simulation may lead to a large variance of the desired quantity. Situations of this kind are usually related to the tails of the energy distribution function, as, for example, in the case of valley repopulation seen above, impact ionization, or escapes of electrons into the gate insulator in a field-effect transistor.

In such cases the variance can be reduced by the following procedure [68]. Within a simulation, when an electron enters a fixed "rare region" of energy, its entering state is recorded and used for generating a fixed number N of different "parallel" simulations; each of these uses different random numbers and ends when the electron exits from the rare region. Then, starting from one of the N exit states chosen at random, the simulation proceeds in the usual way until the rare region is reached again. In the averaging procedures a weight $1/N$ must be given to each parallel simulation.

The variance-reducing techniques above may be very useful in some cases. It must be noted, however, that they are realized somehow by means of a distortion of the original problem, which causes the program to deviate from the strict simulation of possible electron histories.

A strict simulation, in general, has the advantage of yielding a simple and straightforward means of physical interpretation of the phenomenon under investigation. Consequently, as a general rule, a correct balance between computing time, transparency of the simulative procedure, and complexity of the computer program must be found in connection with the particular needs of each single case.

3.13 Comparison with Other Techniques

The MC method is now by far the most extensively used in high-field transport studies and has amply extended the range of problems that can be successfully approached. Other techniques exist, however, that are sometimes still used and deserve consideration since they may be particularly useful under special conditions. In what follows, we shall mention briefly some of these alternative techniques and compare them with the MC method.

3.13.1 Analytical Techniques

Analytical techniques for the solution of the nonlinearized Boltzmann equation were developed mainly during the two decades following the beginning of the study of the hot-electron problem [69, 70] up to the introduction of the numerical techniques, which were suggested at the Kyoto Conference in 1966 [1, 71]. A standard reference for the research performed in that period is Conwell's book [67] and we refer the reader interested in details to this comprehensive review.

Even though analytical techniques require severe approximations in the calculations and can be applied only to extremely simple semiconductor models, they have the merit of having provided a first physical insight into the problem of nonlinear transport. Some important concepts, such as, for example, the electron temperature and the energy and momentum relaxation times, are still valuable heuristic quantities, although it is now known that a rigorous application of these concepts is possible only in a very limited number of actual cases.

Among the analytical techniques developed for the analysis of high-field transport, two main procedures can be recognized. In the first one an a priori analytical form of the distribution function is assumed, with some parameters to be determined by using the Boltzmann equation. A heated and drifted Maxwellian distribution has been widely used:

$$f(\mathbf{k}) \sim \exp[-\hbar^2(\mathbf{k}-\mathbf{k}_d)^2/(2mK_BT_e)], \tag{3.13.1}$$

where $\hbar\mathbf{k}_d$ is the average momentum of the distribution, and T_e the electron temperature, which, under hot electron conditions, is in general higher than the lattice temperature. The two parameters $\mathbf{k}_d$ and T_e are usually determined by satisfying the first two moments of the Boltzmann equation [67]. A distribution function of the type given by Eq. (3.13.1) should be a good approximation when the electron concentration is sufficiently high. Although it is applicable in a limited number of simple cases, it contains most of the essential features of hot-electron physics.

A second technique [72–76] makes use of an expansion of the distribution function in spherical harmonics. For problems with cylindrical symmetry around the direction of the field, the expansion can be made in Legendre polynomials $P_n(\cos\theta)$:

$$f(\mathbf{k}) = f_0(\varepsilon) + f_1(\varepsilon)P_1(\cos\theta) + \cdots, \tag{3.13.2}$$

where θ is the angle between the field direction and $\mathbf{k}$. When the expansion is substituted into the Boltzmann equation, by using the orthogonality property of Legendre polynomials one obtains a system of infinite coupled equations for $f_0, f_1, \ldots$. However, further approximations are required at this stage, and the overall accuracy of the method becomes very questionable.

3.13.2 The Iterative Technique

Other numerical techniques [8, 17, 71, 77–80] for the solution of the Bolyzmann equation were developed more or less at the same time as the MC method. The most important of them is closely related to the MC technique and yields a solution of the Boltzmann equation by means of an iterative procedure. We shall outline it briefly here.

It can be shown [8] that for a uniform system the Boltzmann equation admits the following formal solution:

$$\begin{aligned} f(\mathbf{k}, t) = \frac{V}{(2\pi)^3} \int_0^{\infty} dt'' \exp\left[-\int_0^{t''} \lambda\left[\mathbf{k} - \frac{e\mathbf{E}}{\hbar}t'''\right]dt'''\right] \\ \times \int d\mathbf{k}'\, f(\mathbf{k}', t - t'')P\left[\mathbf{k} - \frac{e\mathbf{E}}{\hbar}t'', \mathbf{k}'\right], \end{aligned} \tag{3.13.3}$$

where $\lambda(\mathbf{k})$ is the total scattering probability.

The iterative technique for the solution of Eq. (3.13.3) consists in substituting an arbitrary function $f_0(\mathbf{k}, t)$ on the right-hand side of Eq. (3.13.3), and calculating $f(\mathbf{k}, t)$ as the right-hand side of Eq. (3.13.3) itself. This function is then resubstituted on the right-hand side, and the procedure is repeated until convergence is achieved.

Numerical iterative solutions of Eq. (3.13.3) have been discussed for stationary conditions, in which the dependence of f upon t vanishes [8, 18, 71, 81]. An important result can be obtained when a self scattering is introduced and the constant scattering rate Γ is made large with respect to the scattering rate λ. Under this condition the following relation holds [82]:

$$f_{n+1}(\mathbf{k}) \simeq f_n(\mathbf{k}) + \frac{1}{\Gamma}\frac{\partial f_n(\mathbf{k}; t)}{\partial t}, \tag{3.13.4}$$

where $f_{n+1}(\mathbf{k})$ and $f_n(\mathbf{k})$ are the distribution functions obtained from the

$(n + 1)$th and the nth iteration, respectively. Each step in the iterative process corresponds, then, to a time increment of $1/\Gamma$ for the evaluation of the physical system. This result enables us to use the iterative method to study the evolution of the distribution function during the transient from a given initial condition $f_0(\mathbf{k})$ to the stationary state.

3.13.3 Comparison of the Different Techniques

When compared with the MC method or other exact techniques, the two analytical procedures sketched above differ markedly in their degree of usefulness. The drifted-maxwellian technique aims at giving a simple physical insight into a hot-electron problem, and, as such, is still a very valuable method when we are not looking for particular details. The Legendre-polynomial expansion, on the other hand, was intended to be the most refined analytical method for an exact as possible solution of the Boltzmann equation and, as such, it has been superseded by the more recent numerical techniques.

As regards a comparison between iterative and MC techniques, we may first comment that both of them given the exact numerical solution for a homogeneous, steady-state, hot-electron phenomenon with noninteracting particles. Both of them can include details of the microscopic model without difficulty and can be extended to time-dependent phenomena.

The iterative technique processes the whole distribution function at each step of the procedure, while with the MC method the distribution function is sampled at each electron flight. For this reason the iterative technique can be more convenient when we deal with physical phenomena that depend on details of the distribution function. On the other hand, some difficulties may arise in the evaluation of the integral in Eq. (3.13.3), and the effect on final results of any numerical approximation may not be easy to control.

The major advantage of the MC technique (besides its easy implementation) arises from its being a direct dynamical simulative procedure, so that any complicated phenomenon can be analyzed in terms of its elementary processes in quite a simple way. Furthermore, fluctuations can be studied with a very straightforward analysis, and, through an ensemble simulation, space-and-time dependent phenomena (as for devices) can be easily studied.

These arguments form the basis for the popularity of the Monte Carlo method, which is at present much more widely used in hot-electron studies than any other technique.

References

[1] Kurosawa, T: Proc. Int. Conf. Phys. of Semiconductors, Kyoto. J. Phys. Soc Japan Suppl. *A49*, 345 (1966).

[2] Fawcett, W., Boardman, D. A., Swain, S.: J. Phys. Chem. Solids *31*, 1963 (1970).

[3] Price, P. J.: Semiconductors and Semimetals *14*, 249 (1979).

[4] Jacoboni, C., Reggiani, L.: Rev. Mod. Phys. *55*, 645 (1983).
[5] Tang, J. Y., Hess, K.: J. Appl. Phys. *54*, 5139 (1983).
[6] Tang, J. Y.: Ph.D. Dissertation, University of Illinois at Urbana-Champain, 1983.
[7] Rees, H. D.: Phys. Lett. *A26*, 416 (1968).
[8] Rees, H. D.: J. Phys. Chem. Solids *30*, 643 (1969).
[9] Sangiorgi, E., Ricco, B., Venturi, F.: IEEE Trans. on CAD 7, 259 (1988).
[10] Hockney, R. W., Eastwood, J. W.: Computer Simulation Using Particles. Maidenhead: Mc Graw-Hill. 1981.
[11] Yorston, R. M.: J. Comp. Phys. *34*, 177 (1986).
[12] Chattopadhyay, D.: J. Appl. Phys. *45*, 4931 (1974).
[13] Takenaka, N., Inoue, M., Inuishi, Y.: J. Phys. Soc. Jap. *47*, 861 (1979).
[14] Lugli, P.: Ph.D. Dissertation, Colorado State University, 1985.
[15] Canali, C., Jacoboni, C., Nava, F., Ottaviani, G., Alberigi-Quaranta, A.: Phys. Rev. *B12*, 2265 (1975).
[16] Jacoboni, C., Nava, F., Canali, C., Ottaviani, G.: Phys. Rev. *B24*, 1014 (1981).
[17] Price, P. J.: In: Proc., 9th Int. Conf. Phys. of Semiconductors (Ryvkin, S. M., ed.), p. 753. Leningrad: Nauka. 1968.
[18] Price, P. J.: IBM J. Res. Dev. *14*, 12 (1970).
[19] Hammarsley, J. M., Handscomb, D. C.: Monte Carlo Methods (Barlett, M. S., ed.). London: Methuen/Chapman and Hall. 1964.
[20] Lebwohl, P. A., Price, P. J.: Solid State Commun. *9*, 1221 (1971).
[21] Lebwohl, P. A., Price, P. J.: Appl. Phys. Lett. *19*, 530 (1971).
[22] Bacchelli, L., Jacoboni C.: Solid State Commun. *10*, 71 (1972).
[23] Matulionis, A., Pozhela, J., Reklaitis, A.: Solid State Commun. *16*, 1133 (1975).
[24] Lugli, P., Ferry, D. K.: Physica *117B*, 251 (1983).
[25] Lugli, P., Ferry, D. K.: Appl. Phys. Lett. *46*, 594 (1985).
[26] Brunetti, R., Jacoboni, C., Matulionis, A., Dienis, V.: Physica *134B*, 369 (1985).
[27] Lugli, P., Ferry, K. D.: IEEE El. Dev. Lett. *EDL6*, 25 (1985).
[28] Jacoboni, C.: In: Proc., 13th Int. Conf. Phys. of Semiconductors (Fumi, G., ed.), p. 1195. Roma: Marves. 1976.
[29] Platzman, P. M., Wolff, P. A.: Waves and Interactions in Solid State Plasmas. New York: Academic Press. 1973.
[30] Potter, D.: Computational Physics. London: Wiley. 1973.
[31] Lugli, P., Ferry, D. K.: Phys. Rev. Lett. *56*, 1295 (1986).
[32] Lugli, P., Reggiani, L.: In: Noise in Physical Systems and 1/f Noise-1985 (D'Amico, A., Mazzetti, P., eds.). Amsterdam-Oxford-New York-Tokyo: North-Holland. 1986.
[33] Bosi, S., Jacoboni, C.: J. Phys. *C9*, 315 (1976).
[34] Lugli, P., Ferry, D. K.: IEEE Trans. El. Dev. *ED32*, 2431 (1985).
[35] Goodnick, S. M., Lugli, P.: Phys. Rev. B *37*, 2578 (1988).
[36] Lebwohl, P. A.: J. Appl. Phys. *44*, 1744 (1973).
[37] Zimmermann, J., Leroy, Y., Constant, E.: J. Appl. Phys. *49*, 3378 (1978).
[38] Fawcett, W., Rees, H. D.: Phys. Letters *A29*, 578 (1969).
[39] Jacoboni, C., Brunetti, R., Reggiani, L.: Proc., 3th Int. Conf. Hot Carriers in Semiconductors. J. Phys. Colloq. *42*, C7-123 (1981).
[40] Brunetti, R., Jacoboni, C., Nava, F., Reggiani, R., Bosman, G., Zijlstra, R. J. J.: J. Appl. Phys. *52*, 6713 (1981).
[41] Boittiaux, B., Constant, E., Reggiani, L., Brunetti, R., Jacoboni, C.: Appl. Phys. Lett. *40*, 407 (1982).
[42] Brunetti, R., Jacoboni, C.: In: Semiconductors Probed by Ultrafast Laser Spectroscopy (Alfano, R. R., ed.), Vol. 1, chapter 12, p. 367. New York: Academic Press. 1984.
[43] Fauquembergue, R., Zimmermann, J., Kaszynski, A., Constant, E., Microondes, G.: J. Appl. Phys. *51*, 1065 (1980).
[44] Hill, G., Robson, P. N., Fawcett, W.: J. Appl. Phys. *50*, 356 (1979).
[45] Brunetti, R., Jacoboni, C.: Phys. Rev. Letters *50*, 1164 (1983).
[46] Brunetti, R., Jacoboni, C.: Phys. Rev. *B29*, 5739 (1984).

[47] Curby, R. C., Ferry, D. K.: Phys. Status Solidi *A15*, 319 (1973).
[48] Shichijo, H., Hess, K., Stillman, G. E.: Appl. Phys. Lett. *38*, 89 (1981).
[49] Shichijo, H., Hess, K.: Phys. Rev. *B23*, 4197 (1981).
[50] Kim, K., Hess, K.: J. Appl. Phys. *60*, 2626 (1986).
[51] Robbins, W. M., Wang, T., Brennan, K. F., Hess, K., Stillman, G. E.: J. Appl. Phys. *58*, 4614 (1985).
[52] Boardman, A. D., Fawcett, W., Ruch, G.: Phys. Status Solidi *A4*, 133 (1971).
[53] Chattopadhyay, D.: J. Appl. Phys. *45*, 4931 (1974).
[54] Seeger, K.: Semiconductor Physics. Wien-New York: Springer. 1973.
[55] Fork, R. L.: Physica *134B*, 381 (1985).
[56] Shah, J., Damen, T. C., Deveaud, B., Block, D.: Appl. Phys. Lett. *50*, 1307 (1987).
[57] Lugli, P.: Solid State Electron. *31*, 667 (1988).
[58] Shah, J., Deveaud, B., Damen, T. C., Tsang, W. T., Gossard, A. C., Lugli, P.: Phys. Rev. Letters *59*, 2222 (1987).
[59] Lugli, P., Goodnick, S. M.: Phys. Rev. Letters *59*, 716 (1987).
[60] Reggiani, L., Lugli, P., Jauho, A. P.: Phys. Rev. *B36*, 6602 (1987).
[61] Chang, Y., Ting, D. Z., Tang, J. Y., Hess, K.: Appl. Phys. Lett. *42*, 76 (1983).
[62] Brunetti, R., Jacoboni, C., Lugli, P., Reggiani, L.: Proc., 18th Int. Conf. Phys. Semiconductors (Engstrom, O., ed.), p. 1527. World Scientific. 1987.
[63] Brunetti, R., Jacoboni, C.: Solid State Electron. *31*, 527 (1988).
[64] Kalos, M. H. (ed.): Monte Carlo Methods in Quantum Problems. Dordrecht: D. Reidel. 1984.
[65] Jacoboni, C., Poli, P., Rota, L.: Solid State Electron. *31*, 523 (1988).
[66] Hammar, C.: Phys. Rev. *B4*, 417 (1971).
[67] Conwell, E. M.: High Field Transport in Semiconductors. Solid State Phys., Suppl. 9. New York: Academic Press. 1967.
[68] Philips, A. Jr., Price, P. J.: Appl. Phys. Lett. *30*, 528 (1977).
[69] Froehlich, H.: Proc. R. Soc. (London) *A188*, 521 and 532 (1947).
[70] Shockley, W.: Bell Syst. Tech. J. *30*, 990 (1951).
[71] Budd, H.: In: Proc., Int. Conf. Phys. Semiconductors, Kyoto. J. Phys. Soc. Japan Suppl. *21*, 420 (1966).
[72] Yamashita, J., Watanabe, M.: Prog. Theor. Phys. *12*, 443 (1954).
[73] Stratton, R.: J. Electron. *5*, 157 (1958).
[74] Stratton, R.: Proc. R. Soc. (London) *A246*, 406 (1958).
[75] Yamashita, J., Inoue, K.: J. Phys. Chem. Solids *12*, 1 (1959).
[76] Reik, H. G., Risken, H.: Phys. Rev. *126*, 1737 (1962).
[77] Hammar, C.: Phys. Status Solidi *A11*, 495 (1972).
[78] Hammar, C.: J. Phys. *C6*, 70 (1973).
[79] Nougier, J. P., Rolland, M.: Phys. Rev. *B8*, 5728 (1973).
[80] Vinter, B.: Doctoral Dissertation, University of Lyngby, 1973.
[81] Vassel, M. O.: J. Math. Phys. *11*, 408 (1970).
[82] Fawcett, W.: In: Electrons in Crystalline Solids (Salam, A., ed.), p. 531. Vienna: IAEA. 1973.

signed to operate at very high frequency (above 1 GHz), power devices, used for their capacity to handle large voltages and large currents, and switching devices, which should be able to change their state from a high-voltage, low-current (off) condition to a low-voltage high-current (on) state in a very short time. On the basis of their function, those devices can also be characterized as small-signal or large-signal amplification devices. An additional class of devices are the ones that are used to detect, generate or convert optical energy to electrical energy and vice versa.

We have chosen here to follow closely the historical evolution of semiconductor devices pointing out, as we proceed, the characteristics of specific structures in comparison with others. A thorough discussion will be given in section 4.3. Since they generally involve different transport physical properties, technological features and electronics applications, we will distinguish between semiconductor devices based on covalent materials (mainly Si), and on polar materials (mainly GaAs and GaAs alloys).

4.2.1 *Evolution of Si Devices*

The first Si transistor, the bipolar transistor, based on two adjacent *p-n* junction, was invented by Bardeen, Brattain and Shockley in 1947/48. The following years were characterized by extensive technological advances, from the control of the growth process to the development of diffusion and mask techniques. These advances allowed a considerable improvement of the device performance and the realization in 1955, that is only seven years after the invention of the transistor, of an all-transistor computer.

As will be seen in the following section, the bipolar transistors are "vertical" devices, since the current flows in the direction perpendicular to the semiconductor surface. Both type of carriers contribute to the current.

A completely different type of device (the junction field-effect transistor, or JFET) was proposed by Shockley in 1951 and realized a few years after. The JFET is a "horizontal" device, where carriers move in the direction parallel to the surface. The conduction process involves predominantly one type of carrier, preferably electrons because of their higher mobility. The name of the device comes from the fact that the current is controlled by an electric field perpendicular to the current flow. The control electrode (gate) is formed by a *p-n* junction (or, in the case of metal-semiconductor field effect transistor, or MESFET, by a Metal-Semiconductor junction).

Much more successful for Si devices has been another type of field effect transistor, namely the MOSFET (Metal-Oxide-Semiconductor Field Effect Transistor) which was first constructed in 1962. Some important features of this transistor (reduced area, rather simple processing, high imput impedence, low power dissipation) make it an ideal element for integrated circuits (I.C.). The years following the introduction of the MOSFET have witnessed an outstanding technological growth that has led to the development of complementary MOS (CMOS) circuits that use p and n-type channels (respectively n-MOS and p-MOS transistors). Because of the high integration level and the extremely low power dissipation CMOS circuits are one of the building blocks of VLSI electronics.

4.2.2 *Evolution of GaAs Devices*

Although the outstanding properties of GaAs for semiconductor device applications have been known since 1950, the first breakthrough was obtained with the invention of the Gunn diode in 1963, followed by the MESFET in 1966.

The major advantages of GaAs in comparison with Si are the following:

— The electron mobility in GaAs is much higher than Si. The peak velocity (that is the maximum velocity reached in the stationary velocity-field curve) is also larger.

— the band gap at room temperature is 1.43 eV for GaAs vs. 1.12 eV for Si. Therefore, it is possible to obtain semi-insulating substrates necessary to achieve an adequate isolation of devices on the same chip. The larger band gap also allows a wider range of temperatures for the operation of the device (typically in the range $-200\ ^\circ C \div 200\ ^\circ C$)

— Enhanced hardness against radiation.

On the other side, GaAs suffers serious disadvantages with respect to Si. The most important are:

— GaAs does not form a stable native oxide on the surface, as the SiO_2 on Si. Therefore, mask processes are more difficult on GaAs, and devices whose properties are based on transport at the oxide-semiconductor interface (as the MOSFET) are not very common.

— Doping via the diffusion of dopants is not possible in GaAs. Therefore, more elaborate and expensive techniques have to be used to introduce dopants in the material.

— GaAs tends to decompose above 500 C, releasing toxic Arsenic in the environment, therefore posing strict safety requirements.

— The cost of GaAs wafers is much higher than Si wafers, because of the larger technological complexity as well as the smaller scale of production. Furthermore, Si is a very abundant element in nature.

For several years after its introduction, the MESFET was used as single device for analog applications, with performances inferior to Si devices. Major technological improvements in the quality of the material and the geometrical control of the device have led to outstanding performance in the microwave field, both for low noise as well as power transistors. High resolution photo-lithographic techniques and ion implantation (see below) have allowed the realization of E-MESFET (Enhancement mode), that is transistors that are normally-off, and that can be turned on with positive bias on the gate. With the introduction of E-MESFET next to the more

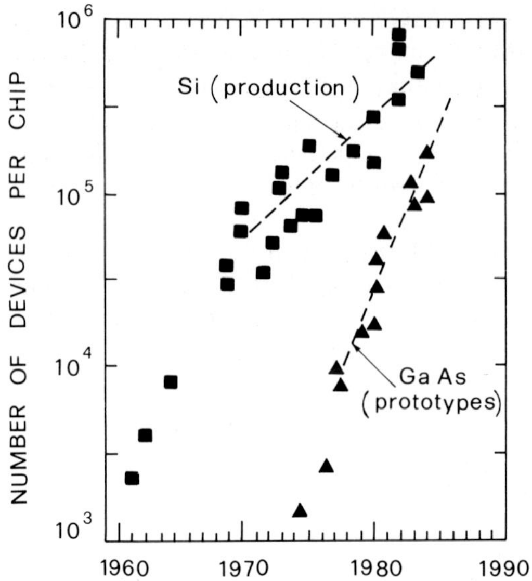

Fig. 4.1. Comparison of the integration level for GaAs (prototype) and Si (production) logic circuits [11]

standard, normally-on, D-MESFET (Depletion mode), the first GaAs I.C.'s were constructed around 1972.

The level of integration of GaAs logic circuits (prototypes) is compared with the one of Si (production) in Fig. 4.1 [11]. Despite the wide gap between the two technologies, the outstanding advances over the most recent years of GaAs digital circuits is evident.

The possibility to use GaAs or other III-V compounds for microelectronics applications has been enhanced by two new techniques for epitaxial growth, namely the Molecolar Beam Epitaxy (MBE) [12] and the Metal-Organic Chemical Vapour-phase Deposition (MOCVD) [13]. These techniques enabled the construction of heterostructures with two different semiconductors that offer great advantages for device applications. Due to the outstanding control over the growth process, it is possible to selectively dope different layers or to change the composition of a given compound with a precision of a single monolayer. Various structures can be tailored at will simply by controlling the growth speed and the elements used in the growth process. A typical single heterostructure $Al_xGa_{1-x}As$-GaAs was shown in Fig. 2.38. As pointed out there, one of the characteristics of the structure is the confinement of the carriers into the quantum well formed at the interface between the two materials. By doping exclusively the large band-gap material (AlGaAs) [14], the carriers are spatially separated from the impurities, leading to very high mobilities.

Heterostructures have been used extensively in the design of novel devices over the last two decades. The two most successful types are the Heterojunction Bipolar Transistor (HBT) [9] and the High Electron Mobility Transistor (HEMT) [8]. The HBT was devised following an idea by Kroemer [15] to improve the performance of bipolar transistors. It keeps the same structure of the Si bipolars, using a large band gap (AlGaAs) emitter and a highly-doped GaAs base. The HEMT is a field effect transistor similar to the MESFET, where the quantum well at the interface between AlGaAs and GaAs constitutes the channel between source and drain. Such a device has been studied by several industrial and university laboratories, and has received many acronyms. As for the GaAs devices, technological advances have brought heterostructure devices into the microelectronic scene as very competitive and innovative components for both analog as well as digital applications.

4.2.3 Technological Features

Even from the short description of the previous sections, it should be clear that technology has been the most important factor in the evolution of semiconductor devices. Here we will offer a brief outline of the processing steps involved in the construction of a *p-n* junction in Si technology [4] (see Fig. 4.2). Although this is just a particular example, it illustrates nicely how different methods can be used for the production of a given structure, and it also traces the historical evolution of fabrication techniques.

In the alloy method, aluminum is placed on an *n*-type silicon wafer as shown in Fig. 4.2a. By heating the system to a temperature slightly higher than the eutectic temperature ($\sim 580°C$ for the Al-Si system) a molten Al-Si mixture is formed. When the temperature is lowered, a recrystallized portion, which is saturated with the acceptor impurities, forms a heavily doped *p*-type region (p^+) on the *n*-type substrate. The aluminum layer on top provides an ohmic contact for the *p*-type region.

An ohmic contact [16, 17] is generally defined as one which supplys the required current with a voltage drop sufficiently small in comparison to the voltage drop in the active region of the device. Ideally, an ohmic contact does not perturb the device performance. In order to have an ohmic contact between a metal and a semiconductor, it is necessary to reduce either the height or the width of the barrier that forms at the interface between the two materials (see section 4.3.4). This is done by appropriately choosing the metal or by doping heavily the semiconductor surface. For instance, on the bottom of the *n*-Si layer an ohmic contact is formed by evaporating a Au-Sb alloy onto the wafer, and alloying it at 400°C to form a heavily doped *n*-type region. The junction location obtained by the alloy method very much depends on the temperature-time alloying cycle and is difficult to control precisely.

The solid-state diffusion method (Fig. 4.2b) was developed in 1956 to give more precise control of the impurity profile. In the diffused mesa method shown in the figure, *p*-type impurities (e.g. boron) are diffused into the *n*-type

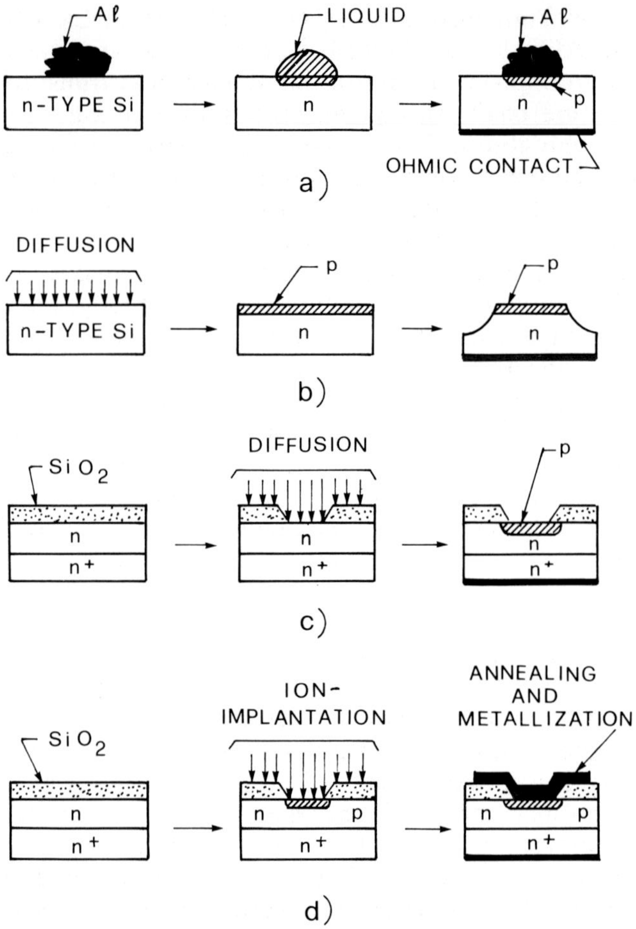

Fig. 4.2. Processing steps involved in the fabrication of a *p-n* junction in Si technology, using *a* alloy method; *b* diffused mesa method; *c* diffused planar junction; *d* ion implanation [4]

substrate. After diffusion, portions of the surface are protected (e.g., by wax or metal contacts), and the rest are etched out to form the mesa structures. A new degree of control over the lateral geometry of the diffused junction is achieved by using an insulating layer that can prevent most donor and acceptor impurities from diffusing through it. A typical example is shown in Fig. 4.2c. A thin layer of silicon dioxide ($\sim 1\ \mu$m) is thermally grown on silicon. With the help of lithographic techniques (e.g., photolithography, x-ray, or electron-beam lithography) portions of the oxide can be removed and windows (or patterns) cut in the oxide. The impurities will diffuse only through the exposed silicon surface, and *p-n* junctions will be formed in the oxide windows. This process, the planar process, has become the principal method of fabricating semiconductor devices and integrated circuits.

Also shown in Fig. 4.2c is the epitaxial substrate, for example, an *n*-layer on

n^+ substrate. Such a substrate is generally used in the planar process to reduce the series resistance. Epitaxy is a technique of crystal growth by chemical reaction used to form, on the surface of crystals, thin layers of semiconductor materials with lattice structures identical to those of the crystal. In this method, lightly doped high-resistivity epitaxial layers are grown on and supported by a heavily doped low-resistivity substrate thus ensuring both the desired electrical properties and mechanical strength.
Figure 4.2d shows a *p-n* junction formed by ion implantation. Ion implantation [18] is one way of changing the metallurgical, electrical or chemical properties of a substrate by introducing energetic charge atomic particles into a given portion of it. For instance, the implantation of dopants atoms into a seminsulating material creates conductive areas which are electrically isolated. A precise control of the total dose, depth profile and area uniformity can be achieved. Ion implantation can be done at room temperature, and the implantation-induced lattice damages can be removed by annealing at about 700°C or less. Therefore, ion implantation is a relatively low-temperature process compared to diffusion, which is generally done at 1000°C or higher.

4.2.4 Scaling and Miniaturization

Since the advent of integrated circuits, the number of transistors on a single chip has increased up to several millions. As a result, the performance of integrated circuits has improved greatly.
How far such a trend can continue, depends on several factors, which are related to the choice of materials, to the type of devices and circuits that are employed, to the overall system (that is the architecture and packaging of integrated circuits on a chip), and, to a certain degree, to the fundamental limits inposed by the physics of the microscopic processes involved in carrier transport.
One key question is how far a single device can be scaled down, and which effect this reduction of the device dimensions has on the performance of an integrated circuit. We present here the scaling laws that apply to the MOSFET [19]. Figure 4.3 shows a traditional device (a) and the scaled-down counter part (b), in which all lateral and vertical dimensions are reduced by a so called "scaling factor" K. This reduction includes oxide thickness, channel length, channel width and junction depth. The doping level is increased by K, and the supply voltage is reduced by the same factor, in order to keep the strength of the electric field constant and maintain an equal safety margin in the operation of the device. Such constant electric field scaling offers great advantages. Clearly, there is a lower limit to scaling, in that the supply voltage has to be kept higher than the noise level to guarantee the proper operation of the device. The time required to switch the device, which depends on the length of the channel, decreases by a factor K. The power dissipated per unit chip remains constant, while the packing

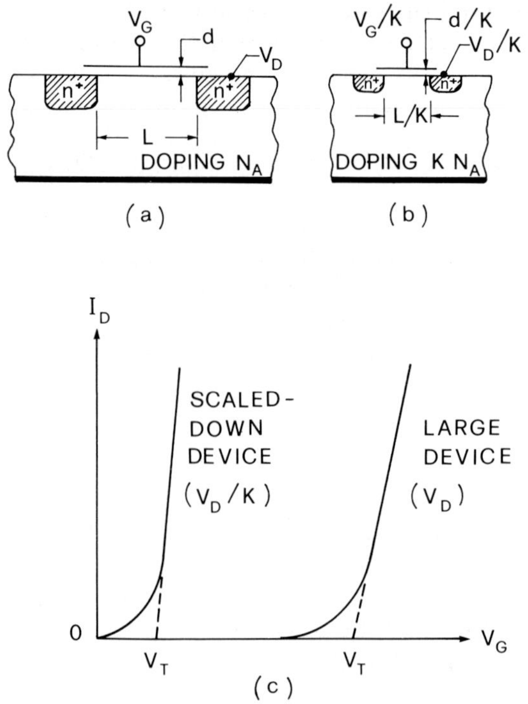

Fig. 4.3. Traditional (*a*) and scaled-down (*b*) structure, and transfer characteristics (*c*) of a Si-MOSFET [19]

density increases by K^2. Last, but certainly not least, the power-delay product (that is the energy consumed in each switching operation) decreases by a factor of K^3. The previous scaling rule can be summarized by saying that, as a result of scaling, a chip contains more devices, each one of which switches faster and uses less power.

If, from the device point of view, scaling is greatly beneficial, the opposite occurs for the interconnecting lines between devices [20]. In part this is due to the tendency of using larger and larger chips. Furthermore, the interconnect time constant RC does not scale as $1/K$, following the MOSFET delay, it tends rather to stay constant. What is important is that, once the switching time of the devices has been optimized by scaling, the limiting factor to the circuit performance will be set by the interconnection time constants.

We will not continue further in the discussion of scaling and miniaturization of integrated circuits, since it goes beyond the scope of the present book. It is though necessary to stress the importance of the miniaturization of semiconductor devices both in the contest of very-large scale integration as well as for single device performance. In the following we will find many examples of submicron devices, where the active length of the device has been reduced well below the micron limit. Such devices will exhibit improved performances and confirm what has been briefly discussed in this section. Furthermore, novel effects such as overshoot and ballistic transport will be observed.

4.3 Physical Basis of Semiconductor Devices

In the present section, we will discuss the most important semiconductor devices with particular attention to their physical characteristics and ways of operation. The discussion will be kept qualitative as far as possible, in order to stress how carrier transport influences the device performance, rather than going into a detailed treatment of the equations that govern the device behavior.

4.3.1 p-n Junction

The *p-n* junction is a fundamental part of most semiconductor devices. It is formed when a *p*-type and an *n*-type region of a semiconductor are in intimate contact. Two idealized models for *p-n* junction are usually considered. Figure 4.4 refers to the case of an abrupt junction, where the dopant concentration changes in a step-wise fashion from a region with acceptor impurity N_A to one of donor impurity N_D. As a result of the diffusion of free carriers on both sides of the junction, a space-charge depleted region is formed at the interface (Fig. 4.4a). By solving Poisson's equation for the one-dimensional structure, the electric field and potential distribution consistent with this charge distribution can be calculated (Figs. 4.4b and 4.4c). The equilibrium energy-band diagram is also displayed in Fig. 4.4d.

The second idealized model of a *p-n* junction describes the interface through a continuous gradient in doping between the *n*-type and the *p*-type regions. This case, referred to as a linearly-graded junction, is illustrated in Fig. 4.5. In both cases, the built-in potential acts as a barrier between the two regions, opposing and balancing the tendency of electrons and holes to diffuse into the regions of lower concentration.

The abrupt approximation seen before is adequate for alloyed, shallowly diffused and ion-implanted junctions. The linearly-graded approximation is more appropriate to deeply diffused junctions.

When a bias is applied to a *p-n* junction, the current-voltage I-V characteristics shown (on a semilog plot) in Fig. 4.6 are obtained [22]. The solid curves refer to the ideal characteristics calculated in the Shockley model [22], while the dashed curves show the real I-V characteristics of an Si *p-n* junction. The ideal curves can be understood starting from the energy diagram shown in Fig. 4.7, which are calculated under the following assumptions:

i) ohmic contacts connect the *p* and the *n* regions to the external voltage source so that negligible voltage is dropped at the contacts;

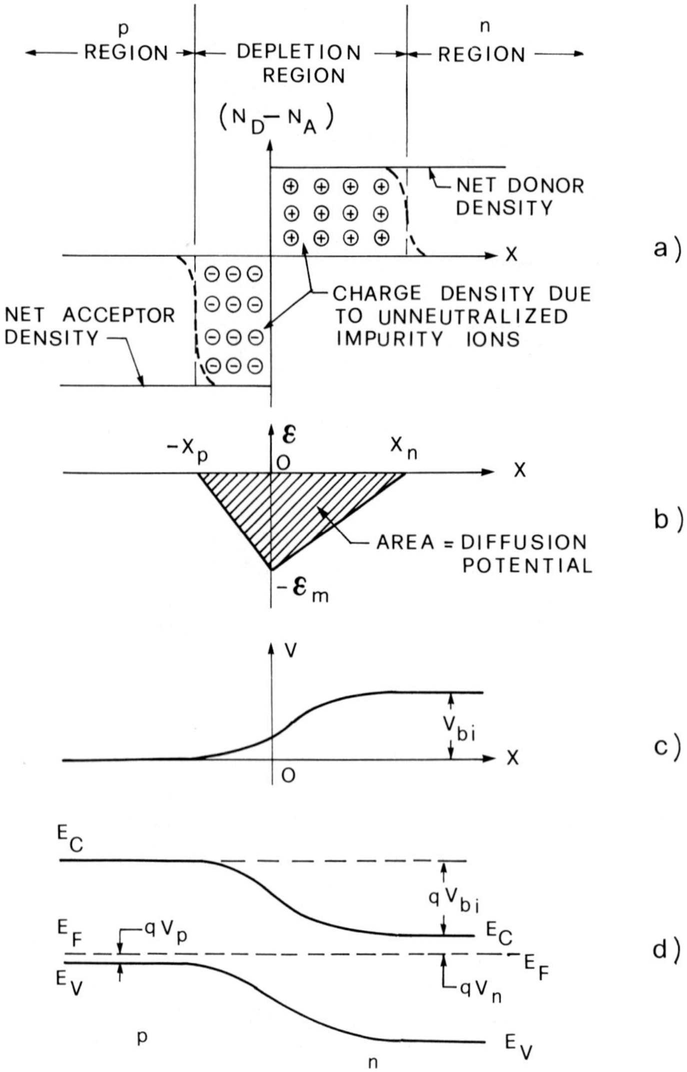

Fig. 4.4. Abrupt *p-n* junction in thermal equilibrium, showing (*a*) the charge distribution; (*b*) the electric field distribution; (*c*) the potential profile with built-in potential V_{bi}; (*d*) the energy band diagram [4]

ii) the boundaries of the depletion regions are abrupt, and outside these boundaries the semiconductor is neutral;
iii) the injected minority-carrier densities are small compared with the majority-carrier densities;
iv) electron and hole currents are constant through the depletion layer, with no contribution from generation current (see below).

Under these assumptions, if we consider the *n*-region to be grounded and a

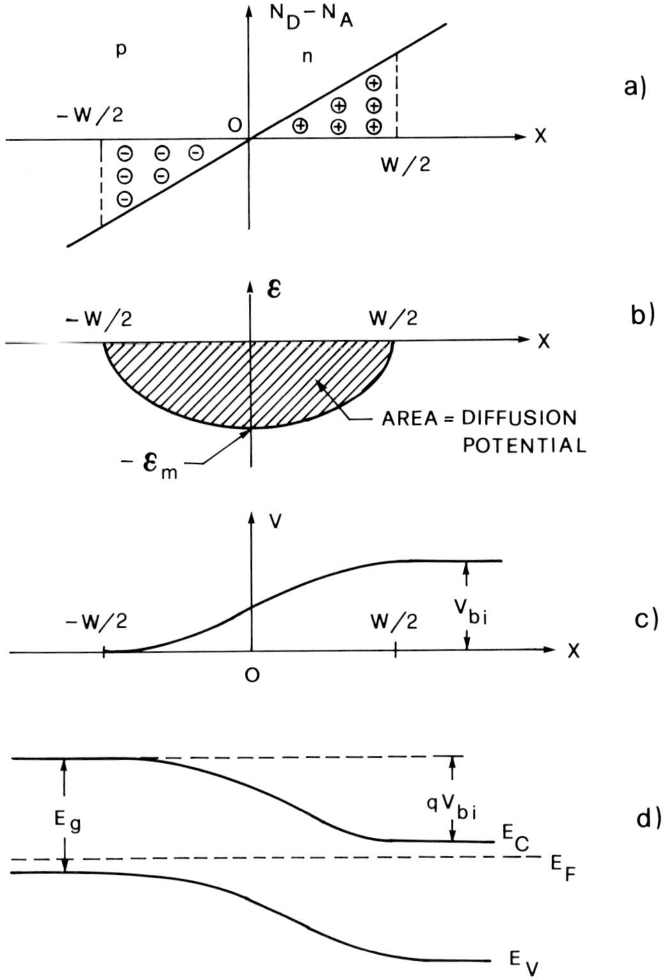

Fig. 4.5. Linearly *p-n* junction in thermal equilibrium, showing (*a*) the charge distribution; (*b*) the electric field distribution; (*c*) the potential profile with built-in potential V_{bi}; (*d*) the energy band diagram [4]

voltage V to be applied to the *p*-region, the entire applied voltage is dropped across the junction. The solution of Poisson's equation found for the thermal equilibrium still applies to the biased junction, with the only change of the potential drop across the junction which goes from V_{bi} to $V_{bi} - V$.

If V is positive (forward bias), the barrier to majority carriers at the junction is reduced, and the depletion region is narrowed (Fig. 4.7a). In simple terms, we can think that majority carriers are pushed by the applied voltage towards the depletion regions, where they neutralize some of the space charge, thus reducing the overall depletion-region width. Under forward bias, appreciable currents can flow even for small applied potential, with

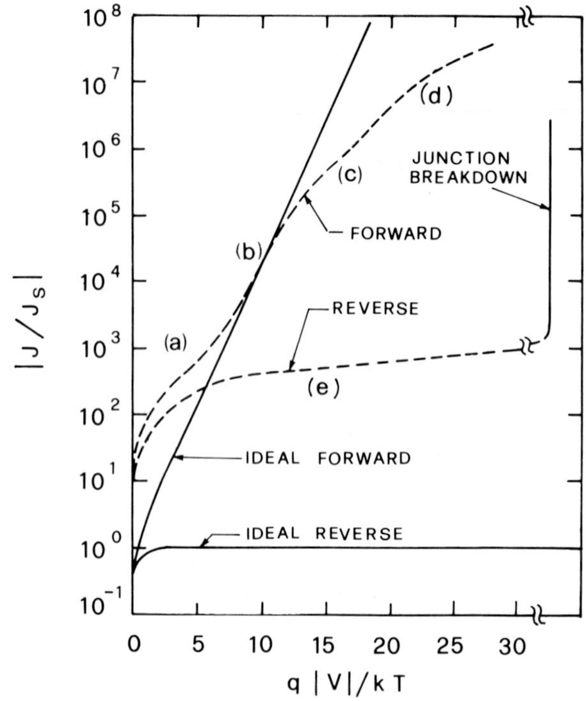

Fig. 4.6. I-V characteristics for a Si *p-n* junction. The solid line refers to the ideal case, the dashed line to a real diode. The different regions in the dashed line are due to the effect of: (*a*) generation-recombination; (*b*) diffusion; (*c*) high-injection; (*d*) series resistance; (*e*) reverse leakage [22]

the law

$$J = J_S(e^{qV/K_BT} - 1)$$

which is known as Shockley equation.

This equation also holds when the junction is under "reverse bias", that is when the applied voltage V is negative. In this case, the applied voltage pulls away carriers from the depletion regions, which consequently widen. The polarity applied to the junction favours the transfer of minority carriers from one region to the other, and because of the low concentrations involved the current levels are very low compared with the forward bias case. Under reverse bias, the current saturates at a value J_S.

The carrier distributions under forward and reverse bias conditions are also shown in Fig. 4.7 (solid curves in the bottom pictures). There the label *o* refers to the equilibrium values of the current concentrations, while n_i indicates the intrinsic carrier concentration. The current flow in a *p-n* junction is determined mainly by the minority carriers. The majority carriers act mainly as a supplier for the injection of the minority carriers, and as charge neutralizers in the neutral regions.

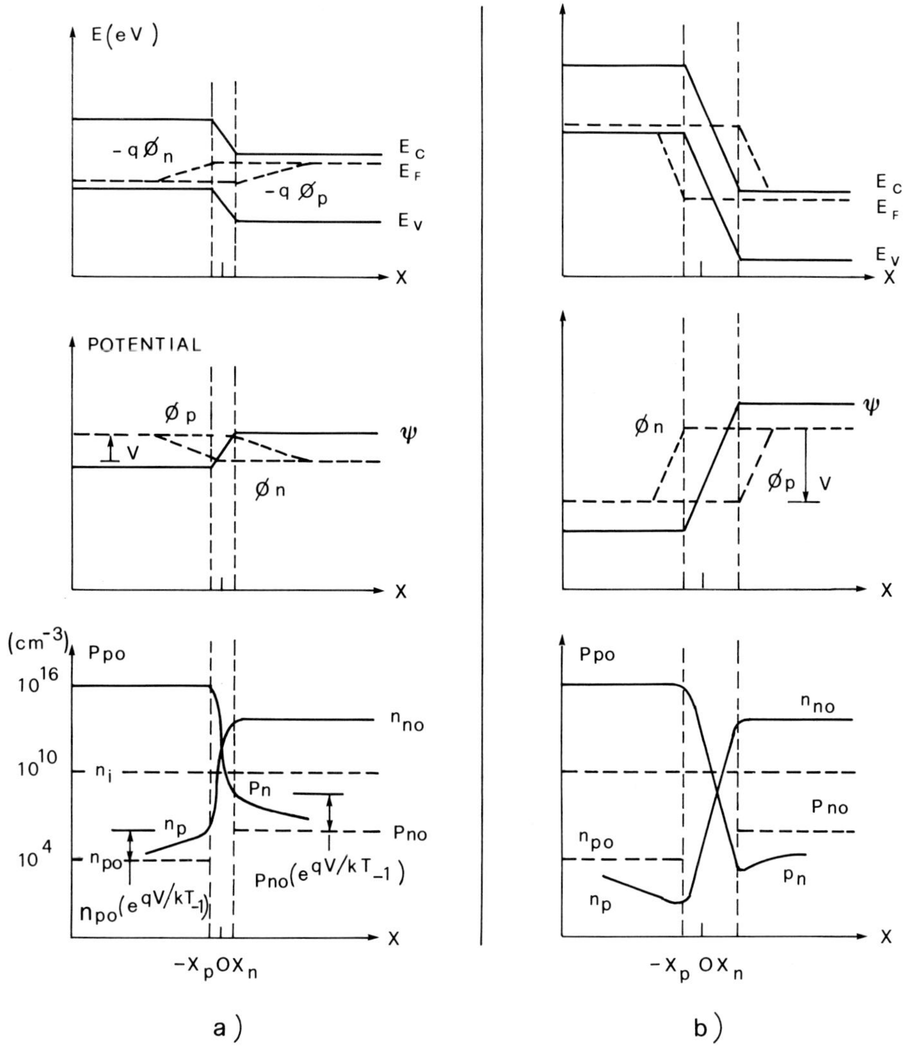

Fig. 4.7. Energy band diagram, intrinsic Fermi level Ψ, and carrier distribution under forward (*a*) and reverse (*b*) bias conditions [4]

While the Shockley model adequately describes the I-V characteristics in germanium *p-n* junctions at low current densities, this is not the case for Si and GaAs. The deviations of real curves from the ideal ones shown in Fig. 4.6 are due to various physical processes that take place in real *p-n* junctions. One of the most important processes is the generation and recombination of carriers in the depletion layer, which cause additional currents to the diffusion currents of the idealized model. In the reverse bias case, generation processes are dominant since the carrier concentrations in the depletion layer are such that $np \ll n_i^2$. The generation currents are strong in

Si and account for the difference in reverse current of Fig. 4.6. On the other hand, at forward bias we have an additional recombination current, which brings the ideal curve in Fig. 4.6 up to the value shown by the dashed curve (b). Additional effects related to high-energy injection (with minority-carrier densities comparable with majority-carrier ones) and series resistance effects (due to the finite resistivity of the neutral regions) are responsible for the deviation from the ideal behaviour at high forward bias (curve c).

As pointed out earlier, the width of the depletion layer decreases at increasing reverse bias. Therefore, the maximum electric field at the junction can reach very high values and lead to the junction breakdown. The breakdown is due to the internal generation of extra-free carriers in the semiconductor, which suddenly becomes highly conductive. The physical mechanisms at the origin of the breakdown phenomenon are tunneling effect and avalanche multiplication (due to impact ionization).

In the former case, the electric field is so strong as to break one covalent bond that binds the valence electrons. This process can be represented as a band-to-band tunneling, in which an electron-hole pair is created as an electron makes a transition from the valence into the conduction band.

In the latter case free carriers can gain enough energy from the field to break a covalent bond and produce an electron-hole pair. The three carriers resulting from the ionization process can in turn cause further ionization, thus leading to a sudden multiplication of carriers in the space-charge region, if the field is sufficiently high.

The *p-n* junction is an essential component of many semiconductor devices that will be dealt with in the following section. It is also a two terminal device, that can perform various functions depending on its specific design (i.e., doping profile, device geometry) and on the bias conditions. Among the various applications of *p-n* junction diodes we can remember the following:

— rectifier; from the I-V characteristics it is clear that *p-n* junctions exhibit a very high resistance for current flowing in one direction and a low resistance for the other direction. They can therefore be used to rectify alternating current,
— voltage regulators,
— variable resistor (varistor),
— variable reactor (varactor), used for harmonic generation, mixing detection and voltage-variable tuning,
— fast recovery diode,
— charge storage diode;

and for microwave applications:

— Tunnel diode,
— IMPATT diode.

Due to the simple technological process involved in its fabrication, *p-n* junctions are used as isolating elements in integrated circuits. The various devices in an integrated circuit are surrounded by *p-n* junctions which are constantly kept under reverse bias, thus guaranteeing the complete isolation of each device.

4.3.2 Bipolar Transistors

The bipolar transistor consists of two adjacent *p-n* junctions, defined respectively the emitter-base and base-collector junctions. Figure 4.8 shows a prototype transistor, with the bias configuration for the three terminals of the device and thc respective currents. The shaded areas indicate the space charge regions around the two junctions. The two junctions can be arranged to give either a *p-n-p* or an *n-p-n* bipolar transistor. The operation of such device is based on the possibility of modulating the current flow in one junction by changing the bias in the neighbouring one. The main components of the current (which is carried by both electrons and holes) are given in Fig. 4.9 (a), together with the energy band diagram in equilibrium (b) and under operating conditions (c), with one junction reverse-biased and the other forward-biased. Under equilibrium conditions, there are very few electrons in the base region, since the base is *p*-type, and virtually no-transfer

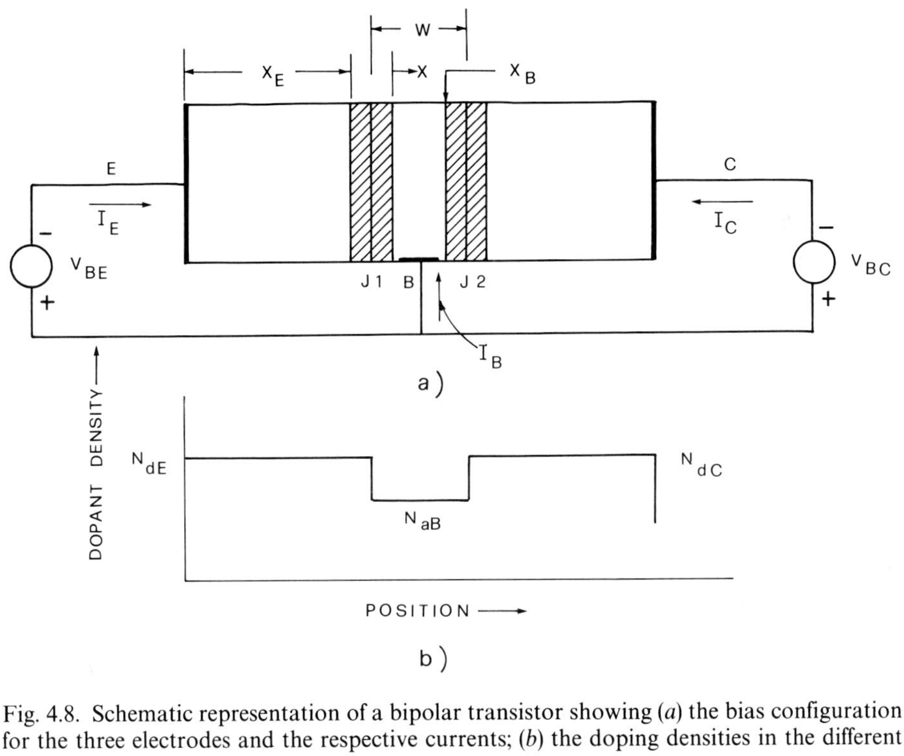

Fig. 4.8. Schematic representation of a bipolar transistor showing (*a*) the bias configuration for the three electrodes and the respective currents; (*b*) the doping densities in the different regions

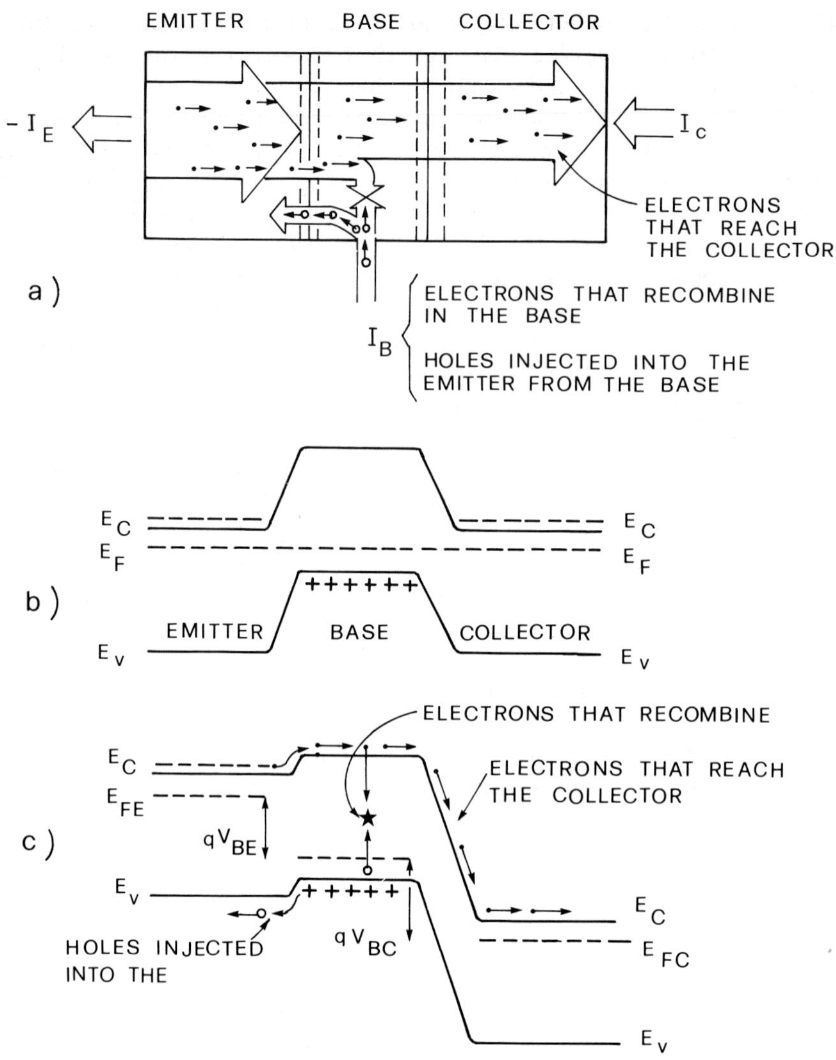

Fig. 4.9. Main current components (*a*); energy band diagram in equilibrium (*b*) and under operating condition (*c*) for a *n-p-n* transistor

from the emitter and collector *n*-type regions occurs because of the energy barrier seen by the electrons.

A negative bias applied to the junctions increases these barriers and depletes further the base region from the few electrons present in equilibrium. A positive bias applied to the junctions has the opposite effect of lowering the barrier height and allowing the flow of large currents through the device.

The most interesting configuration is the one depicted in Fig. 4.9c, since it can lead to signal amplification. There, electrons are injected from the emitter into the base, where they diffuse (as minority carriers) towards the collector. As soon as the electrons enter the base-collector junction, the high electric field at the junction drives them quickly to the collector contact. The collector current (given by electrons flowing from the emitter to the collector) represents the output current of the transistor in the active-biased region. Since the electron injection into the base depends exponentially on the emitter-base forward bias, the voltage V_{BE} applied between base and emitter controls the collector current. The amplification efficiency of the bipolar transistor depends on how small the current is between base and emitter. Figure 4.9a shows the main contributions to the current for the configuration which is usually referred to as common-base configuration. The base current I_B is mainly due to the recombination of electrons and holes in the base, and to hole injection into the emitter through the forward-biased emitter-base junction.

Three important parameters that characterize the static performance of a bipolar transistor are the emitter efficiency γ_E, the common-base current gain α_F and the common-emitter current gain β_F, defined respectively by

$$\gamma = \frac{I_{nE}}{I_E}, \qquad \alpha_F = \frac{I_C}{I_E}, \qquad \beta_F = \frac{I_C}{I_B}.$$

Here, the emitter current I_E is given by the sum of the electron current I_{nE} and the hole current I_{pE} that flow across the emitter-base junction. Since, in steady state conditions, all currents into the transistor must sum to zero (i.e., $I_E + I_B + I_C = 0$), β_F, which gives the current gain when the input current flows between base and emitter while the output current flows into the collector, is related to α_F by

$$\beta_F = \frac{\alpha_F}{1 - \alpha_F}.$$

A reduction of the currents associated respectively with recombination processes in the base and with hole emission into the emitter region, leads to values of α_F close to unity, and, consequently to large values of the current gain β_F. Furthermore, a large emitter efficiency can be obtained with a doping of the emitter region ($N_D \geqslant 10^{20}\ \mathrm{cm}^{-3}$) much higher than the doping of the base region ($N_A \leqslant 10^{19}\ \mathrm{cm}^{-3}$).

A typical structure of a bipolar transistor built on Si planar technology is shown in Fig. 4.10. Planar technology is particularly suitable for integrated circuits, when many devices have to be packed in small regions. In comparison with the single transistor shown schematically in Fig. 4.8, the planar transistor has all electrodes on the top surface of the wafer. Each transistor of an integrated circuit should be electrically isolated to prevent interaction with other devices. The isolation can be obtained either with *p-n* junctions or oxide layers.

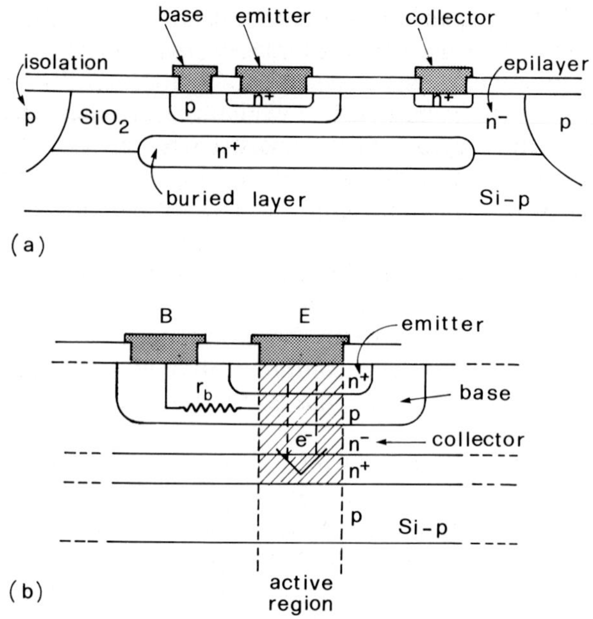

Fig. 4.10. (*a*) Cross section of a vertical *n-p-n* bipolar transistors; (*b*) the enlargement of a portion of the device shows the active region and the base parasitic resistance [11]

Bipolar transistors have found a very wide range of applications such as:

— microwave transistors, operating in the microwave region (above 1 GHz). The main requirement for high frequency operation is that the active areas and the parasitic resistances and capacitances are small;
— power transistors, designed for handling high voltages and large currents in their operation as power amplifiers;
— switching transistors, designed to function as fast switches, able to change the state of the transistor from a high-voltage low-current (off) condition to a low-voltage high-current (on) condition in a very short time.

The main difference between devices designed for amplification (microwave and power transistor) and switching purposes is in the dimension of the various layers in Fig. 4.10. An improvement in the device performances, as well as a higher level of integration, can be obtained by scaling down the dimensions of the device. A reduction of the base thickness has the double effect of reducing the carrier transit time through the base (thus leading to higher cut-off frequencies and faster switching times) and the electron-hole recombination (thus cutting down the recombination current and enhancing the current gain). An additional increase in the amplification of a bipolar transistor can be achieved by cutting down the hole injection into the emitter through a reduction of the base doping concentration. Unfortunately, the improvements listed above have some serious drawbacks that can be understood from Fig. 4.10b. In order to reduce the collector parasitic series resistance a buried n^+ layer is added to the planar structure. Another parasitic resistance (indicated with r_b in the figure) exists between the base contact and the active region of the transistor (shaded area). Such resistance increases when the base thickness and doping level are lowered. The RC circuit formed with the parasitic capacitances associated with the junctions present in the device (emitter-base, base-collector, and isolating junctions) causes a deterioration of the intrinsic speed of the transistor.

As pointed out earlier, the reduction of the horizontal and vertical dimensions of the device leads to improved performances (for example, it diminishes the parasitic capacitances). An alternative way to improve the transistor performance is through the use of heterostructures, as will be shown in the next section.

We have focused so far on *n-p-n* transistors. *p-n-p* transistors are described in the same way as the *n-p-n* ones once type of carriers and polarities are reversed. Since the electron mobility is higher than the hole mobility in Si, more attention in IC technology has been given to the production of *p-n-p* structures. The *p-n-p* transistor is very important in a family of densely packed digital IC's, usually called "integrated-injection logic" or I^2L. The I^2L is formed by a lateral *p-n-p* and a vertical *n-p-n* transistors; the collector of one device functions simultaneously as the base of the other.

This short review on the bipolar transistor has focused on Si technology. Although GaAs have some outstanding advantages over Si (in particular the higher electron mobility), it has not been possible to construct junction bipolars on GaAs which are competitive with the Si ones. The main drawbacks of GaAs concerning bipolar transistors are the impossibility to diffuse dopants into GaAs, and therefore to form *p-n* junctions, and the extremely low mobility of holes (even lower than Si).

Such problems have been overcome, as we will see in the next section, using heterojunction bipolar transistors (HBT).

4.3.3 Heterojunction Bipolar Transistor

We have discussed in chapter 2 some of the properties of $Al_xGa_{1-x}As$-GaAs heterostructures. In particular, we have seen that the different band gaps of the two materials lead to a discontinuity in the band edges at the interface. The two-dimensional nature of the transport along the heterojunction was pointed out, together with the importance of modulation doping. Another great advantage of the AlGaAs-Gas system is found in connection with bipolar transistors. In this case, conduction takes place across the heterojunction, which is formed by *n*-type $Al_xGa_{1-x}As$ and *p*-type GaAs. The band diagram of an *n-p-n* heterojunction bipolar transistor is shown in Fig. 4.11 for the two cases of an abrupt (a) and graded (b) emitter-base junction [23]. The grading is achieved through a gradual

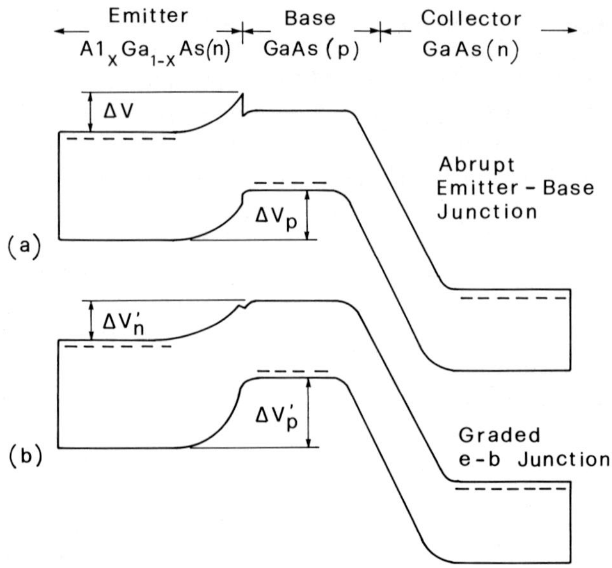

Fig. 4.11. Band diagram of a *n-p-n* heterojunction bipolar transistor with (*a*) abrupt and (*b*) graded emitter-based junction [23]

variation of Al composition in the $Al_xGa_{1-x}As$ emitter. The possibility of spatially controlling the band gap is one of the most outstanding achievements of MBE growth techniques. In the active bias configuration shown in Fig. 4.11, the operation of the transistor is the same as for the homojunction bipolar described in the previous section. The potential advantages of such a structure had already been recognized by Shockley in 1948, but only the modern technological growth techniques have made its construction possible. Several features of the HBT can be noticed from the figure. In the first place, the emitter has a larger band gap than the base. At the emitter-base junction, the potential barrier ΔV_p seen by the holes (of the p^+ base) is higher than the barrier for the electrons (of the n^+ emitter).

The hole motion from the base into the emitter is prevented by the barrier ΔV_p. Therefore, the emitter current through the forward-biased emitter-base junction is entirely due to the electron injection. It is possible at this point to increase the base doping level without suffering from an increased hole current into the emitter. The high doping that can be achieved (even higher than the emitter) leads to a significant reduction of the parasitic resistance r_b, and to an overall enhancement of the device performance. As seen from Fig. 4.1b, the hole confinement is even better for the graded emitter-base junction.

The discontinuity in the conduction band, with the typical spike shown in Fig. 4.11, is another considerable advantage of the HBT. Such discontinuity acts, in fact, as a "launcher" for fast electrons into the base. Electrons that overcome the barrier, are accelerated quasi-ballistically by the junction field, and can cross the base region at very high velocities. Therefore, if the base thickness is sufficiently reduced (≈ 1000 Å), very rapid transit times can be achieved, which lead, in turn, to very high cut-off frequencies and ultrashort switching times. The use of the conduction-band discontinuity as a launcher has several drawbacks, since the discontinuity reduces the electron injection into the emitter, and requires higher biases of the emitter-base junction to sustain a relevant current. A proper balance of all these effects is necessary in order to obtain improved performances of HBT's.

Band-gap engineering [24], that is the possibility to control locally the band-gap variation of certain regions of the device, enables high transistor performances to be achieved. By gradually varying the doping concentration along the base region, it is possible to effectively tilt the conduction band in the way illustrated in Fig. 4.12. Electrons moving in the potential gradient of the base are drifted by an effective electric field across the base. The transit time is reduced with respect to the normal case (no compositional grading in the base), when electrons move in the base only by diffusion.

The actual structure of a HBT is shown in Fig. 4.13; n^+ layers are added to the emitter and collector, respectively, to facilitate ohmic contact and the flow of lateral current from the base to the collector contact. Oxide isolation

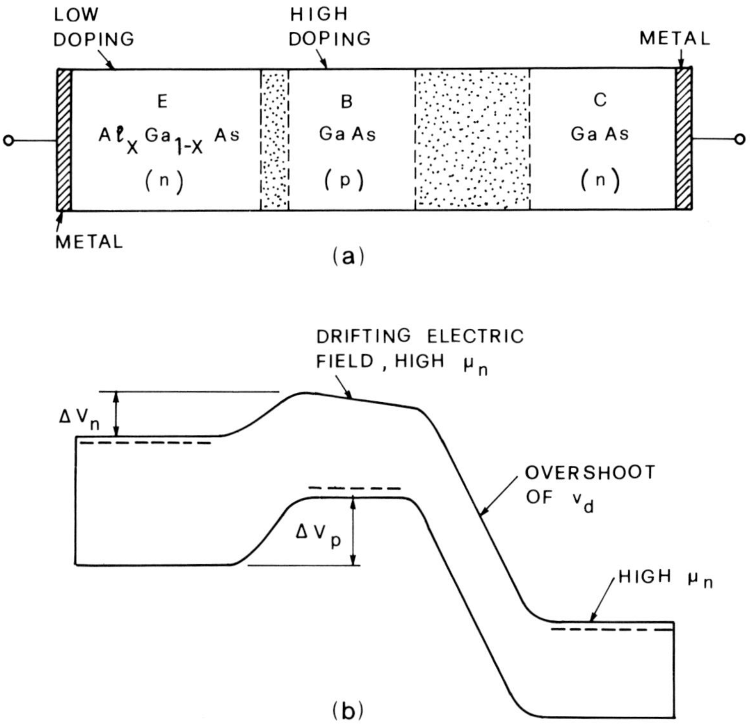

Fig. 4.12. (*a*) Schematic representation and (*b*) energy band diagram for a heterojunction bipolar transistor with gradually varing composition in the base [23]

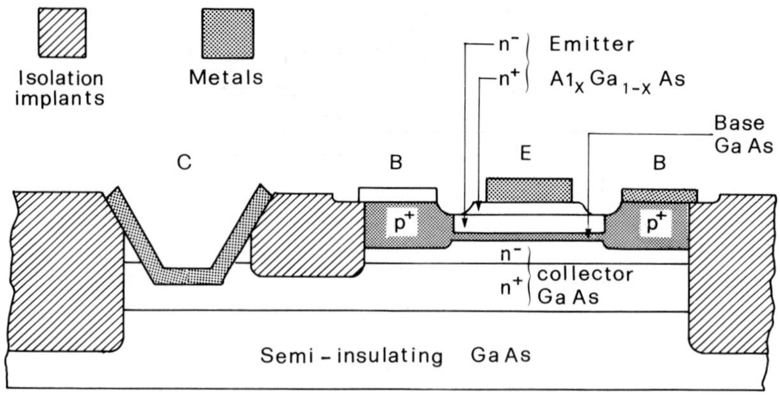

Fig. 4.13. Cross section of a real heterojunction bipolar transistor. The n^+ layers under emitter and collector allow the realization of Ohmic contacts [23]

(obtained by implantation techniques) is used in the structure to separate different devices.

4.3.4 *Metal-Semiconductor Contacts*

Metal-semiconductor contacts (also called Schottky contacts) are another fundamental component of semiconductor devices, as we have already seen in the previous sections. The connection of single devices and integrated circuits with the external environment is mostly achieved through metal-semiconductor contacts. As usual, the energy band diagram of the structure provides much valuable information on the electrical properties of the junction. In equilibrium, a metal is characterized by its work function ϕ_m, and a semiconductor by the work function ϕ_s, the electron affinity χ and the band gap $E_g = E_c - E_v$. Fig. 4.14 illustrates the case of Au and n-type Si [5]. The idealized band diagram of the metal-semiconductor contact is shown in Fig. 4.14a. The requirement of a constant Fermi level across the structure, and of a continuous vacuum level E_0 leads to a band bending on the semiconductor side, with a discontinuity in the conduction band-edge given by $\phi_B = \phi_M - \chi$. In the so-called depletion approximation, a positively-charged region exists on the semiconductor side of the interface, balanced by a negative δ-like distribution in the metal (Fig. 4.14b). The depleted layer is a result of the transfer of electrons from the n-type semiconductor into the metal. The electric field in the idealized structure is presented in Fig. 4.14c. The voltage drop across the δ-function of space-charge in the metal is virtually zero. Thus the total voltage drop across the entire space-charge region (which is indicated by ϕ_i) occurs entirely within the semiconductor. The potential barrier at the interface between metal and semiconductor prevents in general the flow of current across the interface. By applying an external bias to the structure, it is possible to modify the barrier height from

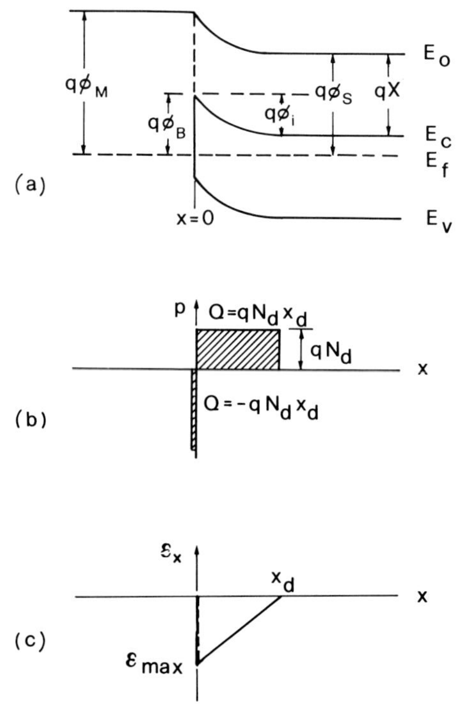

Fig. 4.14. (*a*) Energy band diagram, (*b*) charge distribution, and (*c*) electric field distribution for an ideal metal-semiconductor junction [5]

the semiconductor side (since the total voltage drop is all on this side of the structure). The barrier is reduced when a positive bias is applied to the semiconductor, and it is increased for the opposite polarity (see Fig. 4.15). With the notation indicated in the figure, we will call forward bias the case $V_a > 0$, and reverse bias the case $V_a < 0$.

So far, we have considered only idealized structures. In reality, the metal-semiconductor interface presents many surface states (associated with defects, dislocations etc), which can modify the previous picture. Furthermore, the current flow that is obtained by biasing the structure can modify the structure itself, leading to an effective lowering of the barrier [4]. Nevertheless, for the purpose of our presentation, the idealized model contains all the essential features. The most important processes that characterize current transport in metal-semiconductor interfaces under forward bias are shown in Fig. 4.16 [25]. Contrary to the case of *p-n* junctions, here the current is due mainly to the majority carriers (electrons in the case indicated in the figure). The process labelled 1 in Fig. 4.16 refers to the injection of electrons into the metal over the potential barrier. This is the dominant process for moderate doping of the semiconductor (for example Si with $N_D \leqslant 10^{17}$ cm^{-3}) at room temperature.

The quantum-mechanical tunneling across the barrier (process 2) is very

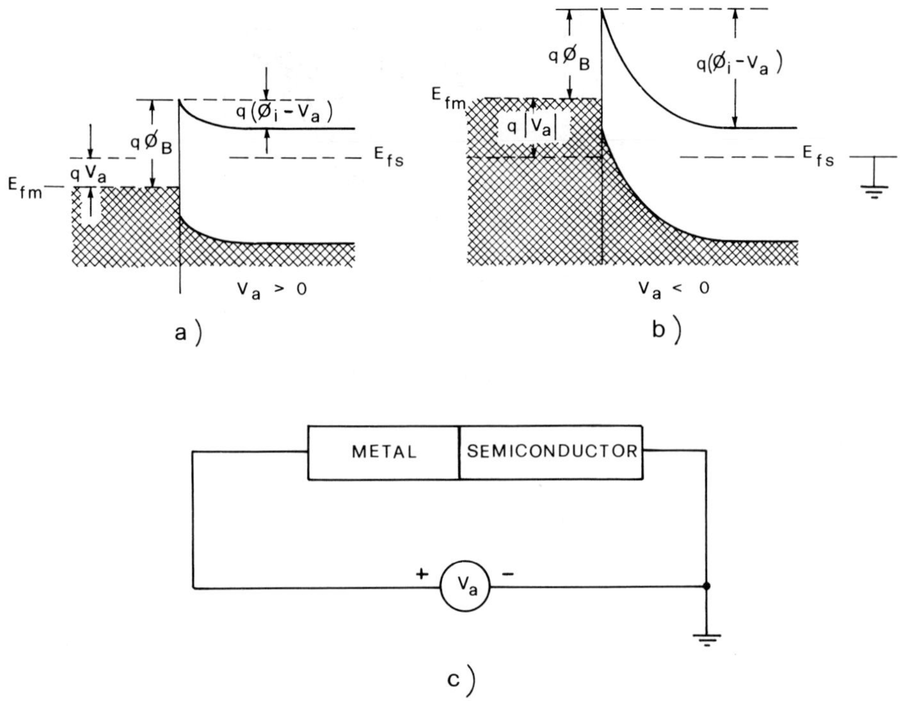

Fig. 4.15. Energy band diagram of an ideal metal-semiconductor junction under applied (*a*) forward bias and (*b*) reverse bias. The semiconductor side is taken as a reference (ground voltage) as shown in (*c*) [5]

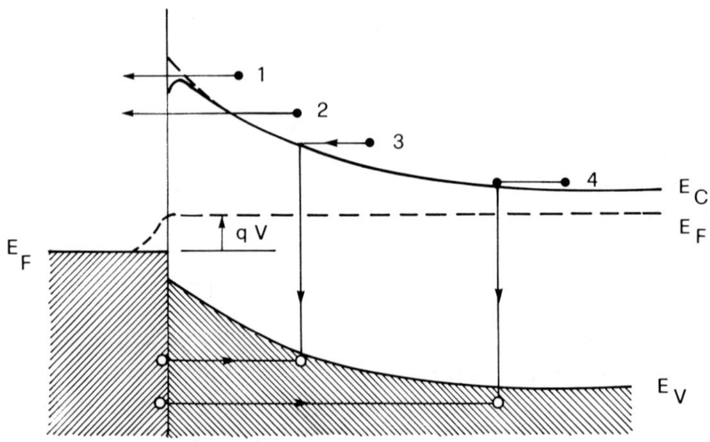

Fig. 4.16. Basic transport process under forward bias: *1* emission over the barrier; *2* tunneling across the barrier; *3* electron-hole recombination in the depletion region and *4* in the neutral region [25]

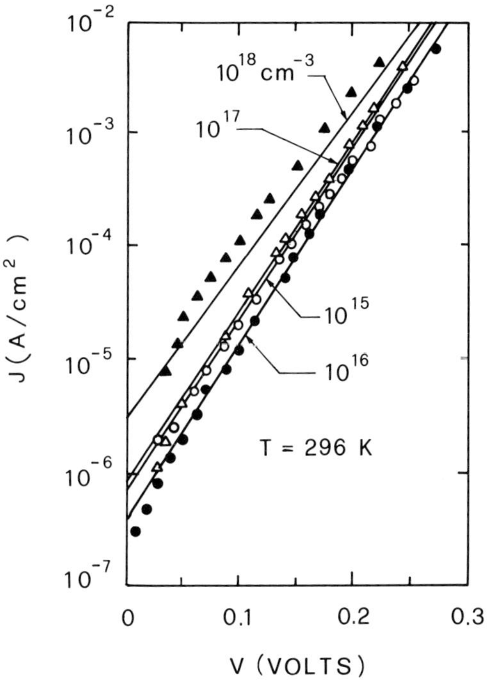

Fig. 4.17. Experimental (symbols) and theoretical (continuous line) I-V characteristics for Au-Si structure at different doping densities [4]

important for heavily-doped semiconductors, since the width of the depletion region (and consequently the barrier width) is inversely proportional to the square root of the doping concentration. Process 3 indicates electron-hole recombination in the space-charge region (and is identical to the recombination process in *p-n* junctions).

The electron-hole recombination in the neutral zone, due to hole injection from the metal into the semiconductor, is represented by process 4.

A set of typical current-voltage characteristics for Au-Si barriers are shown in Fig. 4.17 for various doping densities. Theoretical and experimental values are shown respectively by continuous lines and symbols. The current can be expressed as

$$J = J_s[\exp(qV/nK_BT) - 1],$$

where J_s is the saturation current, and n is the so-called ideality factor. It can be noticed that, for $n = 1$, the previous equation coincides with the one obtained for the *p-n* junction. At low doping, J_s is essentially constant and n is very close to unity. Both J_s and n deviate from such behaviour as the doping increases.

Metal-semiconductor devices are usually referred to as Schottky diodes. Since such diodes are electrically similar to one-sided abrupt *p-n* junctions, they can perform many similar functions. Schottky diodes have a widespread use as fast switches and power rectifiers in integrated circuits. A very important application of the metal-semiconductor interface concerns ohmic contacts. As mentioned in section 4.2.3, a contact is called ohmic when it has negligible contact resistance, that is, under bias conditions, the voltage drop across the contact is negligible compared to the voltage drop in other regions of the device. In such conditions, no power is dissipated at the contact, which can be treated as a system in thermal equilibrium even when current is flowing. Despite their importance, very little is known about ohmic contacts. The best and safest technique of establishing ohmic contacts involves the presence of heavily doped regions of the semiconductor at the interface with the metal. Metal-n^+-n structures can be obtained in a variety of ways (diffusion, implantation, alloying). As seen before, tunneling is the dominant transport mechanism in such case, and the current encounters a minimal resistance from the contact.

4.3.5 *Metal-Semiconductor Field-Effect Transistor*

The Metal-Semiconductor Field-Effect Transistor (MESFET) is the first of a series of field-effect transistors (FET) that will be analyzed in the present and following sections. In FET's, carrier transport involves mainly one type of carrier, contrary to the bipolar devices discussed previously, where contributions to the current come from both electrons and holes. For such reasons, FET's are also referred to as "unipolar" devices.
The schematic structure of a MESFET is shown in Fig. 4.18. The current flows between the two ohmic contacts at the source and drain electrodes. A rectifying Schottky-barrier acts as a gate electrode and controls the current flow in the device.
The MESFET is a particular form of the Junction Field-Effect Transistor (JFET), where the gate electrode is formed by a reverse-biased *p-n* junction.

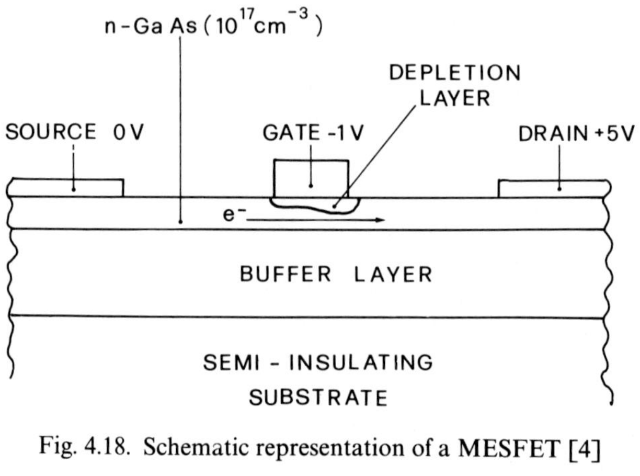

Fig. 4.18. Schematic representation of a MESFET [4]

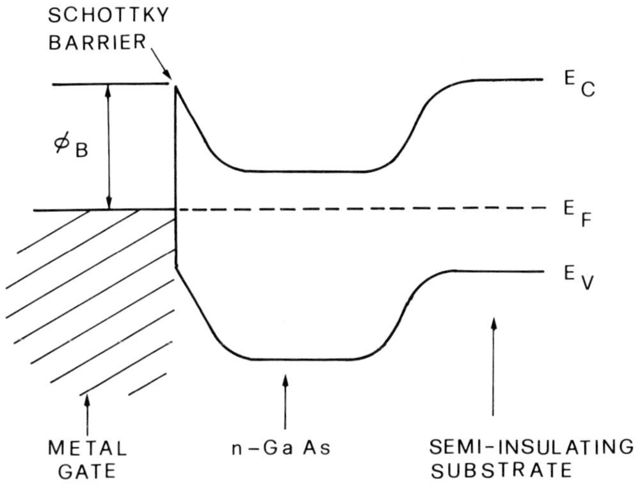

Fig. 4.19. Energy band diagram under the gate of a MESFET for zero bias conditions

Because of the technological and performance advantages of the Schottky barrier versus *p-n* junctions, MESFET's are more frequently used in electronic applications.

MESFET's are usually fabricated by growing an epitaxial layer of GaAs (typically doped around 10^{17} cm^{-3}) on top of a semi-insulating GaAs substrate. The band diagram of an *n*-type unbiased structure under the gate is illustrated in Fig. 4.19. The metal-semiconductor barrier creates a depletion region under the gate electrode. Carriers are then confined in a narrow "channel" near the interface between the epilayer and the semi-insulating substrate.

Current flows through the device when a positive voltage is applied to the drain electrode. Due to the potential drop along the channel, the width of the depletion region under the gate is not uniform. At zero gate voltage, the MESFET considered here is normally-on, since the conductive channel is open. Normally-on FET's are usually referred to as Depletion-mode (D-)MESFET's.

By applying a negative bias on the gate electrode, the width of the depletion layer increases and the conductive section of the channel decreases.

At a gate voltage normally called pinch-off voltage (V_p), the depletion region extends over the epilayer and virtually closes the channel. A sketch of the depletion regions for different bias voltages is shown in Fig. 4.20. The change in the depletion width causes a change in the resistivity of the channel.

The ideal I-V characteristics of a MESFET (and in general of any field-effect transistor) are shown in Fig. 4.21. Three regions can be identified: at low drain voltage V_D, the drain current I_D is proportional to V_D (linear region); at higher values of V_D the drain current remains essentially constant (saturation region); at even higher drain voltage, I_D increases rapidly with a slight increase of V_D (breakdown region). The saturation current (and the corre-

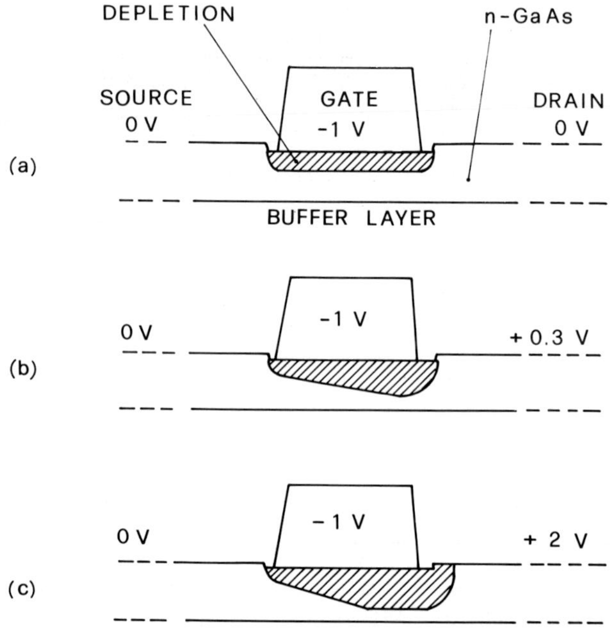

Fig. 4.20. Extension of the depletion region of a MESFET for different bias voltages [5]

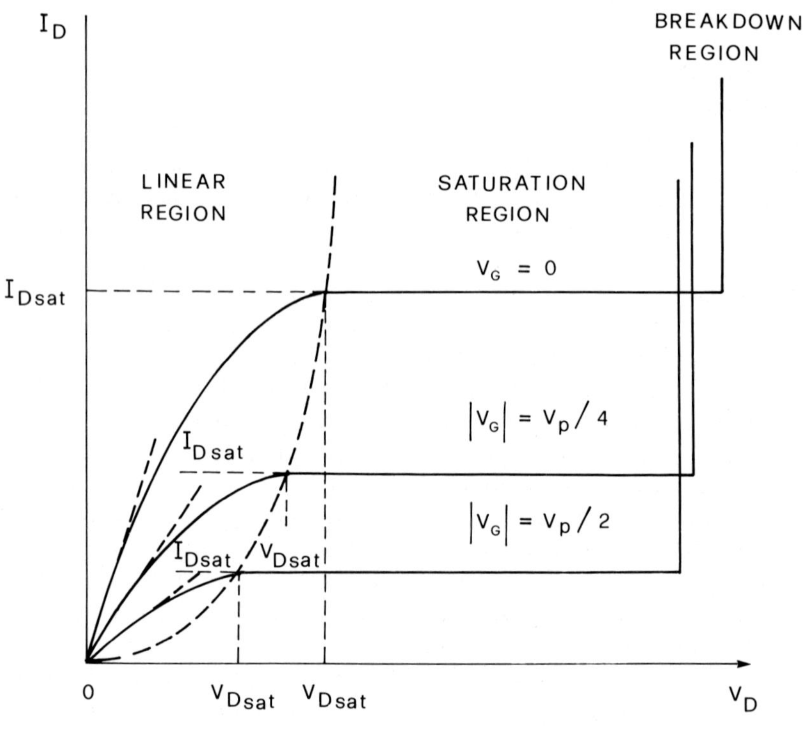

Fig. 4.21. Ideal I-V characteristic for a MESFET

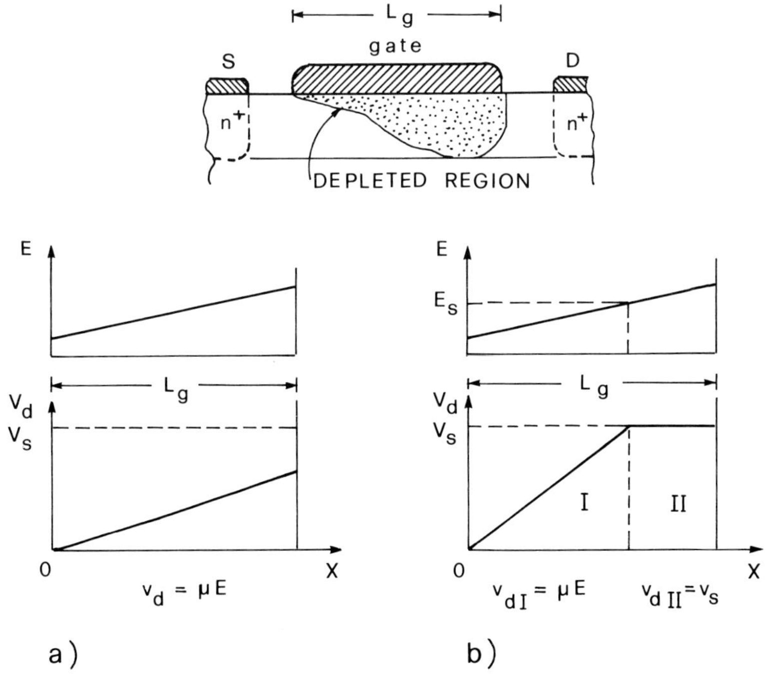

Fig. 4.22. Schematic representation of the variation of electric field and drift velocity according to (*a*) the constant mobility model and (*b*) the saturation-velocity model

sponding voltage) decreases as the gate voltage increases, due to the larger channel resistance.

The physical interpretation of the characteristics of a MESFET is not trivial because of the complicated band structure of GaAs (see section 2.2). The simple models that have been given [4], (based on a constant mobility, or on velocity saturation along the channel) provide meaningful results for Si devices. A schematic representation of the two models is shown in Fig. 4.22. The complications arise from the shape of the GaAs drift velocity versus field curve. As shown in section 2.8, such a curve presents a maximum at fields of around 3 kV/cm at room temperature, then decreases in correspondence with electron transfer into the low-mobility satellite valleys. At high fields, experimental results indicate an increase of the drift velocity and the absence of saturation effects. It is clear that the approximations contained in the model shown in Fig. 4.22 do not adequately represent the essential features of GaAs. A qualitative explanation of the current saturation in GaAs MESFET is illustrated in Fig. 4.23 [26]. Because of the positive drain voltage, the depletion width is wider near the drain. The resulting decrease of the channel opening must be compensated by an increase of either drift velocity or carrier concentration in order to guarantee current continuity through the channel. The electric field peaks at the drain-end of the gate, where most of the potential drop occurs. The stationary drift velocity reaches a maximum at the x_1 point, in correspondence with the

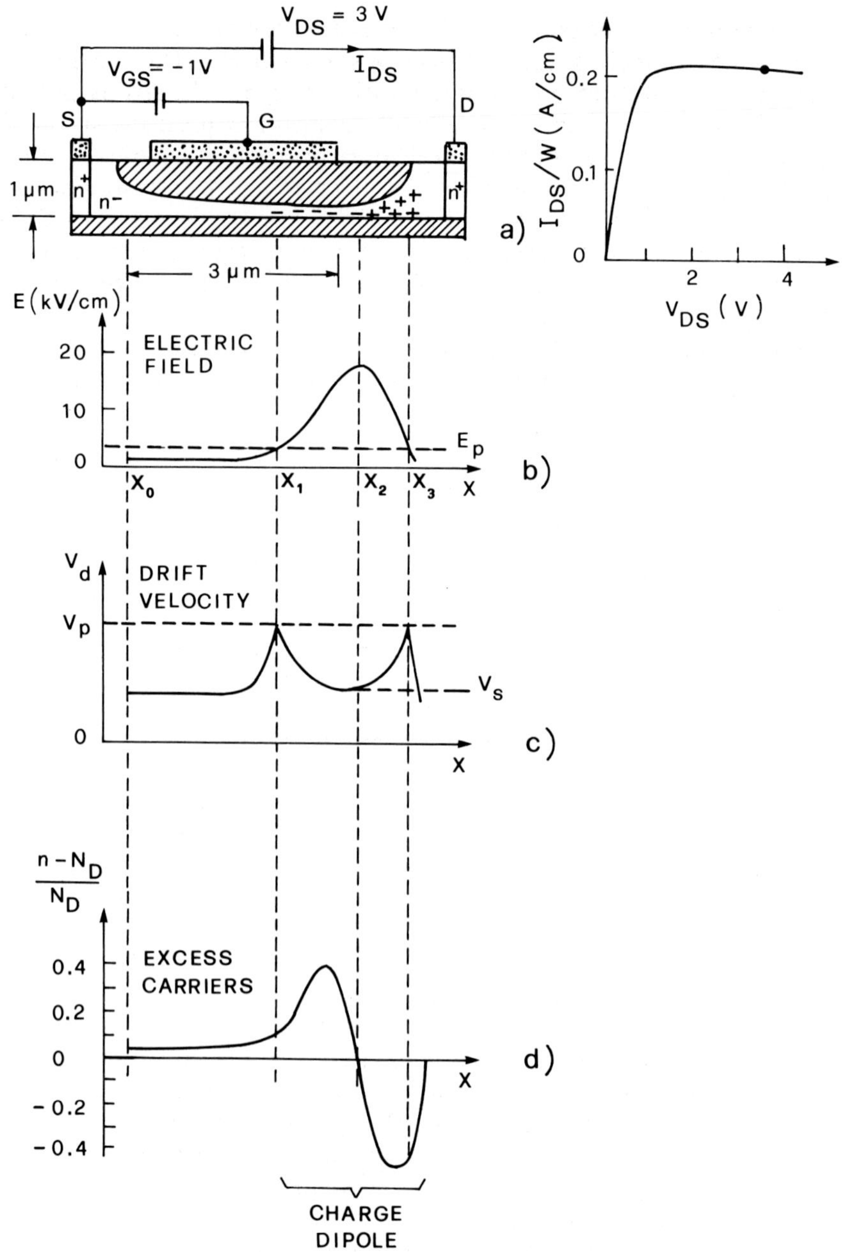

Fig. 4.23. Pictorial description of the current saturation mechanism in a MESFET [4]

critical field for intervalley transfer, then decreases when the electric field reaches its maximum value at x_2. As the field drops to the low values characteristics of the drain region, the velocity goes through another maximum at x_3. To preserve current continuity, a strong electron accumulation has to form between x_1 and x_2, where the channel is narrower and the electrons slower. On the contrary, electron depletion occurs between x_2 and x_3, where the channel widens and electrons move faster. The accumulation and depletion regions form a stationary dipole layer. The strong potential drop in the layer determines the value of the drain saturation current. The model just presented requires the transport process to remain stationary, that is the possibility for the electrons to always be in steady-state with the field. It was shown in section 2.9 that such an assumption is not valid for very short devices, when overshoot phenomena occur. Examples of this situation will be dealt with in chapter 6, in connection with the Monte Carlo simulation of devices.

The performance of a MESFET strongly depends on various technological aspects involved in the construction of the device. The possibility of im-

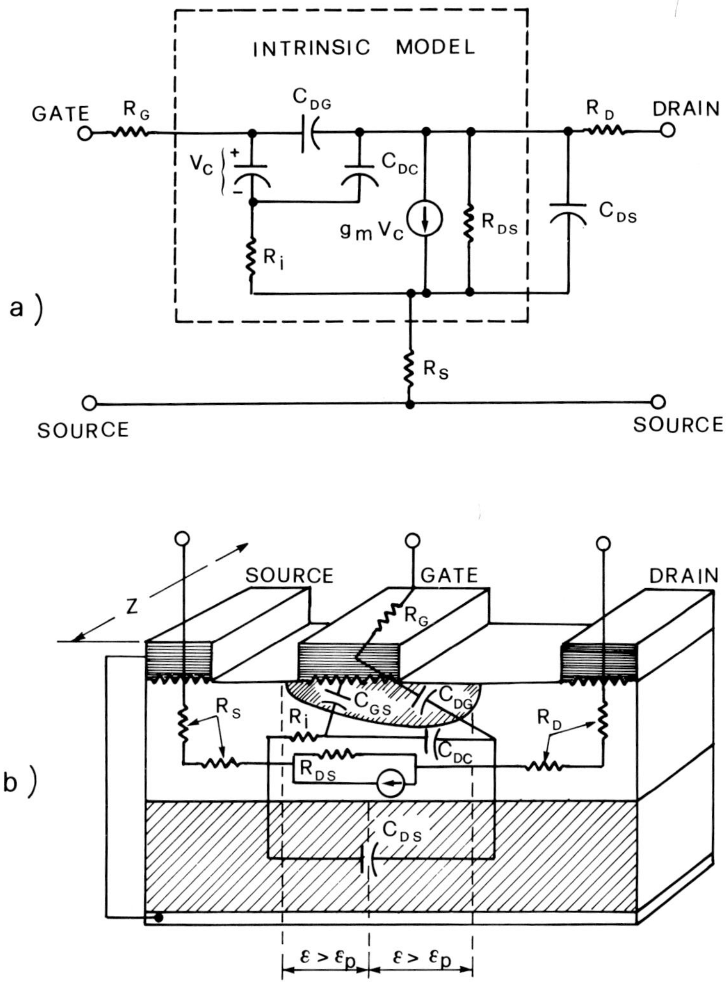

Fig. 4.24. (*a*) Equivalent circuits of a MESFET and (*b*) Physical origin of the circuit elements [26]

proving the device performance can be investigated starting from the small-signal equivalent circuit of a MESFET in the saturation region shown in Fig. 4.24a. The physical origin of the circuit elements is displayed in Fig. 4.24b. The intrinsic elements are represented by the total gate-to-channel capacitance ($C_{DG} + C_{GS}$) and the input resistances R_i and R_{DS} under the gate (which show the effects of the channel resistance). The extrinsic parasitic elements are the source resistance R_S, the drain resistance R_D, and the substrate capacitance C_{DS}. The parasitic resistances are responsible for degradation of the drain conductance and transconductance. Furthermore, they control the frequency response of the transistor, through the RC time constant. Several technological improvements have been introduced to reduce the parasitic elements. One of the most important is certainly provided by the technique of ion implantation, which allows to selectively dope given areas of a chip by introducing a controlled high density of dopants.
Using masking techniques in conjunction with ion implantation, it is possible to accurately control the doping levels and profiles through the choice of the implanted species and of the implantation energy. In this way it is possible to reduce parasitic resistances by heavily doping the regions underneath the source and drain contacts (Fig. 4.25a). Another important advantage of ion implantation is the possibility of building enhancement MESFET, which require a very small channel width in order to have no conduction (completely depleted channel) at zero gate voltage. A normally-off MESFET built using ion implantation is shown in Fig. 4.25b.
The device performance can also be improved by modifying the device geometry. For instance, a reduction of the gate length leads to higher cut-off

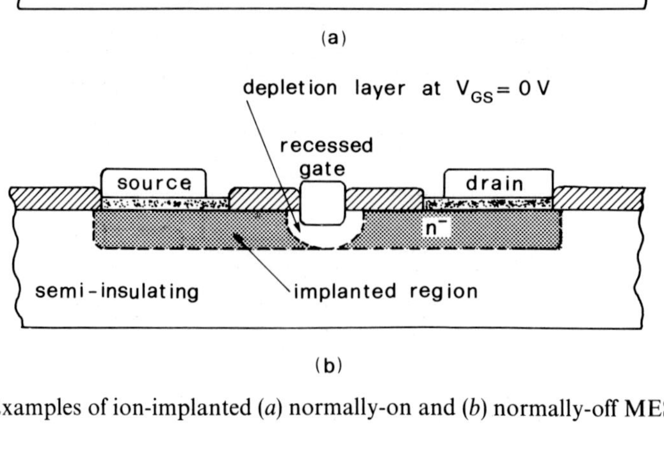

Fig. 4.25. Examples of ion-implanted (*a*) normally-on and (*b*) normally-off MESFET's [11]

frequencies, through the reduction of the gate-to-source capacitance. Self-alignment, that is the realization of source-to-drain spacing which coincide with the gate length, also contributes to the reduction of parasitic elements. GaAs MESFET's have found widespread use as single devices (for low-noise amplification or high-efficient power generation) and in integrated high-speed logic applications. The large scale integration in GaAs has been possible mainly through the use of planar technology combined with ion implantation and new lithographic techniques. The availability of both E-mode and D-mode FET's has allowed the introduction of new logic circuits (such as direct coupled FET logic-DCFL), characterized by low dissipation and high integration, two requirement necessary for integrated circuits.

4.3.6 Metal-Oxide-Semiconductor Field-Effect Transistor

The Metal-Oxide-Semiconductor Field-Effect Transistor (MOSFET) is a type of field-effect-transistor based on the properties of the metal-oxide-semiconductor system (mainly Metal-SiO_2-Si) examined in section 2.10. The typical structure of a *p*-type transistor is shown in Fig. 4.26. Two heavily doped regions are formed under the source and drain contacts. The metal (or polysilicon) contact on the insulator gate controls the carrier conduction in the device. A fourth contact on the back of the transistor is also present. When no bias is applied to the gate with respect to the substrate, the n^+ regions under the source and drain are isolated, as they form two adjacent *n-p* junctions. The depletion regions of the two junctions are indicated by the shaded area in Fig. 4.27. The band structure for flat-band conditions on a plane perpendicular to the gate is also shown. When a sufficiently high positive voltage is applied to the gate, the strong transversal field depletes

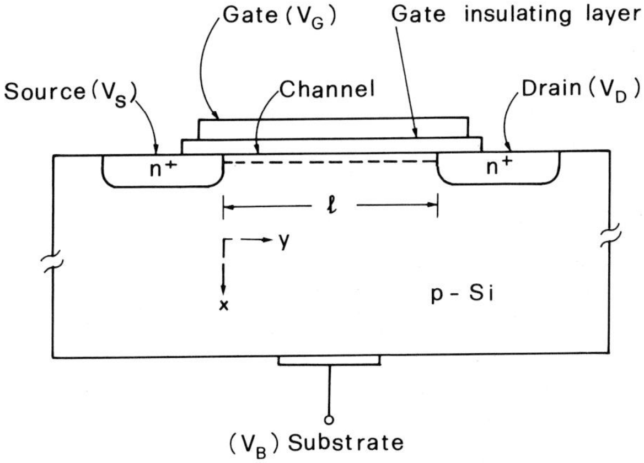

Fig. 4.26. Structure of a typical MOSFET [5]

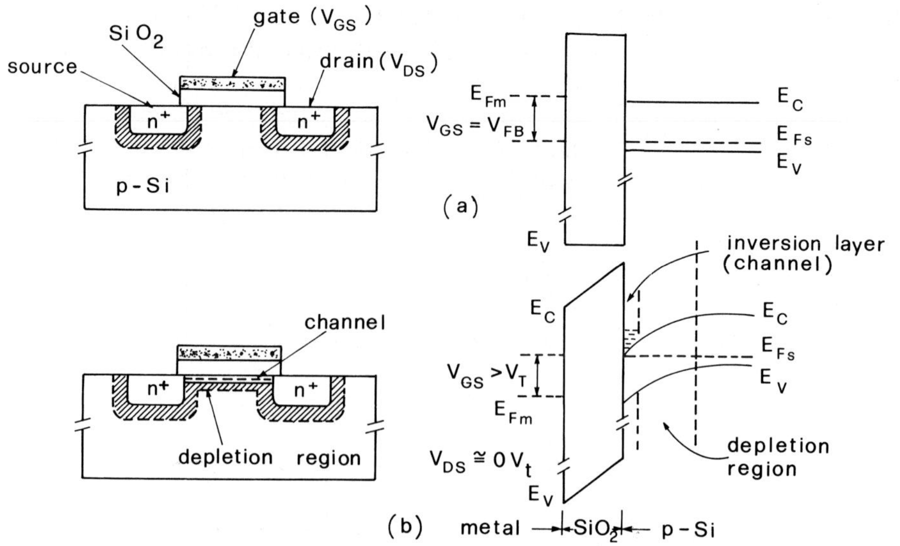

Fig. 4.27. Cross section and energy band diagram of an enhancement-mode MOSFET under (*a*) flat band condition and (*b*) inversion conditions [11]

the Si region near the surface pushing the majority carriers (holes in this case) towards the substrate. An inversion layer is formed, which opens a conductive channel between the doped source and drain regions (Fig. 4.27b). When, in addition to the gate voltage, a drain voltage is applied, current flows from the source to the drain. By looking at the device as a part of an outside circuit, we can say that in such a conductive mode the output current carries a logical signal to the next stage of the circuit. Electrons flow primarily through regions where they are majority carriers, in contrast to the bipolar transistor, where current is carried by injected minority carries. At zero gate voltage, no current flows from source to drain. Such a device is called "enhancement mode" (or normally-off) because the positive gate voltage is used to increase the channel conductance. "Depletion mode" (normally-on) device can be obtained by forming an *n*-type doped layer at the Si-SiO_2 interface. Even at zero gate voltage, the connection through the *n* layer allows current to flow from source to drain. A negative voltage on the gate gradually depletes the conductive channel and its resistance increases up to the point where the transistor is turned off. The ideal current-voltage characteristics of an "enhancement mode" MOSFET are illustrated in Fig. 4.28. Two main features can be identified: at low drain voltage, the current increases linearly with the potential, while at high drain voltage the current is constant. Such behaviour can be understood qualitatively by looking at Fig. 4.29 that shows the effect of bias on the depletion region. Under strong inversion conditions ($V_G > V_T$, where V_T is the threshold voltage, that is the voltage at which the channel becomes conductive), for low drain voltage the depletion region is nearly uniform along the channel. The drain current

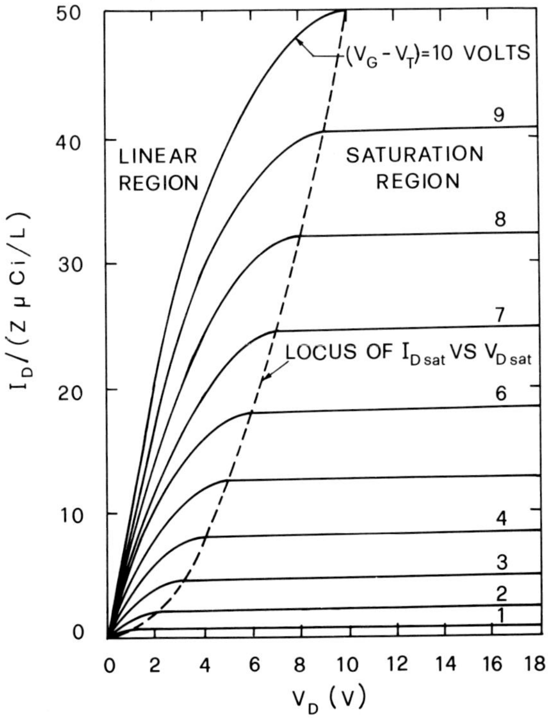

Fig. 4.28. Ideal I-V characteristics for a normally-off (enhancement) MOSFET [4]

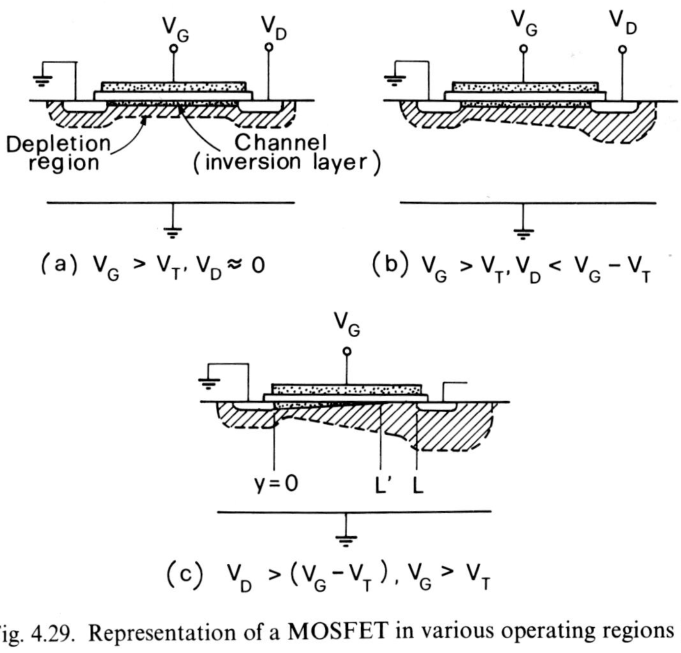

Fig. 4.29. Representation of a MOSFET in various operating regions [4]

varies linearly with the drain voltage, and the conductance between source and drain is proportional to the effective gate bias ($V_G - V_T$). As V_D increases, the thickness of the depletion region varies along the channel, being wider at the drain edge. Because of the variation of the potential drop along the channel, the confining well at the Si-SiO_2 interface is also not uniform. As a result, the drain current starts deviating from the linear behaviour characterizing the low-drain voltage region. If the drain voltage is further increased (with $V_D > V_G - V_T$), a high-field region forms at the drain edge of the gate, with very low electron densities. Here electrons are accelerated up to their saturation velocity. The drain current becomes insensitive to any variation of the drain voltage, being determined only by the rate at which electrons arrive at the edge of the high-field region.

The water analogy of Ref. [5] can explain this situation in a pictorial way (see Fig. 4.30). Source and drain are depicted as deep reservoirs connected by a variable-depth canal. The source to drain voltage is represented by the relative elevation difference between the two reservoirs. The gears that control the channel depth play the role of the gate voltage (Fig. 4.30a). As long as source and drain are at the same level no water flows. When the drain reservoir is lowered with respect to the source, water starts flowing towards the drain. At the drain-end the canal depth decreases as the water velocity increases. For low-level gradient (i.e. small source drain potential), the water flow is controlled by the depth of the canal and by the gradient itself (Fig. 4.30b). This is equivalent to the linear region of the I-V curves. As the drain is lowered further, the flow is completely controlled by the capacity of the canal. Any additional lowering of the drain results in a deeper

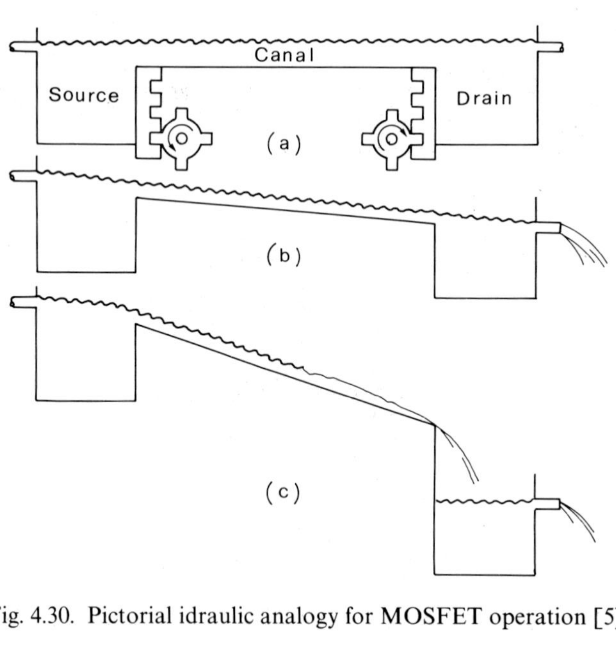

Fig. 4.30. Pictorial idraulic analogy for MOSFET operation [5]

free fall of water over a waterfall. The rate of flow only depends on the delivery rate at the drain-end of the canal, independent of the total drop over the cataract. Such a situation corresponds to the saturation region of the transistor.
The simple theory illustrated so far is based on a number of assumptions that in general limit its applicability. Many of these assumptions have to be questioned when small MOSFET's are considered [4, 5, 27]. For instance, the gradual-channel approximation, which states that the quantity of charge in the channel is completely controlled by the gate, fails in the case of submicron gate lengths. In such situations, it is necessary to consider the influence of drain and source junctions on the channel charge.
Another effect typical of submicron structures is a drain-induced subthreshold current, due to the lowering of the potential barrier between source and drain, caused by the overlap of their respective depletion regions as the device dimension is reduced. Last but not least, hot electron effects will become important, as we will see in detail in chapter 6.

4.3.7 High Electron Mobility Transistor

The High Electron Mobility Transistor (HEMT) is the prototype of the heterojunction field effect transistors. The cross section of such device is shown in Fig. 4.31. The HEMT combines some characteristic features of GaAs FET's and a Si MOSFET, and allows some of the limiting factors of the MESFET to be overcome while still enjoying the high mobility of GaAs. The main features of the HEMT is the spatial separation between the carriers (electrons confined in the two dimensional electron gas) and their parent impurities (confined in the AlGaAs layer), obtained through the

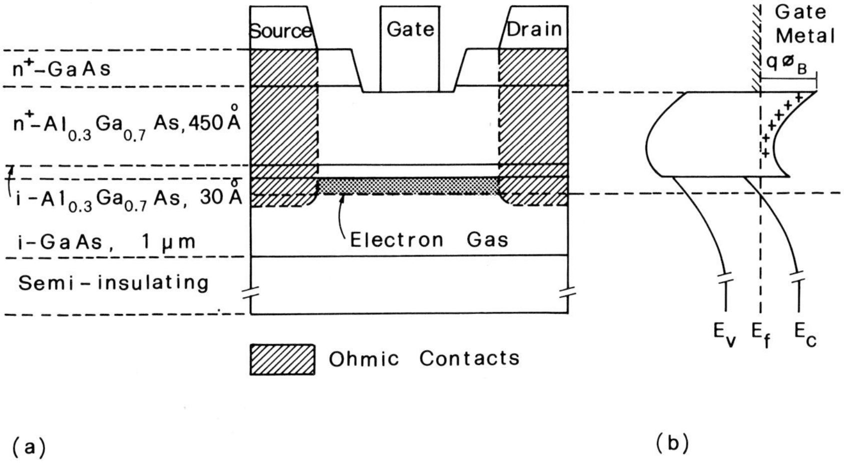

Fig. 4.31. (*a*) Cross section and (*b*) energy band diagram of a HEMT

technique of modulation doping. In a way, the HEMT can be viewed as a MOSFET where the insulator has been substituted by a wider band gap semiconductor. The electron concentration in the quantum well at the GaAs-AlGaAs interface determines the channel conductivity. An undoped AlGaAs layer ("spacer") is usually introduced to help to remove the channel electrons from the impurities. As pointed out in section 2.10 the high transverse field at the interface causes quantization of the electron motion in the direction perpendicular to the interface. The modulation and control of the channel charge, and therefore of the current, is achieved through the Schottky junction (gate) on the AlGaAs layer, in the same way as in the GaAs MESFET. The difference in the electron affinity of the metal gate and the AlGaAs causes a depletion region to be formed at the metal interface. In a normally-on ("depletion mode") device (Fig. 4.32a), a second depletion region occurs at the heterojunction, because of the electron transfer into the quantum well [28]. At zero gate bias, the AlGaAs is totally depleted and the conductive channel is constituted by the 2DEG in the GaAs. By applying a negative bias at the gate, it is possible to reduce the channel conductance by depleting the channel charge. A normally-off ("enhancement mode") device (Fig. 4.32b) can be obtained by reducing the size of the AlGaAs layer and/or by increasing the height of the Schottky barrier. The depletion region then extends over the GaAs, and prevents the formation of the 2DEG. In order to ensure conduction, a positive voltage is applied to the gate contact (dashed line in Fig. 4.32b). From the point of view of charge control and

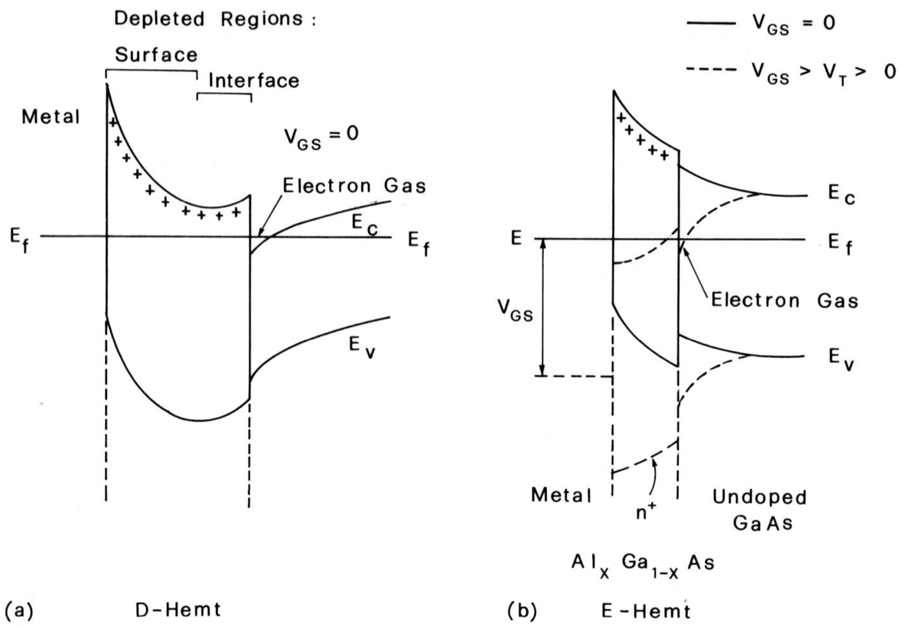

Fig. 4.32. Schematic energy band diagram for (*a*) a depletion-mode HEMT ($V_{GS} = 0$) and (*b*) an enhancement-mode HEMT ($V_{GS} = 0$, continuous line; $0 < V_{GS} < V_T$, dashed line) [23]

current-voltage relations, the HEMT is similar to the MOSFET. The gate voltage V_T below which there is no charge in the channel is called threshold voltage. Its value heavily depends on technological issues such as the doping and Al-concentration control in the growth process of the AlGaAs layer.
The carrier concentration in the channel grows linearly with the applied bias. The AlGaAs layer can be viewed as a parasitic FET parallel to the one characterized by the heterojunction. Since the AlGaAs is heavily doped, it is necessary to prevent carriers from flowing into such a low-mobility channel. For D-HEMT devices, the parasitic FET is turned off as long as the two depletion regions (at the metal-AlGaAs and at the AlGaAs-GaAs interfaces) overlap. For E-HEMT, the positive gate voltage has to be kept below a certain value (typically 0.8 eV), at which the AlGaAs layer becomes conductive. At this critical voltage, the gate can no longer control the channel charge, as a neutral region forms in the AlGaAs. Such voltage can be seen as the threshold voltage for the parasitic FET, and sets an upper limit for the proper operation of the device.
Contrary to the Si-SiO_2 interface (where the barrier height is around 3 eV), the electrons in the AlGaAs-GaAs system are confined within the quantum well by a barrier of around 0.3 eV. Therefore, at moderately high fields (in the direction along the interface) electrons can gain enough energy to surmount the barrier. Such an effect, which was described in section 2.10 is called "Real Space Transfer". The transfer of carriers from the high-mobility channel into the low-mobility AlGaAs layers due to heating effects is a major source of degradation for HEMT's.
The I-V characteristics of a HEMT are qualitatively similar to those of a MOSFET. The drain current can be approximated by [29]

$$I_{DS} = q n_s v(x) w,$$

where q is the electron charge, w the channel width, and $v(x)$ the electron velocity in the channel, x denoting the direction from source to drain. The sheet electron concentration n_s depends on the effective gate voltage $V_G - V_T$ and on the channel voltage $V_c(x)$ through the relation

$$n_s = \frac{\varepsilon}{qd}[V_G - V_T - V_c(x)],$$

where ε is the GaAs permittivity, and d is the thickness of the AlGaAs layer. At small drain-to-source voltages, a linear current-voltage relation is found. At higher voltages, saturation of the drain current is reached. Such behaviour can be qualitatively understood on the basis of the analysis already presented for MOSFET and MESFET. The channel can be divided into two main regions, the first characterized by a constant mobility, up to a critical value E_c of the electric field, the second by a saturated velocity, constant at fields higher than E_c. In the first region, the charge density decreases along the channel, compensated by an increase of the electric field (and hence of the electron velocity), to maintain a constant current from

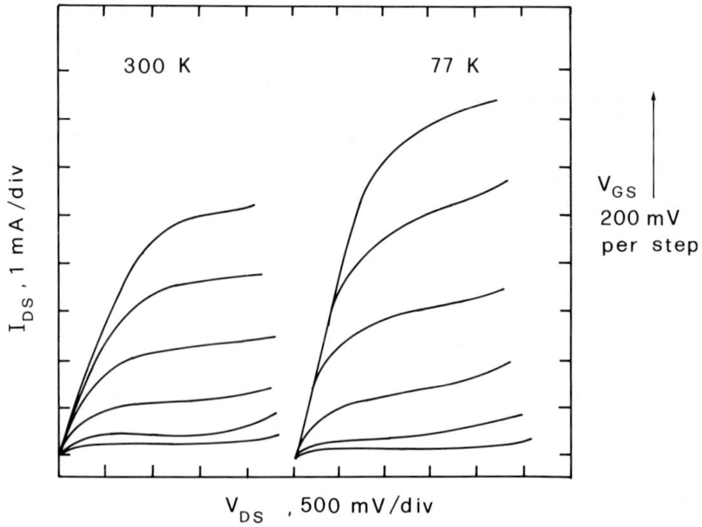

Fig. 4.33. I-V characteristics for a HEMT at room temperature and nitrogen temperature [11]

source to drain. In the second region, as the velocity saturates, a large part of the current is carried by diffusion.

Typical I-V characteristics for a HEMT are shown in Fig. 4.33 [29]. The improvement in the device performance at lower temperatures is a direct consequence of the increased carrier velocity.

4.3.8 *Hot Electron Transistors*

One way of reducing the transit time is to shorten the channel of a FET or the base of a bipolar transistor. A double advantage is reached, in that electrons have to cross shorter distances before being collected, and the potential drop under bias occurs over a smaller region, thus leading to higher electric fields. When the size of the active region is of the order or smaller than the electron mean free path, overshoot phenomena can occur (see section 2.9). The electrons are accelerated by the electric field up to velocities that exceed the saturated velocity of the material, leading to very high effective drift velocities and correspondingly short transit times.

An alternative way of reducing the transit time across a device, is to "launch" electrons into the active region with high kinetic energy and high drift. If the length of the active layer is sufficiently small, electrons will not suffer many collisions in their motion. Therefore, they could maintain their high speed up to the collecting contact, assuming transit times much shorter than in standard transistors.

Such an idea was already put forward by Mead in 1960 [30], in connection with the metal-insulator-metal-insulator-metal (MIMIM) structure shown

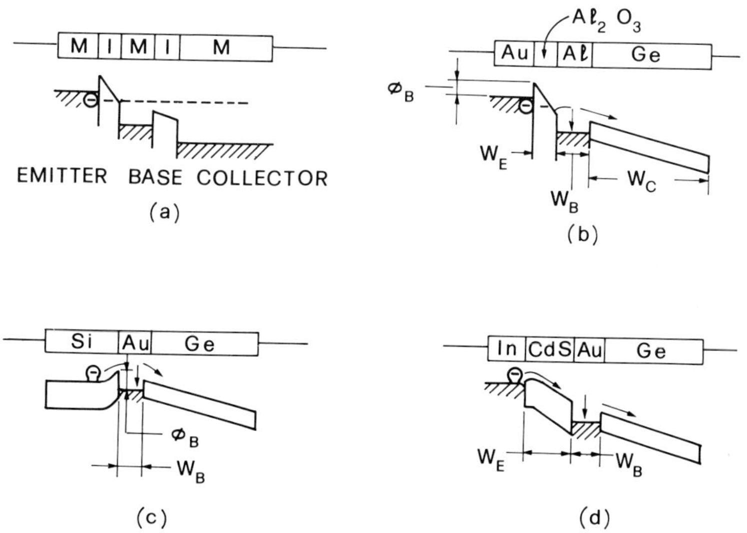

Fig. 4.34. Four different types of hot-electron transistors: (*a*) MIMIM structure; (*b*) tunnel transistor; (*c*) metal-base transistor; (*d*) space-charge-limited transistor [31]

in Fig. 4.34a. There, electrons tunnel through the first thin insulator layer into the metal base. Since their excess energy is well above the thermal energy $K_B T$, the device was called a "hot electron" transistor. The second "MIM" structure, between base and collector, confines the base electrons. The emitter-base voltage controls the tunneling current into the base, while the base-collector voltage controls the number of electrons that reach the collector. Clearly, in order to achieve a high transfer coefficient (that is a high ratio of collected to injected current) it is necessary to reduce energy and momentum losses in the base. Only those electrons that have a sufficient momentum in the direction perpendicular to the collector barrier contribute to the collector current. The others, which are trapped in the base, contribute to the base current. Other hot-electron transistors designed to improve the device performance are illustrated in Fig. 4.43d-b [31]. The main difference in these transistors is the method in which electrons are injected into the base. The most promising structure is probably the metal-base transistor (Fig. 4.34c), since, in principle, the hot electrons should suffer little scattering in transversing the thin Au base. The transit time should therefore be of the order of a few femtoseconds. In addition, the low resistance of the metal would allow a fast modulation of the base voltage and very high current densities. Unfortunately, the transmission of the electrons through films of gold as thin as 10 nm was found to be very small, due to technological reasons connected with the metal-semiconductor interface, and to more fundamental questions associated with the quantum reflection at that interface. More recently, the metal base has been substituted by a silicide base, which should guarantee a better uniformity of the interfaces [32].

Two hot-electron transistors based on GaAs are particularly important since they allowed for the first time the observation of ballistic electrons

[33, 34]. They were called respectively Planar Doped Barrier (PDB) transistor [35], and Tunneling Hot Electron Transfer Amplifier (THETA) [36]. As in the previous cases, they differ mainly in the way electrons are launched into a doped base, and are collected. In one case, the emitter-base and base collector barriers are formed by a plane of berillium atoms embedded into highly doped GaAs regions. The berillium atoms (Fig. 4.35) attract electrons,

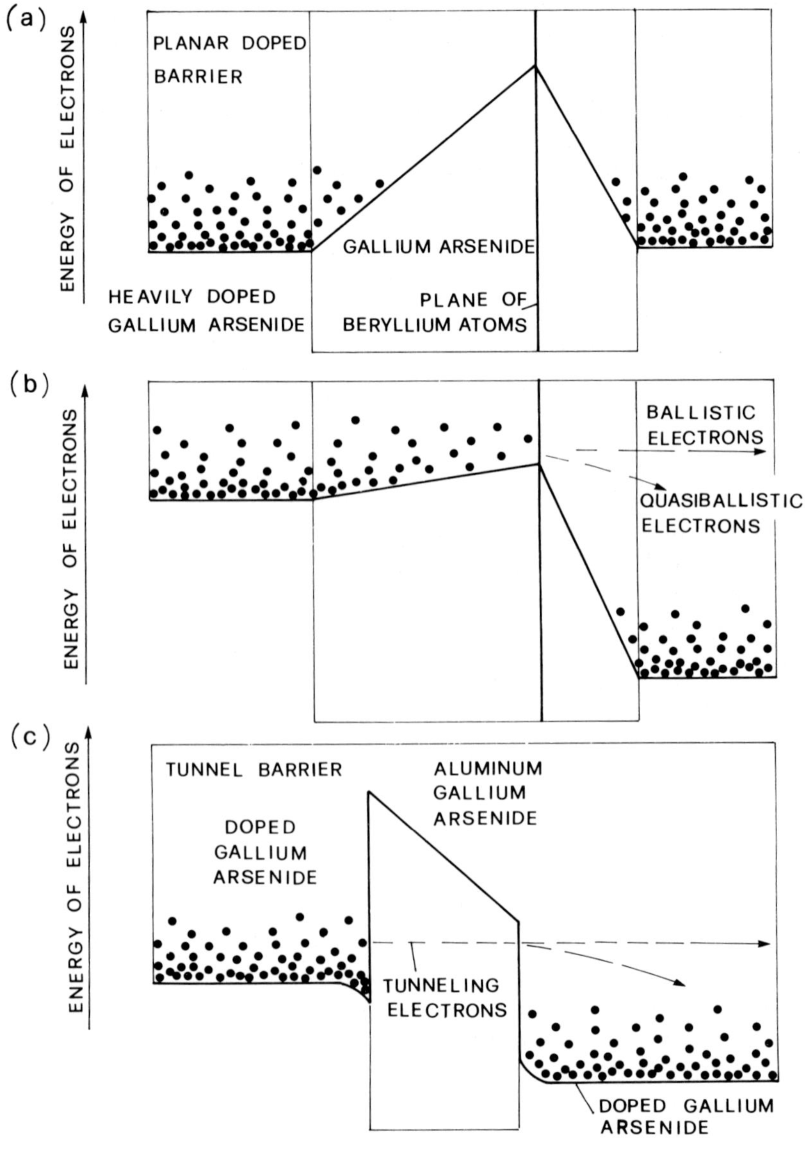

Fig. 4.35. Two different types of ballistic injection: a planar doped barrier (*a*) without and (*b*) with applied bias, and (*c*) tunnel barrier with applied bias [37]

forming a plane of negative charge, which gives rise to a triangular potential which opposes the motions of the electrons from the heavily doped region [37]. The application of a given potential lowers the barrier for the electrons on one side of the berillium layer, allowing current flow across the barrier. In the second case, tunneling emitter and collector barriers are created using a GaAs-AlGaAs heterojunction. If the AlGaAs barrier layer (which is sandwiched between two heavily doped regions of GaAs) is thin enough, electrons can tunnel through the barrier. At zero bias, there is no net current flow. The application of a bias rises the potential energy of the electrons on one side of the barrier. Those electrons tunnel through the barrier and enter the base region with high energy.
The full band diagrams for the PDB and the THETA devices are shown in Fig. 4.36. In both cases, emitter-collector current is given by the number of injected electrons that can surmount the collector barrier after crossing the

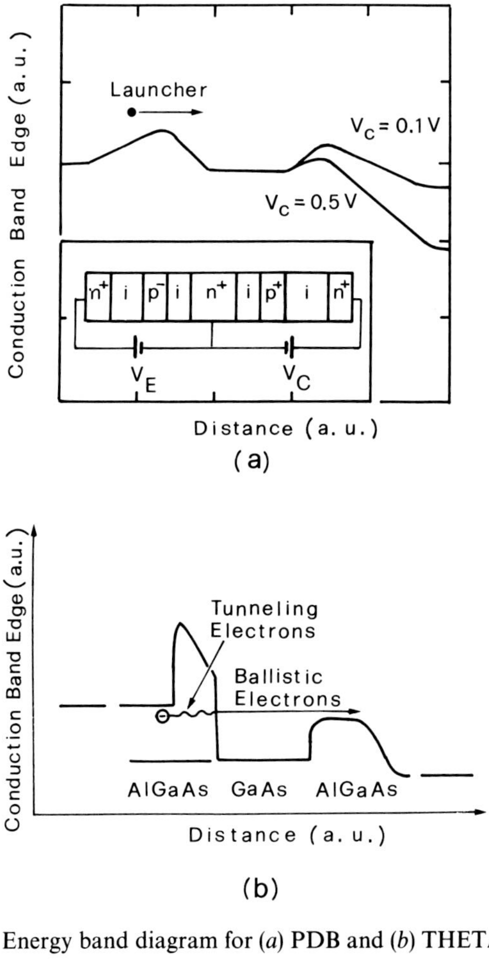

Fig. 4.36. Energy band diagram for (*a*) PDB and (*b*) THETA devices

base. Such a number is dependent on the base-collector bias, and on the details of the collector barrier. It was found for example that grading of the edge of the GaAs-AlGaAs base collector interface was necessary to reduce the quantum reflection of the barrier and improve the transfer ratio of the THETA device.

Despite their great potential, the two devices just presented have not yet been able to offer a valuable alternative to more traditional bipolar and field-effect transistors. The main reason lies in the very high scattering rate that the hot electrons suffer in the base. As was shown in section 2.4.4, when fast carriers move into doped regions, several phenomena such as carrier-carrier and carrier-plasmon interaction become very important. In the case of the PDB and the THETA devices, those are two major sources of degradation, together with the electron reflection at the collector barrier. It has been shown that the use of InGaAs in the base leads to improved performances over GaAs, due to the smaller electron mass of the former material.

4.3.9 Permeable Base Transistor

Another device in which ballistic electron transport is being sought is one in which a metal base is patterned into a permeable grid, defined lithographically with finger-like structures having widths and spacings of the order of 100 to 1000 nm [38]. The device, called Permeable Base Transistor (PBT), consists of four layers; the n^+ GaAs substrate, the *n*-type GaAs emitter layer, the patterned metal film (usually tungsten) and the *n*-type GaAs collector. An alternative to the metal base is the (platinum or cobalt) silicide base. A schematic diagram of the PBT is shown in Fig. 4.37. The device can be considered the solid-state counterpart of the vacuum tube diode. The PBT can be viewed as a vertical FET, having an ultrashort gate determined by the thickness *L* of the grating. The principles of behavior of a permeable metal-base transistor can be understood from Fig. 4.38, where the calculated

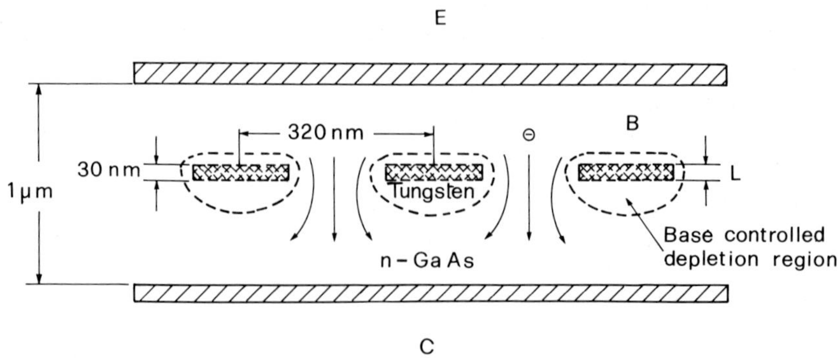

Fig. 4.37. Schematic diagram of a permeable base transistor

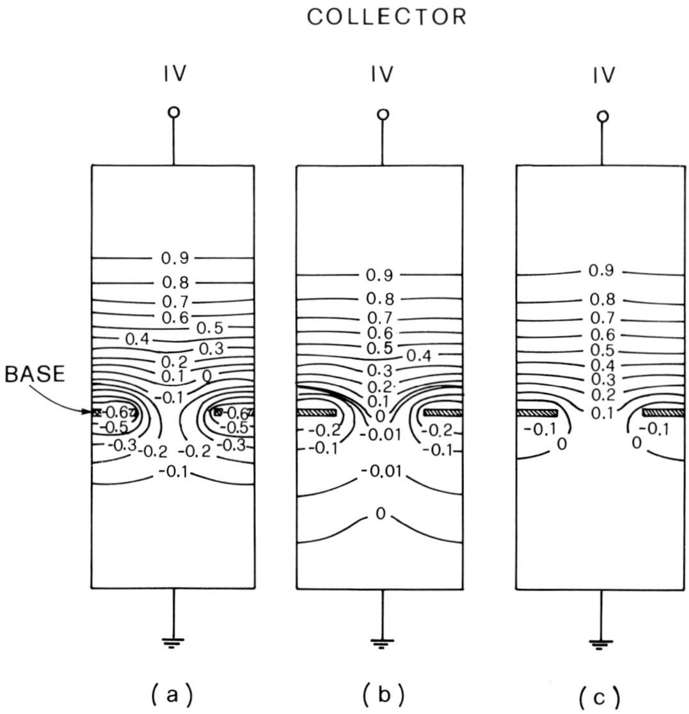

Fig. 4.38. Calculated equipotential lines for the unit cell of a PBT at three different base bias conditions (*a*) $V_{BE} = 0$ V, (*b*) $V_{BE} = 0.3$ V, (*c*) $V_{BE} = 0.5$ V [38]

equipotential lines are shown for a cross section of a unit cell of the device. In Fig. 4.38, 1 volt is applied to the collector, with no bias on base and emitter. The electrons moving from the emitter to the collector must surmount a potential barrier created by the Schottky barrier of the metal base (negative equipotential lines in the plot). Such a barrier is at its lowest in the center of the opening between the metal grids, and highest at the metal-semiconductor interface. Under those bias conditions, the collector current is very small. However, a forward bias on the base lowers the barrier and favours the current flow toward the collection (Fig. 4.38b). At even higher base bias, carrier accumulation develops, and the current starts being space-charge-limited (Fig. 4.38c).

Depending on the base grid spacing and on the carrier concentration, barrier-limited carrier flow can exist at high current densities, resulting in outstanding performances of the device.

4.4 Comparison of Semiconductor Devices

In the previous section we focused on various semiconductor devices from a physical point of view, trying to clarify the principles of operation of such devices. Here we want to look at the ways used to characterize a device and quantify its performance. Therefore the most important parameters will be introduced and briefly discussed. On the basis of these parameters, a comparison between different devices will be given.

4.4.1 *Device Parameters*

In general terms, the speed of operation for a semiconductor device depends on the type of logic circuit that it belongs to. Let us take as an example a field effect transistor.

For digital logic circuits, as found for example in computers, transistors act as switches. The speed of a transistor is then defined as the time required to be pinched off, that is to reach the stage when no current flows from source to drain. The most basic logic circuit is the "inverter", which converts a low voltage input (a "0" or "off" logic state) into a high voltage output (a "1" or "on" state) and does the opposite as well. An inverter circuit that uses complementary MOSFET (CMOS) technology (discussed in section 4.2.1) is shown in Fig. 4.39.

Unless the circuit is switching, one transistor is always "off" (that is, it does not conduct any current) and the other is "on". The transfer curve for an ideal and a real inverter is illustrated in Fig. 4.40. Ideally, such a curve should be a step function, with infinite slope at the transition point. In practice this cannot be achieved. The speed of the circuit is given by the time it takes to switch from the "0" to the "1" state, or viceversa. In the inverter circuit shown in Fig. 4.39, the switching operation corresponds to turning on one device and turning off the other.

In analog circuits, such as those found in receivers, radios, and televisions, the role of the transistor is to be able to transform small changes of the control electrode (gate) into corresponding large changes in the drain current. Such a function is called "current gain", and is defined as the ratio of the drain to the gate current. Such gain drops at increasing frequencies.

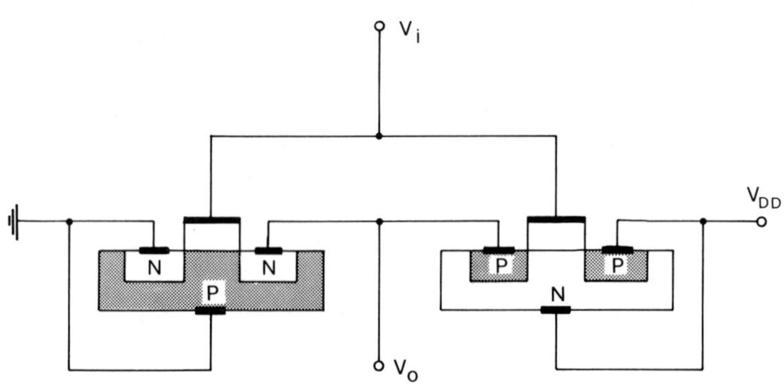

Fig. 4.39. Schematic representation of an inverter circuit based on CMOS technology

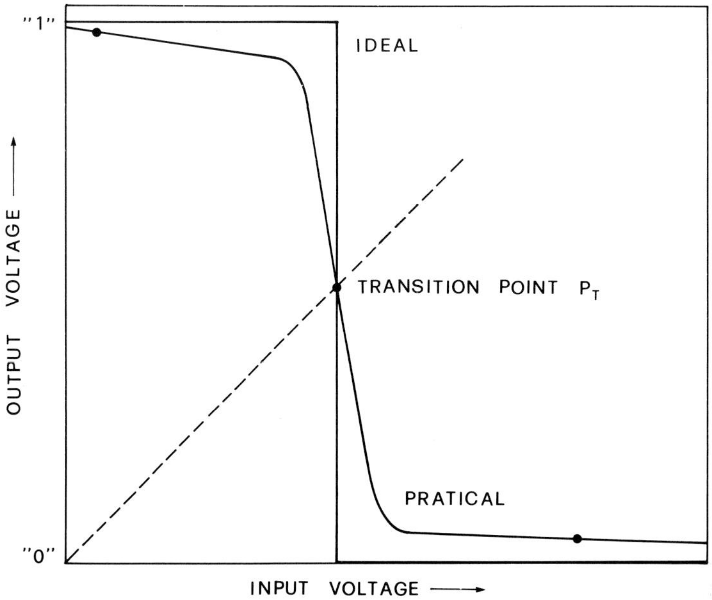

Fig. 4.40. Transfer curve for an ideal and a pratical inverter circuits [20]

The speed of a transistor for analog applications is therefore associated with the ability to maintain high gain at high frequencies.

The previous concepts can be better defined through a series of parameters that will be now introduced.

For high-speed logic applications, the parameter that controls the speed of a device in a circuit is the transconductance, defined as the variation of the output current in response to a changing input voltage. For example, in a MOSFET the transconductance g_m is formally given by

$$g_m \equiv \frac{I_D}{V_G},$$

where I_D and V_G are respectively the drain current and the gate voltage. The role of the transconductance can be understood by looking at the schematic picture of Fig. 4.41, where a transistor is represented as a current generator, with a current I given by the product of the input voltage V_{in} and the transconductance g_m. The transistor is coupled to a capacitive load. For a given input bias, a device with higher g_m provides more current to charge and discharge the capacitances connected to it. Therefore, switching times are reduced and higher operating frequencies can be reached. Furthermore, a high transconductance allows an equal output current for a lower input voltage, leading to a reduction of the power dissipated in switching operations.

The major limitation to the speed of the response of a transistor in a circuit (such as the inverter shown in Fig. 4.39) is associated with the time needed

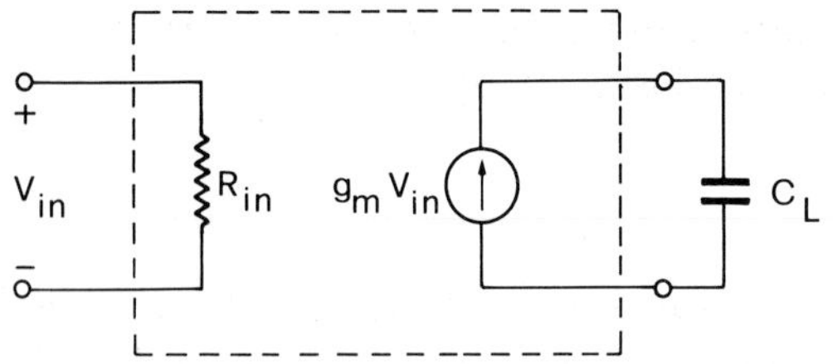

Fig. 4.41. Elementary equivalent circuit for a device with capacitive load

to charge the capacitances associated with the device and the elements to which it is connected in the circuit. Therefore, the ultimate speed of a switching device is determined by the ratio of the transconductance to the sum of the device and interconnection capacitances. The higher the transconductance, the faster the device. A measure of the intrinsic speed of a device is the transit time, that is the time required for carriers to cross the active region of the device (the channel for a field-effect transistor and the base for a bipolar transistor). The transit time depends on the nature of the carrier transport. In a FET, for example, carriers move under the influence of an electric field that drags them from source to drain, while, in a bipolar, carriers diffuse through the base. In both cases, the transit time depends on the length of the active region, although the type of behavior depends on the occurrence of hot-electron phenomena such as velocity saturation, velocity overshoot, ballistic injection, etc., ..., as previously discussed.

Another intrinsic limit, especially for high-frequency operations, is given by the *RC* time constant, that is the time required to charge the capacitances inherent to the device structure. An example is given by the elements of the equivalent circuit of the MESFET shown in Fig. 4.24. In general, it can be said that *RC* time constants are usually larger than the transit time, and therefore constitute the major obstacle to device speed.

Another parameter which is often cited in connection with ultra-high speed integrated circuits is the power-delay product $P\tau$, given by the product of the power dissipated (per gate) in a given circuit and the propagation delay (or gate switching speed). The power P depends on the global environment and sets a limit to the density of transistors that can be placed on a chip before heat dissipation in the circuit becomes a problem. The power-delay product gives the dynamical switching energy for a circuit, and should be kept as low as possible when high scale integration (that is a large number —$10^5 \div 10^7$—of devices on a chip) has to be achieved for ultra-high speed logic applications. The power-delay product depends mainly on the type of material and technology used in the single device and in the circuit and on the biasing voltages.

For high frequency applications (such as in microwave transistors) the most important parameters are the current-gain cut-off frequency f_T, power-gain cut-off frequency $f_{\max}$ and the noise figure NF. The two frequencies f_T and $f_{\max}$ denote the maximum frequencies at which a transistor has, respectively, current gain and power gain. The former, f_T, depends only on the physics of carrier transport, and is given by

$$f_T = \frac{1}{2\pi\tau},$$

where τ is the carrier transit time. The latter, f_{max}, depends both on f_T and on the parasitic resistance and capacitance of the device. Transistors, properly matched with appropriate impedances, amplify up to the power gain cut-off frequency f_{max}. Such frequency depends substantially on the transistors design and construction, and for field-effect transistors is normally higher than f_T.

The noise figure NF is defined as the ratio of the total mean-square noise voltage at the output of a transistor to the mean-square noise-voltage from thermal noise in the source resistance.

4.4.2 Comparison of Semiconductor Devices

We are now in the position to attempt a comparison between the various devices that have been presented earlier. We will try to classify and compare the devices in a homogeneous way, according to their use in microelectronics.

It was previously pointed out that transistors can be built in two alternative ways, with a horizontal or a vertical geometry. In the first case, carriers flow parallel to the planes of crystal growth, whereas in the second they flow perpendicular to those planes. The most common type of horizontal device is the field-effect transistor (FET), in which source, gate and drain contacts lie in one horizontal plane on the top surface of the wafer. Such geometry does not present particular fabrication difficulties and fabrication costs can be kept low. Vertical devices (such as bipolar transistors) are on the contrary more costly because of the crystal growth technique employed. The advantage over horizontal structures is in the possibility of controlling the thickness of each layer down to a few tenths of a nanometer. Submicron dimensions are therefore easily obtained for vertical structures, while for horizontal devices the limiting factor are the lithography techniques that are used to "write" the contacts and spacings between them on the wafer.

Technology and cost are two fundamental factors involved in determining which device should be used for a given application.

From the point of view of the materials, we have seen in section 4.1 that silicon has some unique properties that make it the best semiconductor material for integrated circuits. One fundamental factor in favour of Si is given by the complementary MOSFET (CMOS) technology. Such technology has became the dominant technology for processing information because it allows a low-power-consumption and high-speed inverter to be introduced simply by coupling an n-type and a p-type channel MOSFET. Gallium Arsenide on the contrary offers advantages from the point of view of electron velocity, which immediately allows reduced transit times and higher operation frequencies. For these reasons, GaAs technology has already made a substantial impact on analog circuits, dominating the scene for applications over most of the microwave band (up to 30 GHz). The reduction of gate length and the use of heterojunction devices will probably become established in the near future as the dominant technology also for the millimeter-wave band, which extends up to 300 GHz and is still largely unexploited.

A useful first comparison can be made between different field-effect transistors. Figure 4.42 shows the drain current-gate voltage characteristics for MESFET, HEMT and MOSFET at various operating conditions [23]. Several features are of importance. Within the same class of transistors, the transconductance increases at decreasing temperatures (in HEMT g_m goes from a value of 196 mS/mm at 300 K to 494 mS/mm at 77 K, while for MESFET the enhancement is from 120 mS/mm to 220 mS/mm, although

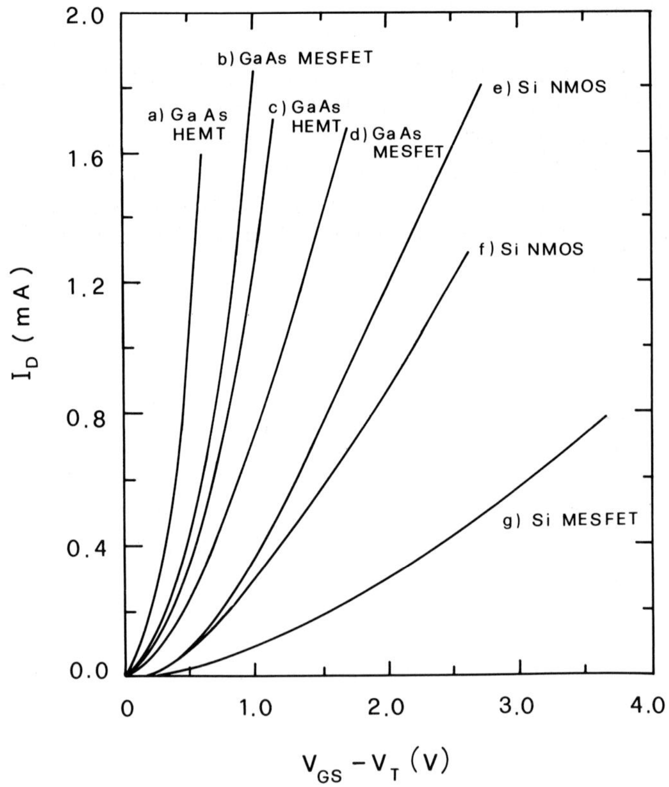

Fig. 4.42. Drain current-gate voltage characteristics for various FET's. The abscissa represents the difference $V_{GS} - V_T$. (*a*) HEMT, L_g = gate length = 0.7 μm, at 77 K; (*b*) GaAs MESFET, $L_g = 0.5$ μm, at 77 K; (*c*) HEMT, $L_g = 0.7$ μm, at 300 K; (*d*) GaAs MESFET, L_g = 1 μm, at 300 K; (*e*) Si *n*-MOS, $L_g = 0.7$ μm, at 300 K; (*f*) Si *n*-MOS, $L_g = 1.3$ μm, at 300 K; (*g*) Si MESFET, $L_g = 1$ μm, at 300 K [23]

in the latter case the reduction of the gate length also plays an important role). The improved device performance at lower temperatures is a direct consequence of the higher electron velocity. The comparison between the transconductance of different devices shows a considerable advantage of the HEMT compared to MESFET and MOSFET (the latter exhibiting a relatively small g_m of 40–45 mS/mm at room temperature). Such a leading position can be attributed to various factors:

a) the higher electron velocity of electrons in the channel of a HEMT with respect to the two other structures. The difference increases at low temperatures due to the suppression of impurity scattering achieved via modulation doping;

b) the parasitic source resistance (which has a very strong influence on the transconductance) is reduced in the HEMT;

c) the control of the channel charge is increased by the fact that the two-dimensional electron gas is displaced only few hundred Å from the gate of the HEMT. This implies that higher current densities can be modulated by smaller gate biases.

It should be remembered that a high value of g_m has remarkable consequences, in that it allows the realization of logic circuits characterized by low power dissipation and high switching speed.

Figure 4.43 shows the delay-power curve for logic gates in a large scale integrated circuit for various technologies [8]. The diagonal lines indicate constant values of the power-delay product. The optimal technology is the one that occupies the lower left corner of the graph, since it would guarantee

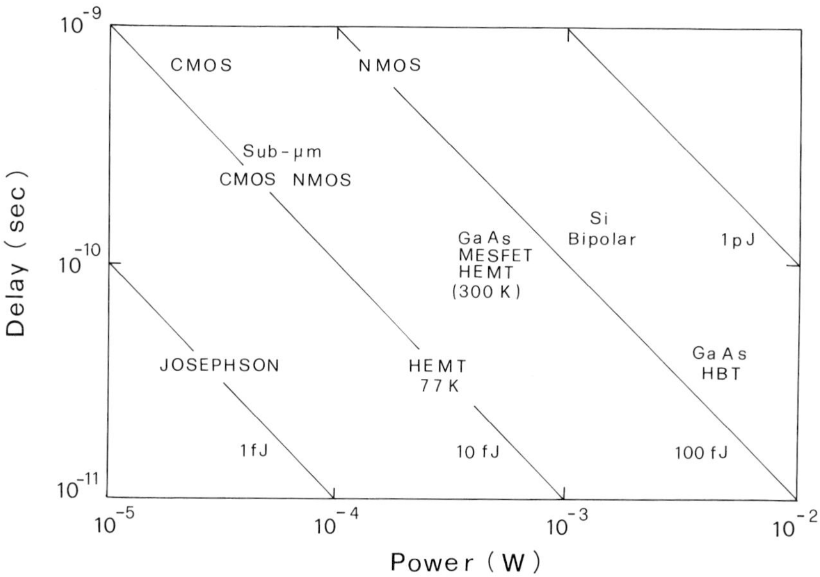

Fig. 4.43. Comparison of delay-power curve for logic gates in a large-scale integrated circuit for various technologies [23]

small delay times (i.e. high speed) as well as low power consumption (i.e. high levels of integration). The Josephson Junction technology offers by far the best combination of high speed and low power, but it is has been abandoned for logic applications because of technological difficulties. Silicon technology presents the best figure for power dissipation, with CMOS giving the optimal performance for VLSI applications (due to the fact that CMOS circuits only dissipate while switching). Despite the advantages achieved in terms of speed by going into submicron dimensions, Si MOS circuits are much slower than the GaAs counterparts. For room temperature applications, HEMT and MESFET technologies are practically equivalent. HEMT circuits gain both in speed and power when cooled down at nitrogen temperature. Silicon bipolars, and even more heterojunction bipolars, have competitive delay times (HBT offering the lowest delay for room-temperature operations) but suffer from very high power dissipation.

The previous two figures set the potential for the realization of integrated circuits based on a given technology. However, other factors have to be accounted for. When high integration levels are sought, a great degree of uniformity of the threshold voltage across the wafer is required. This is, for example, a very stringent constraint for HEMT circuits, where, in order to assure an acceptable range of variation of the threshold voltage (within few percent of the logic swing of the circuit) it is necessary to control the width of the AlGaAs layer by few Å and the aluminum concentration within one percent. Furthermore, the complexity of the construction process, and the yield and reproducibility of the final product are parameters that will determine the success of a given technology.

Table 4.1 gives an overall comparison of the technologies available for high speed applications (on a scale from 1 = best to 5 = worst), combining the speed-power considerations of Fig. 4.42 with more technological issues. It is clear that the most promising candidate for high speed integrated circuits are HEMT structures, once material problems of the AlGaAs are solved and a better doping control achieved. At present, and certainly for the near future, Si technology dominates and will dominate the field.

Table 4.1. *Comparison of various devices for high speed applications, graded from* 1 (*best*) *to* 5 (*worst*) [8]

	GaAs			Si	
	HEMT	MESFET	HBT	MOSFET [CMOS]	BJT
Speed	1	3	2	5	4
Power delay product	1	2	4	1	4
Lithographic requirement	2	3	1	4	1
Doping control	4	4	1	2	1
Processing complexity	2	1	5	3	5
Materials problems	4	3	4	1	2

As anticipated previously, a different situation is found for high-frequency applications. There, GaAs is the dominant material. In this case, uniformity requirements, that limit the possibility of high integration in GaAs, are somewhat relaxed. Furthermore, heat removal is easier to accomplish. The emphasis in the devices goes towards high frequencies of operation, with high gain, low noise, and high power capabilities.

Figures 4.44 and 4.45 show respectively a comparison of the current-gain cut-off frequency f_T and the minimum noise figure NF for MESFET's and HEMT's as a function of gate length [39]. We can see that both quantities

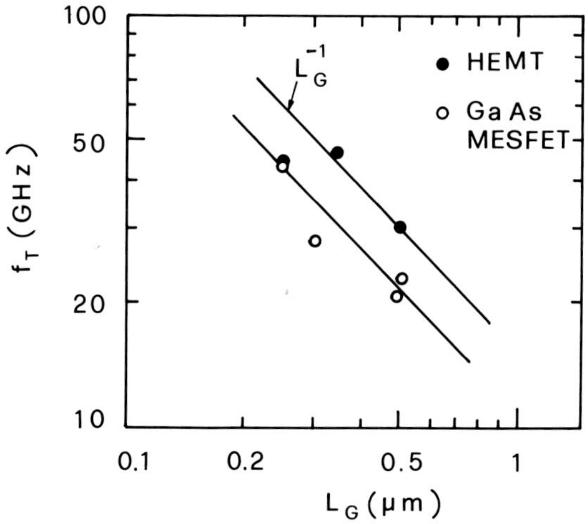

Fig. 4.44. Current gain cut-off frequency versus gate length for experimental GaAs MESFET's end HEMT's at room temperature [39]

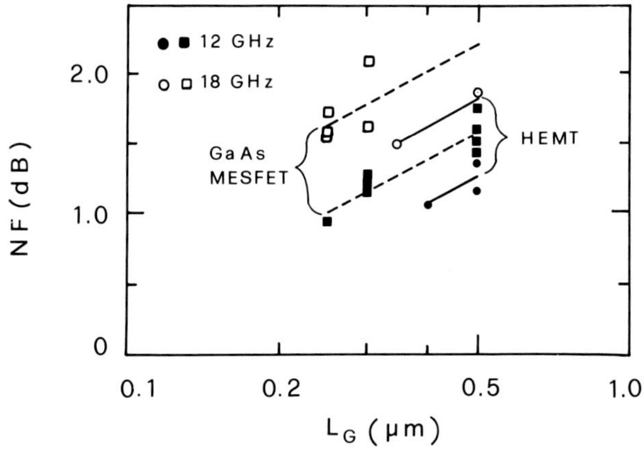

Fig. 4.45. Minimum noise figure versus gate length of low-noise HEMT's compared with GaAs MESFET's [39]

Table 4.2. *Comparison of various devices for high frequency applications* [40]

Transistor Parameter	HBT	MESFET	HEMT	PBT
Affiliation	Rockwell	GE	GE	MIT
Gain, dB	12.1; 25 GHz	9.5; 60 GHz	5.3; 90 GHz	15; 40 GHz
f_{max}, GHz	105	180	170	223
f_T, GHz	55	55	70	50
Power density, W/mm	2.5; 12 GHz	0.24; 60 GHz	0.4; 60 GHz	—
Efficiency, %	35; 12 GHz	9; 60 GHz	13; 60 GHz	—
Noise figure, dB	—	3.4; 60 GHz	2.5; 60 GHz	2.8; 18 GHz
Ass. gain, dB	—	3.8; 60 GHz	4.4; 60 GHz	8; 18 GHz
g_m, mS/mm	10.000	420	600	120
Linewidth, μm	1.3	0.25	0.25	0.16
Lithography	Optical	*E*-beam	*E*-beam	X-ray
Structure	Vertical	Horizontal	Horizontal	Vertical

are sensitive to the gate length, because of their dependence on the inverse transit time across the channel. Very high frequencies are obtained with submicron structures (where overshoot effects start playing an important role) up to values of 40 – 50 GHz for gates below the half-micron limit. The HEMT is far superior to the MESFET due to the higher carrier velocity in the channel. An area where the MESFET appears to have an advantage over the HEMT is in high-power microwave applications, where the HEMT is limited by the maximum drain voltage [39].

An overall comparison of the most suitable candidates for high-frequency operation (up to the mm-wave band) is given in Table 4.2 [40]. It is clear that the different FET's are the best devices for low power, low noise performance at such frequencies. The vertical transistors, such as HBT and PBT, might be more successful when high power is needed at high frequencies. For all devices illustrated in the table, monolithical integration in microwave circuits (MMIC's) is feasible.

References

[1] Mead, C. A., Conway, L.: Introduction to VLSI Systems. Reading, Mass.: Addison-Wesley. 1980.
[2] Moore, G.: Proc. IEDM 1975, p. 11. New York: IEEE Press. 1975.
[3] Ferry, D. K., Akers, L. A., Greeneich, E. W.: Ultra-Large-Scale Integrated Microelectronics. Englewood Cliffs, N.J.: Prentice-Hall. 1988.
[4] Sze, M.: Physics of Semiconductor Devices, 2nd edn. New York: Wiley. 1981.
[5] Muller, R. S., Kamins, T. I.: Device Electronics for Integrated Circuits. New York: Wiley. 1986.
[6] Solomon, P. M.: Proc. IEEE *70*, 489 (1982).
[7] Eden, R. C.: Proc. IEEE *70*, 5 (1982).
[8] Solomon, P. M., Morkoc, H.: IEEE Trans. El. Devices *ED-31*, 1015 (1984).
[9] Kroemer, H.: Proc. IEEE *70*, 13 (1982).
[10] Scavennee, A.: J. de Physique *C4* [Suppl. 9], 115 (1988).

[11] Zanoni, E., Lugli, P., Canali, C., Alberigi. Quaranta, A.: Fisica e Tecnologia *9*, 75 (1986).
[12] Pamlin, B. R. (ed.): Molecular Beam Epitaxy. Oxford: Pergamon Press. 1980; Ploog, K., Graf, K.: Molecular Beam Epitaxy of III-V Compounds. Berlin-Heidelberg-New York: Springer 1984; Cho, A. Y.: J. Vac. Sci. Technol. *16*, 275 (1979).
[13] Razeghi, M., Duchemin, J. P.: In: Two-Dimensional Systems, Heterostructures and Superlattices (Baver, G., Kuchar, F., Heinrich, H., eds.), p. 100. Berlin-Heidelberg-New York: Springer. 1984; Jones, M. W.: In: Two-Dimensional Systems, Heterostructures and Superlattices (Baver, G., Kuchar, F., Heinrich, H., eds.), p. 115. Berlin-Heidelberg-New York: Springer. 1984.
[14] Esaki, L., Tsu, R.: IBM Res. Rep. RC-2418 (1969).
[15] Kroemer, H.: Proc. IRE *45*, 1535 (1957).
[16] Lepselter, M. P., Andrews, J. M.: In: Ohmic Contacts in Semiconductors (Schwartz, B., ed.), The Electrochemical Society Symposium Series, p. 159. New York: Electrochemical Society 1969.
[17] Rideout, V. L.: Solid State Electron. *18*, (541 (1975).
[18] Pickar, K. A.: Ion Implantation in Silicon-Physics, Processing and Microelectronics Devices. In: Applied Solid State Science (Wolfe, R., ed.), Vol. 5. New York: Academic Press. 1975.
[19] Dennard, R. H., Gaensslen, F. H., Yu, H., Rideout, V. L., Bassons, E., Blanc, A. R.: IEEE J. Solid State Circuits *SC-9*, 256 (1974).
[20] Meindl, J. D.: Scientific American *257*, 54 (1987).
[21] Keyes, R.: Proc. IEEE, *63*, 740 (1975).
[22] Moll, J. L.: Proc. IRE *46*, 1076 (1958).
[23] Zanoni, E., Lugli, P., Canali, C., Alberigi-Quaranta, A.: Fisica e Tecnologia *10*, 83 (1987).
[24] Capasso, F.: In: Semiconductors and Semimetals, Vol. 24, p. 319. New York: Academic Press. 1988.
[25] Rhoderick, E. H.: Metal-Semiconductor Contacts. Oxford: Clarendon Press. 1978.
[26] Liechti, C. A.: IEEE Trans. Microwave Theory Tech. *MTT-24*, 279 (1976).
[27] Ferry, D. K.: In: VLSI Electronics: Microstructure Science, Vol. 1, p. 231. New York: Academic Press. 1981.
[28] Hill, A. J., Ladbrooke, P. M.: GEC J. Res. *4*, 1 (1986).
[29] Drummond, T. J., Masselink, W. T., Morkoc, H.: Proc. IEEE *74*, 773 (1986).
[30] Mead, C. A.: Proc. IRE *48*, 359 (1960).
[31] Moll, J. L.: IEEE Trans. El. Dev. *ED-10*, 299 (1963).
[32] Poate, J. M., Dynes, R. C.: IEEE Spectrum, p. 38. February 1986.
[33] Levi, A. F. J., Hayes, J. R., Platzman, P. M., Wiegmann, W.: Phys. Rev. Lett. *55*, 2071 (1985).
[34] Heiblum, M., Nathan, M. I., Thomas, D. C., Knodler, C. M.: Phys. Rev. Lett. *55*, 2200 (1985).
[35] Malik, R. J., Hollis, M. A., Eastman, L. F., Woodard, D. J., Wood, C. E. C., Aucoins, T. R.: In: Proc., Conf. Active Microwave Devices, Cornell University, Ithaca, N.Y., 1981.
[36] Heiblum, M.: Solid-State Electron. *24*, 343 (1981).
[37] Heiblum, M., Eastman, L. F.: Scientific American *256*, 64 (1987).
[38] Bozler, C. O., Alley, G. D., Murphy, R. A., Flanders, D. C., Lindley, W. T.: IEEE Tech. Dig. Int. Electron. Dev. Meet., p. 384 (1979).
[39] Mimura, T., Nishiuchi, N., Abe, M., Shibatomi, A., Kobayanshi, M.: Superlattice and Microstr. *1*, 365 (1985).
[40] Art, W. H.: Microwellen Magazin *14*, 106 (1988).

5 Monte Carlo Simulation of Semiconductor Devices

5.1 Introduction

After introducing the general methodology of the Monte Carlo algorithm and reviewing some of the essential features of semiconductor devices, we are now ready to discuss in detail the application of the Monte Carlo simulation to semiconductor devices. Although every specific device presents peculiar aspects that make its modelling not easily extendable to other devices, it is nevertheless possible to give some general guidelines and define specific blocks that are common to all simulations [1].

In the following sections we will refer to self-consistent calculations, where the Monte Carlo particles (or superparticles) move in a field distribution that is determined self-consistently (through the solution of Poisson's equation) in accordance with the particle distribution itself.

Other possibilities exist, as we will see by looking at specific applications in the next chapter. In general, the MC simulations of devices can be classified as follows:

(1) self-consistent simulations [2–7];
(2) one particle simulations on a fixed potential distribution [8–9];
(3) Ensemble simulations on a fixed potential [10];
(4) hybrid self-consistent simulations [11–13].

In the second and third cases, a potential distribution is determined at the beginning of the simulation. Such distribution can be the result of a simulation performed with traditional methods (in general drift-diffusion), or can be a piece by piece constant approximation of a known distribution. Although not as accurate as the self consistent simulations, those algorithms can be very useful for studying special physical effects in devices.

The so called hybrid method is based on the idea of coupling the MC simulators with traditional ones, trying to determine the regions where each can be successfully applied. The idea is very interesting. There are in fact regions of devices where the MC dynamics requires very long computation times. Low field regions and retarding barriers belong to this category.

There, electrons are in thermal equilibrium with the thermal bath, and their main features could be effectively reproduced by simpler simulators than the Monte Carlo. The problems arises at the boundary between regions. The main problem, not yet solved, is that it is very hard to reproduce the injection mechanism into the active region of devices from low field zones. If this interface could be modelled accurately, the two simulators would run in parallel, with enormous advantages. Examples will be given in the next chapter.

As a prototype for the simulated device we will take the GaAs MESFET. The choice has been made for both historical reasons (since the MESFET was the first device selfconsistently simulated with a MC code [14]) as well as for reasons of simplicity, since the MESFET can be described by a fairly simple physical model in comparison to other devices. Therefore, we will be able to discuss the main features of the simulation without worrying about the complications of the physical system.

The modifications required by the simulation of other devices will be introduced along the way. The next chapter will present specific applications to a wide range of devices.

As pointed out earlier, the Monte Carlo analysis of a device consists in the simulation of the real and k-space dynamics of charge carriers in a well defined potential distribution. The physical system is defined as described in chapter 3, through the energy band diagram and the scattering mechanisms that characterize the material of interest.

The geometry of the system under investigation is specified at the beginning of the simulation, by defining the region of interest, the doping profile, and the contact regions.

The carrier dynamics obeys the classical equations of motion

$$\frac{d\mathbf{r}}{dt} = \mathbf{v}(\mathbf{r}), \tag{5.1.1}$$

$$\frac{d\mathbf{v}(\mathbf{r})}{dt} = \frac{e}{m}\nabla\phi(\mathbf{r}) = -\frac{e}{m}\mathbf{E}(\mathbf{r}), \tag{5.1.2}$$

where $\mathbf{r}$ and $\mathbf{v}$ are, respectively, the carrier position and velocity, m the effective mass, $\mathbf{E}$ the electric field, and ϕ the electrostatic potential which is related to the charge distribution by Poisson's equation:

$$-\nabla\cdot(\kappa\,\nabla\phi) = 4\pi\rho\,. \tag{5.1.3}$$

The permittivity κ might also be position dependent when different materials form the device. The charge distribution ρ is given, in the most general case, by

$$\rho = -e(n - p - N_D^+ + N_A^-), \tag{5.1.4}$$

where n and p are, respectively, the electron and hole densities, N_D^+ and N_A^- the concentration of ionized donors and acceptors, and e the electron charge.

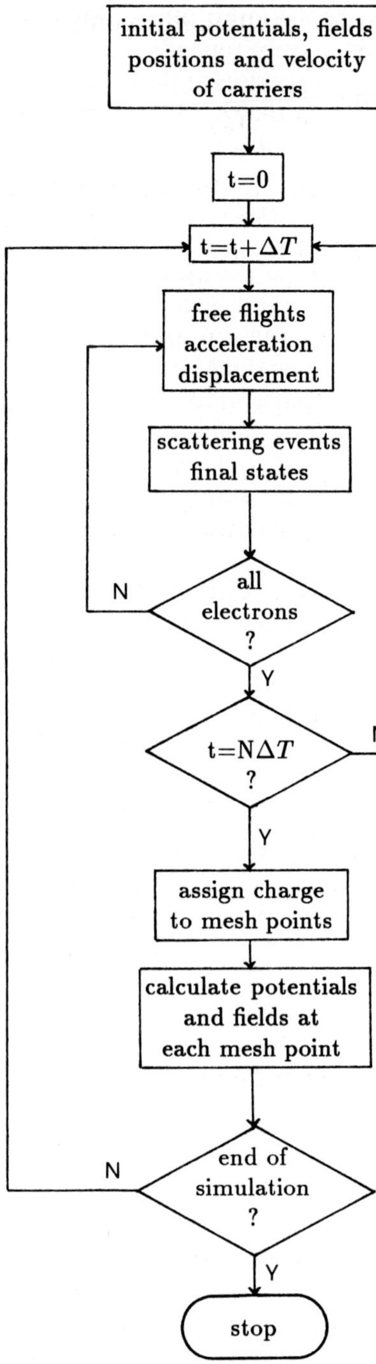

Fig. 5.1. Flow chart of a typical Monte Carlo simulation of semiconductor devices [1]

For the case of homogeneous material, when only electrons and donors are considered, Eq. (5.1.3) reduces to

$$\nabla^2\phi = -\frac{4\pi}{\kappa}\rho = \frac{4\pi}{\kappa}e(n - N_D^+). \tag{5.1.5}$$

The mobile charge is given directly by the Monte Carlo simulation. It is therefore "granular" in nature. On the contrary, the potential ϕ (and the electric field **E**) are represented by values on a regular array of mesh points. It is then necessary to map discrete values defined at the grid points onto the corpuscolar distribution of carriers, and viceversa. This is done by mediating from plasma studies the so-called "Particle Mesh (PM) Method" [15].

In the PM method, differential operators, such as the laplacian ∇^2, are replaced by finite-difference approximations on the mesh. Potentials and forces at particle positions are obtained by interpolating on the array of mesh-defined values. Mesh-defined densities are obtained by the opposite process of assigning the particle charge to nearby mesh-points in order to create a mesh-defined charge density.

The basic algorithm for the PM force calculation consist of the following steps:

(1) assign the charge to the mesh points;
(2) solve Poisson's equation on the grid;
(3) compute the components of the electric field from the mesh-defined potential;
(4) interpolate the field at the particle positions.

The input of such algorithm is the Monte Carlo distribution of mobile charges, the output are the components of the electric field that pushes the carriers. The full self-consistent Monte Carlo simulation (as shown in Fig. 5.1) of the device is therefore made up of an alternate cycle of Monte Carlo particle dynamics and PM force calculation, repeated until convergence is reached.

The following paragraphs will deal with those issues in detail. Most of the material of Sec. 5.2–5.4 is taken from Refs. [15] and [16] (and references therein).

5.2 Geometry of the System

The first step in the Monte Carlo simulation of a device is the definition of a simplified and tractable model of the actual device. Figure 5.2 shows a top view of a MESFET, where the three main regions define the contact (source, gate and drain) pads. Typically, source and drain contacts extend over areas of few thousand μm^2. The separation between source and drain is around few μm. In the perpendicular y direction in the figure, just under the contacts, a GaAs epitaxial layer of few tenths of a micron (usually n-type at 10^{17} –

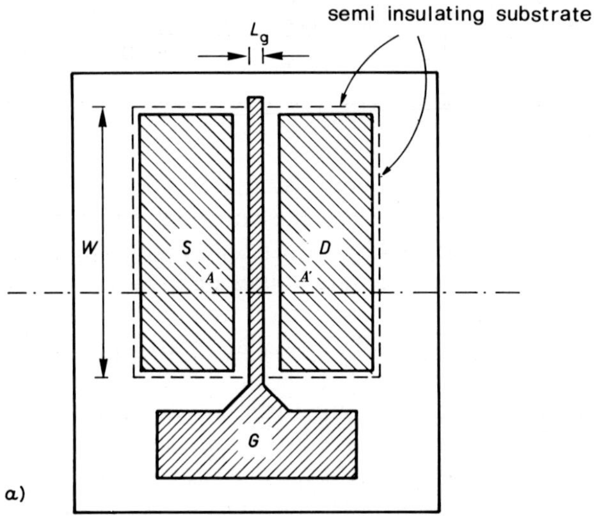

Fig. 5.2. Top view of a GaAs MESFET, showing the source, gate and drain contact pads

10^{18} cm^{-3}) is grown on a semi-insulating GaAs substrate. Source and drain contacts are low resistance ohmic contacts, while the gate contact forms a Schottky barrier between the metal and the GaAs epilayer.

The necessity to create a simple model for the MESFET is clear from the following considerations. In the first place, the volume of the region shown in Fig. 5.2 (taking for simplicity a conservative figure of 0.1 μm for the epilayer depth and 1,000 μm^2 for the top area) is roughly 100 μm^3 (10^{-10} cm^3). Assuming a carrier concentration of 10^{17} cm^{-3}, we find in such volume 10^7 carriers. It is clearly impossible to simulate that number of carriers in a reasonable time even with the fastest of today's computers. Furthermore, the knowledge of the physical mechanisms that govern the behavior of a contact is still very poor. To overcome such difficulties, three main assumptions are made (notice that the following assumptions are common also to more traditional simulators not based on the MC method):

(i) There is no significant variation of physical quantity along the direction perpendicular to the x-y plane of Fig. 5.2 thus a two-dimensional (2D) model can be used.

(ii) It is not necessary to consider the whole extension of source and drain contacts, but it is possible to simulate a reduced active part of it.

(iii) Source and drain contacts are treated as ideal, zero-resistance ohmic contacts, and the gate contact is represented as a perfectly absorbing boundary.

The geometry of the simulated region is shown in Fig. 5.3. The cross section is taken between the points **A** and **A′** of Fig. 5.2. It turns out that most of the current flows through the area considered here. Therefore, as long as spatial inhomogenities along the z direction can be neglected, assumptions

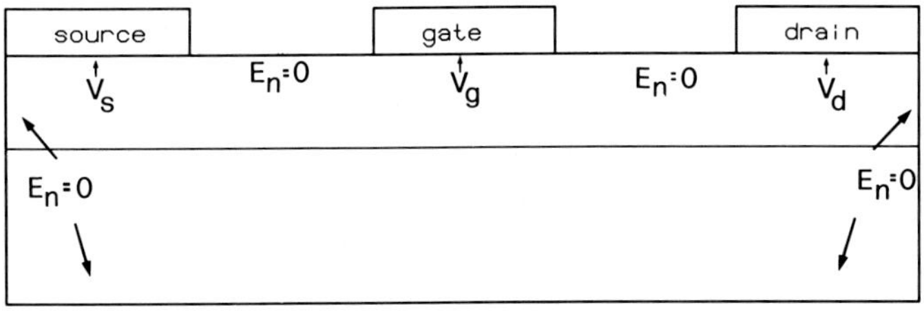

Fig. 5.3. Cross section of the simulated devices, taken on a plane perpendicular to the point *A-A'* of Fig. 5.2. The boundary conditions on the external surfaces are also show

(i) and (ii) above are well justified. Care should be taken in restricting the area of interest in the device. Since the boundary conditions play a very important role in the simulation, the size of the simulated area should be checked for different applications. We will come back to this point below. It should be pointed out that the MC program is inherently three-dimensional (3D). A 2D model is used mainly to the computational requirements of a 3D solutions of Poisson's equation, and to limit the number of simulated superparticles, by projecting them onto the x-y plane.
The interface between doped and undoped regions is usually assumed to be abrupt. All donors in the epilayer are assumed to be ionized. We will see in chapter 6 how these assumptions can be easily relaxed, and how more sophisticated models can be used.

5.2.1 *Boundary Conditions*

Boundary conditions (BC) enter the simulation at two different stages: in the solution of Poisson's equation, and during the Monte Carlo dynamics. Under operating conditions, external voltages are applied to the device at the contacts. It is clear that the value of the potential in the simulation must take on given values on the electrode boundaries (Dirichlet conditions). Elsewhere on the boundary the gradient of the potential is taken equal to zero corresponding to a vanishing electric field (Neumann conditions). Figure 5.3 shows the boundary conditions imposed on the prototype device. On the free internal boundaries the zero-field condition is related to the natural tendency of the free charge to neutralize any field variation within a Debye length of any imposed charge (in this case the edge of the electrode). On the top surface (apart from the electrode regions), the semiconductor is in contact with the air. The continuity of the displacement $D = \kappa E$ imposes that $\kappa_s E_s = \kappa_a E_a$, where the subscripts s and a refer respectively to semiconductor and air. Since $\kappa_s \gg \kappa_a$, and E_a has a finite value, it follows that E_s just inside the semiconductor must be very small.
The carrier motion is also subject to boundary conditions, which should be consistent with those imposed on the potential on the field. On those boundaries where Neuman (zero electric field in the direction normal to the

surface) conditions apply, two choices are available: reflecting boundaries, where the component of the velocity normal to the interface is reflected, and periodic boundaries, where an electron that exits through an interface is reinjected with the same velocity at the other end of the device.

As it will clearly appear from the discussions of BC that will follow, the two BC's are equivalent only as long as carriers are thermalized at the interface (that is no net drift, and average energy is equal to $\frac{3}{2}K_BT$). When such a condition cannot be maintained, the consistency between MC and potential BC's has to be restored, either by enlarging the simulated region of the devices, or by redefining other appropriate field conditions at the interfaces.

The source and drain electrodes are modelled as ideal ohmic contacts.

As mentioned in section 4.3.4, an ohmic contact can be treated as a system in thermal equilibrium even when current is flowing. The voltage drop at the contact is negligible and no power is dissipated. Such a condition can be realized in the simulation by maintaining a neutrally-charged region (that is a layer of cells when the number of mobile charges is always equal to the number of impurities) in the vicinity of the ohmic contact. During the particle dynamics, some of them will exit the simulated area through the source and drain contacts. The neutrality condition is achieved by injecting, at each time step, a sufficient number of electrons to equalize the doping density. Spatially, the injection takes place in a uniform way along the cells most adjacent to the ohmic contact (that is the new coordinates can be generated randomly with a flat distribution). In k-space the injected electrons are distributed with a hemi-maxwellian distribution at the lattice temperature.

Although not very satisfactory the modelling of the source and drain electrodes reflect the scarce knowledge of the microscopic behaviour of Ohmic's contacts. (See also section 6.2).

The gate electrode is modelled as a ideal Schottky contact, with a known barrier height. Then, the potential is given by the difference of the applied voltage and the barrier height. The contact is simulated as a perfectly absorbing boundary. For the MESFET, under normal operating conditions, the region directly underneath the gate is completely depleted and very few electrons do actually reach the gate. A simulation of the effect of reflecting Schottky barrier has been performed by Arnold and Hess [17] using the transmission matrix technique.

In the simulation of heterostructure devices addditional care is required, both in the boundary conditions for the potential and for the MC particle dynamics. For the potential, a back contact can be used to allign the quasi fermi level of the different regions across the heterojunction. The energy gap discontinuity is forced on the otherwise continuous potential at the interface. For the MC simulation, the conservation of energy and momentum parallel to the heterojunction is imposed on particles crossing the heterojunction. The same effective mass for electrons on the two materials forming the heterostructure is usually assumed.

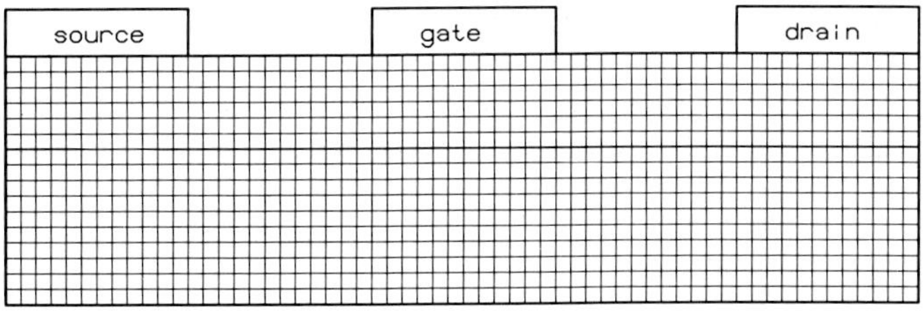

Fig. 5.4. Typical discretization of the simulated area, with uniform size grid

5.2.2 Grid Definition

A typical discretization of the simulated device is shown in Fig. 5.4. There, a uniform rectangular grid is used. Several alternatives exist, from a non-uniform rectangular grid [18–19] to a triangularly-shaped mesh as used in finite elements calculations [20–21]. The definition of the grid (concerning both type and size of the cells) is mostly a matter of experience, depending on the details of the simulation. The guidelines that should be followed are set by the following physical considerations.

The resolution of the potential and force fields is limited by the size of the mesh. Only those field variations which have a wavelength longer than the spacing of the mesh can be accurately represented by mesh values. While the limited resolution of the mesh proves unacceptable for studying correlated systems (plasmas etc.), the same is not true for devices where the limited mesh resolution introduces the properties required to destroy the unphysical correlations between the limited number of computational superparticles. Provided that the mesh spacing is smaller than the wavelength of importance in the physical system (for example the Debye or the diffusion length) and the number of computer superparticles in each cell of the mesh is large enough (~ 10) to maintain low levels of fluctuations, then the PM model will accurately represent the physical system. All physical phenomena whose characteristic wavelength is shorter than the mesh spacing will not be resolved in the simulation.

There are devices (for example the MOSFET or HEMT) where most of the charge transport takes place in a very narrow region. In such conditions, the use of a very fine uniform mesh spacing would not be affordable, since it would lead to a huge number of mesh points (and therefore to huge matrices). Non uniform mesh sizes are used in this case [5, 22].

5.2.3 Superparticles

As it is impossible to simulate all the actual carriers of a device, we are forced to select a subsection of the electron population, containing N_e particles, that is representative of the overall distribution.

During the Monte Carlo dynamics (free acceleration and scattering), the N simulated particles are then regarded as a sample from the distribution in the real device, whereas during the field calculation the particles are regarded as superparticles. Each simulated particle represents a rod of charge with effective charge per unit length

$$q = -e\frac{nl_x l_y}{N} = -\frac{eA}{V},$$

where l_x, l_y are the dimension of the active area, V the volume of the device and A is the cross-sectional area. The charge of the superparticles is then assigned to the grid points in accordance to the rules that will be given in the next section, and the potential on each grid point is calculated by solving Poisson's equation.

We will see later that, if necessary, it is possible to use different types of superparticles in the same simulation [5, 8, 23]. Problems arise when inter-particle scattering is included in the simulation, due to the use of super-particles (and consequent loss of information on the z direction). A discussion of this point can be found in [5].

The following paragraphs will deal with some of those issues in detail.

5.3 Particle-Mesh Force Calculation

Each of the four steps outlined for the Particle-Mesh (PM) algorithm in section 5.1 introduces errors, due to truncation, interpolation or other sources. It is necessary to see how the error of each stage influences the other stages, and how their combination affects the final output of the procedure, that is the force (electric field) acting on the particles (carriers). In fact, quite large numerical errors in single steps of the algorithm might be acceptable if their outcome is an accurate value of the force.

We will first focus on the one-dimensional PM algorithm, which despite its simplicity contains most of the important physical features of the method. We will then extend it to the two-dimensional case that applies to our prototype device.

5.3.1 Particle-Mesh Calculation in One Dimension

The main emphasis of this part will be given to the algorithms for charge assignment (that is the way the charge of the simulated particles can be distributed to the grid points) and force interpolation (that is how the values of the force calculated on the grid points can be assigned to the particle positions).

The computational domain (see Fig. 5.5 [15]) is divided into cells of width Δx. The mesh points (at which the values of charge density, potential, and fields are stored) are located at the center of each cell. Periodic boundary

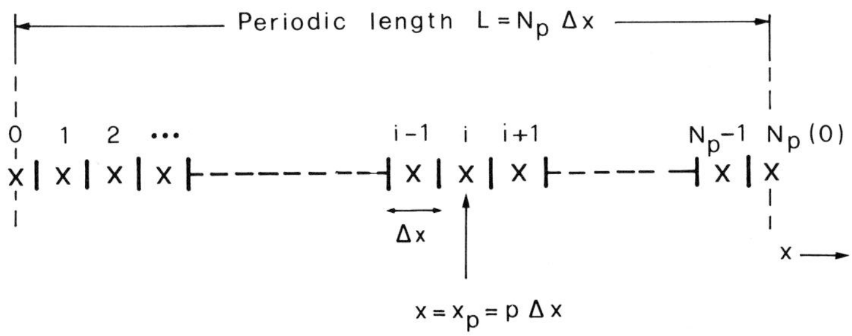

Fig. 5.5. Computational domain for the Particle-Mesh calculation in one-dimension. Periodic boundary conditions apply, so that mesh point N_p is the periodic image of mesh point 0 [15]

conditions $\phi(x) = \phi(x + L)$ are appropriate for the present case, so the domain may be imagined as periodically repeated.
There is a hierarchy of levels for the charge assignment algorithm, depending on the numbers of nodes that are involved. The same hierarchy works for the force interpolation algorithm. We will examine the two lowest order methods, called respectively Nearest-Grid-Point scheme [24] and Cloud-In-Cell [25] scheme.

Nearest-Grid-Point (NGP) Scheme

In the NGP scheme all the charge from a given particle is assigned to the nearest grid point. The force on the particle is also taken as the one calculated on the nearest grid point.
The charge density at a mesh point p is given by the total charge in the cell surrounding the mesh point p divided by the cell volume:

$$\rho_p = \frac{1}{\Delta x} \sum N_S q + \rho_0 , \tag{5.3.1}$$

where the sum extends over all particles in cell p. In Eq. (5.3.1), N_S is the sheet carrier density, q the superparticle charge, and ρ_0 the background charge density. The compact notation $\rho_p \equiv \rho(x_p)$, etc., ... , is used for all position dependent quantities. Introducing a charge assignment function ω we can write

$$\rho_p = \frac{q N_S}{\Delta x} \sum_{i=1}^{N_p} \omega(x_i - x_p) + \rho_0 , \tag{5.3.2}$$

where, in this case,

$$\omega(x) = \begin{cases} 1, & |x| \leqslant \Delta x/2, \\ 0, & \text{otherwise}. \end{cases} \tag{5.3.3}$$

Within a finite difference scheme, the field equations are:

$$\frac{\phi_{p-1} - 2\phi_p + \phi_{p+1}}{\Delta x^2} = \rho_p\,, \tag{5.3.4}$$

$$E_p = \frac{\phi_{p-1} - \phi_{p+1}}{2\,\Delta x}\,. \tag{5.3.5}$$

The NGP force interpolation follows the same scheme as the particle assignment. The force field in any point of a cell takes the value of the mesh point at the center of the cell. In terms of the assignment function, this condition is written as

$$F_i = F(x_i) = N_S q \sum_{p=0}^{N_g-1} \omega(x_i - x_p) E_p\,. \tag{5.3.6}$$

The sum is here over the mesh points (with periodic boundary conditions), while it was over the particles in Eq. (5.3.2).
There is a discontinuity in the force when particles cross a cell boundary. If we look for example at the force between two particles within the NGP scheme, we find that the forces depend both on particle separations as well as on the position of the particles inside the mesh. If the two particles are displaced on the grid keeping their distance fixed, then the interparticle force fluctuates with a period given by the cell width Δx. This lack of translational invariance is the strongest limitation of the model in the attempt to reproduce the physical reality of the system. The largest fluctuation in the interparticle forces occurs when two particles are separated by less than the cell width Δx, and the force switches from or to their maximum as a charge crosses the cell boundary.

Cloud-in-Cell (CIC) Scheme

In the CIC scheme a better approximation can be obtained assigning the charge of a superparticle to the two nearest neighbour cell points. In this case the interpolation function ω is given by

$$\omega(x) = \begin{cases} 1 - |x|/\Delta x\,, & |x| \leqslant \Delta x\,, \\ 0\,, & \text{otherwise}\,, \end{cases} \tag{5.3.7}$$

we then have

$$\rho(x_p) = \frac{qN_S}{\Delta x} \sum_{i=1}^{N_p} \omega(x_i - x_p) + \rho_0\,, \tag{5.3.8}$$

$$F_i = F(x_i) = \sum_p \omega(x_i - x_p) F(x_p)\,, \tag{5.3.9}$$

for charge assignment and force interpolation, respectively.
In comparing the two methods, we can note that:

1. CIC gives a much smoother force than the NGP scheme (piecewise linear rather than piecewise constant)

2. CIC reduces the amplitude of fluctuations in the interparticle forces as the particles are displaced with respect to the mesh.
3. The residual errors in the forces are spatially more localized for the CIC scheme.
4. In the NGP the forces are discontinuous in value, while the CIC gives forces that are continuous, but with discontinuous derivatives.

The hierarchy in the charge assignment schemes can be interpreted physically by ascribing a finite width to the distribution of charge of each particle. The various levels correspond to different cloud shapes $S(x)$ of a particle with unit charge, defined as the particle charge density as a function of the distance x from the center of the particle itself. As shown pictorially in Fig. 5.6 [15], a particle carries with it the cloud shape appropriate to the assignment scheme. The overlap area between the cloud shape and the cell determines the fraction of charge to be assigned to the mesh point of that cell. In addition to the NGP and CIC schemes, the figure also presents the third level of the hierarchy of charge assignment, the so-called "Triangularly Shaped Cloud" (TSC) scheme. In 1D such scheme uses three mesh points and is characterized by a piecewise quadratic assignment-interpolation function. As a result, the forces and their first derivatives are continuous [15].

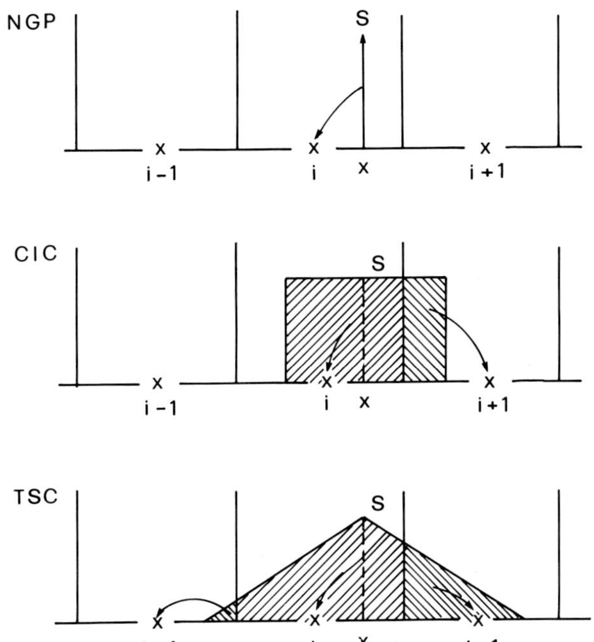

Fig. 5.6. Pictorial description of the cloud shape interpretation of the charge assignment [15]

Mixed Schemes

The choice of the assignment-interpolation scheme depends on several considerations, which range from the desired accuracy of the result to the cost effectiveness of the algorithm. It is possible to use different schemes for charge assignments and force interpolation, respectively. While in the case of both CIC and NGP, the forces between a pair of particles was equal and opposite (guaranteeing momentum conservation) and the force of a particle upon itself (self force) was zero, in the mixed scheme this is no longer assured. In order to avoid instabilities in the numerical algorithm, it is necessary to restrict the force interpolation to, at most, the same order as the charge assignment [15, 26].

5.3.2 Charge Assignment Schemes in Two Dimensions

The charge assignment in two dimensions is a straightforward generalization of the algorithms described earlier. Figure 5.7 [16] illustrates the CIC and NGP methods in 2D. As for the previous case, the NGP scheme assigns the charge the nearest grid point. The CIC shares the particle charge between the four closest nodes, with a distribution that is weighted by the overlap area of the cloud shape and the cell.

The lowest order charge assignment function for a particle at position $\mathbf{r} = (x, y)$ is

$$\omega(\mathbf{r}) = \begin{cases} 1, & -\Delta x/2 \leqslant x \leqslant \Delta x/2,\ -\Delta y/2 \leqslant y \leqslant \Delta y/2, \\ 0, & \text{otherwise}, \end{cases} \tag{5.3.10}$$

where $\mathbf{R} = (\Delta x, \Delta y)$ is the vector describing the cell dimensions of the mesh used.

The CIC scheme corresponds in 2D to a four-point scheme, with charge assignment function equal to

$$\omega(\mathbf{r}) = \begin{cases} (1-|x|/\Delta x)(1-|y|/\Delta y), & -\Delta x/2 \leqslant x \leqslant \Delta x/2,\ -\Delta y/2 \leqslant y \leqslant \Delta y/2, \\ 0, & \text{otherwise}. \end{cases} \tag{5.3.11}$$

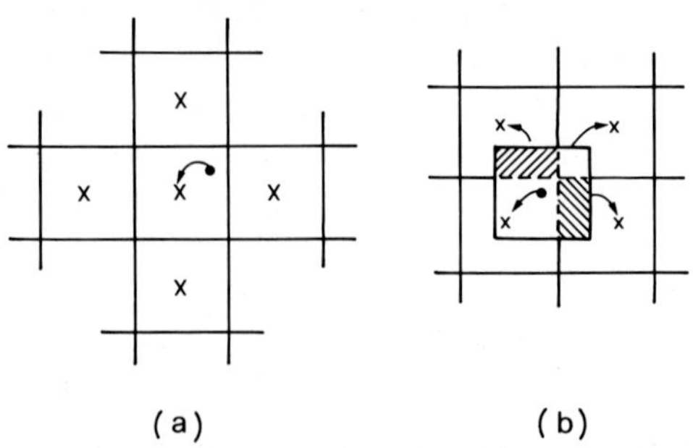

Fig. 5.7. Charge assignment scheme in two-dimension: *a* nearest-grid-point method; *b* Cloud-in-cell method [16]

As for the 1D case, the list can be extended indefinitely, going up with the number of points involved in the scheme. The compromise has to be found between the constraints that need to be satisfied by the calculation (in terms of smoothness, momentum conservation, etc.,...) and the complications (also in terms of costs) that are introduced by the level of complexity of the algorithm.
For the case of non-uniform grids, the CIC scheme should be slightly modified, defining a local cloud shape for the charge assignment [5]. Therefore, when the mesh is refined in some regions of the simulation domain, the particle charge only spreads over a small distance, which grows as the mesh size relaxes.

5.4 Poisson Solver and Field Distribution

The simulation of semiconductor devices with Monte Carlo techniques requires the solution of Poisson's equation to obtain the spatial distribution of potential and electric field. MC simulations avoid many of the limitations that affect more conventional techniques based on drift-diffusion equations, when submicron dimensions are considered. However, an accurate solution of Poisson's equation must be provided, to guarantee the correct treatment of the particle dynamics. As seen earlier, the charge density ρ is given by the difference of the mobile charge (obtained directly from the MC via the particle-mesh calculation) and the fixed impurity profile (assigned as an input of the simulation).
Several methods exist for the numerical solution of Poisson's equation. In the following sections we will focus on the finite different formulation for one- and two-dimensional problems. For treatment of Poisson's equation in three dimensions see [27]. Alternative approaches based on finite element methods [28] will be not be dealt with since they do not easily adapt to the MC simulation (an example of a Monte Carlo device simulation based on a triangular grid can be found in [8]). An interesting pseudo-spectral collocation method, so far applied only to 1D simulations [29], will also be discussed.
In the following discussion, we will always refers to cases where boundary conditions are assigned on the borders of the domain that defines our physical system. This corresponds to the situation illustrated in Fig. 5.3. In the case where electrodes have also to be defined inside the region, a general method called "Capacity Matrix Method" can be used [15].

5.4.1 Finite Difference Scheme

We will consider here the form of the matrices which arise from the finite difference algorithm applied to Poisson's equation.
In one dimension, Poisson's equation (5.1.5) can be written as

$$\frac{d^2\Phi}{dx^2} = \rho\,, \tag{5.4.1}$$

where $\rho(x)$ is the known source function, while $\Phi(x)$ is the unknown potential. On an equally-spaced mesh (with space step Δx) the functions $\rho(x)$ and $\Phi(x)$ are replaced by vectors of elements ρ_j and Φ_j, defined at the mesh points x_j, with $1 \leqslant j \leqslant J$. The total length L of the region of interest is $L = J\,\Delta x$. Within a finite difference scheme, the second derivative $d^2\Phi/dx^2$ is replaced by the difference operator [16]

$$\Delta_x'' = \frac{\Phi_{j+1} - 2\Phi_j + \Phi_{j-1}}{\Delta x^2}\,, \tag{5.4.2}$$

and Poisson's equation (5.4.1) is approximated by the set of equations,

$$\Phi_{j+1} - 2\Phi_j + \Phi_{j-1} = \Delta x^2 \rho_j\,, \tag{5.4.3}$$

which is to apply at every internal point j on the mesh. Assuming for example that the boundary conditions specify the value of the potential at the end points, $\Phi_1 = w_1$ and $\Phi_J = w_J$, then Eq. (5.4.3) forms a set of simultaneous linear equations,

$$\begin{aligned} \Phi_1 &= w_1\,, \\ \Phi_1 - 2\Phi_2 + \Phi_3 &= \Delta x^2 \rho_2\,, \\ \Phi_2 - 2\Phi_3 + \Phi_4 &= \Delta x^2 \rho_3\,, \\ &\cdots \\ \Phi_{J-2} - 2\Phi_{J-1} + \Phi_J &= \Delta x^2 \rho_{J-1}\,, \\ \Phi_J &= w_J\,. \end{aligned} \tag{5.4.4}$$

In matrix notation we can write

$$\mathbf{A\Phi} = \mathbf{w}\,, \tag{5.4.5}$$

where the unknown variables Φ_j for all j are represented as a vector $\mathbf{\Phi}$, while the right-hand side of the set of Eqs. (5.4.4) consists of a known vector $\mathbf{w}$. The matrix $\mathbf{A}$ has the form,

$$\mathbf{A} = \begin{bmatrix} 1 & 0 & 0 & \ldots & 0 \\ 1 & -2 & 1 & 0 & \ldots \\ 0 & 1 & -2 & 1 & \\ & & \ldots & & \\ & & 1 & -2 & 1 \\ 0 & \ldots & 0 & 0 & 1 \end{bmatrix}. \tag{5.4.6}$$

$\mathbf{A}$ has non-zero elements only along the central three diagonals of the matrix (tridiagonal matrix). Because of the large number of zeros the matrix is called "sparse".

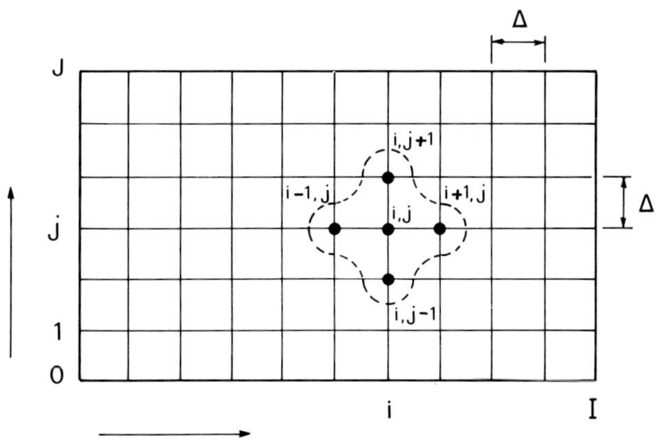

Fig. 5.8. Five-point difference scheme for the solution of Poisson's equation on a square grid (with $\Delta x = \Delta y = \Delta$) [16]

In two dimensions, Poisson's equation is written as

$$\frac{\partial^2 \Phi}{\partial x^2} + \frac{\partial^2 \Phi}{\partial y^2} = \rho . \tag{5.4.7}$$

In solving this equation on the rectangular grid shown in Fig. 5.8 (with $0 \leqslant x \leqslant X, 0 \leqslant y \leqslant Y$), both differential operators are replaced by the correspondent difference operator. Thus, when ρ_{ij} is defined at every lattice point (ij), with $0 \leqslant i \leqslant I, 0 \leqslant j \leqslant J$, the potential Φ_{ij} satisfies the equation

$$\frac{\Phi_{i+1,j} - 2\Phi_{i,j} + \Phi_{i-1,j}}{\Delta x^2} + \frac{\Phi_{i,j+1} - 2\Phi_{ij} + \Phi_{i,j-1}}{\Delta y^2} = \rho_{ij} . \tag{5.4.8}$$

The equation is greatly simplified for a square grid ($\Delta x = \Delta y = \Delta$):

$$(\Phi_{i+1,j} + \Phi_{i-1,j} + \Phi_{i,j+1} + \Phi_{i,j-1}) - 4\Phi_{ij} = \Delta^2 \rho_{ij} . \tag{5.4.9}$$

This equation is usually called the five-point difference equation, since it connects every grid point to the four nearest neighbours. Higher level approximations that take into account a larger number of neighbours can be adopted.

In the more general case of non-uniform mesh spacing, Eq. (5.4.9) takes the form:

$$(a_{i,j}\Phi_{i+1,j} + b_{i,j}\Phi_{i-1,j} + c_{i,j}\Phi_{i,j+1} + d_{i,j}\Phi_{i,j-1}) - e_{i,j}\Phi_{ij} = \Delta^2 \rho_{ij} , \tag{5.4.10}$$

where the coefficients a, b, c, d, e vary on the grid depending on the chosen discretization. We will limit the discussion to the uniform mesh case, with $a = b = c = d = 1$ and $e = -4$.

The discretized Poisson's equation consists of a system of JI linear equations. It is useful to replace all the variables Φ_{ij} by singly subscripted

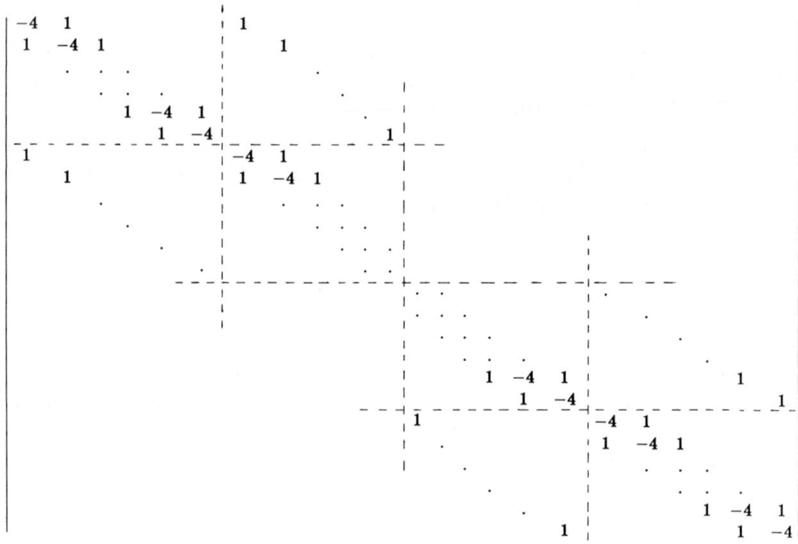

Fig. 5.9. Five-point difference matrix for Poisson's equation in two dimension

elements of the vector **u** (which has dimension IJ),

$$u_k = u_{iJ+j} = \Phi_{i,j} \tag{5.4.11}$$

and similarly to define the vector **w** from ρ_{ij}

$$w_k = w_{iJ+j} = \Delta^2 \rho_{ij} \tag{5.4.12}$$

The elements of the vector **u** are obtained as solutions to the simultaneous equations (5.4.8) and therefore, they satisfy, as for the one dimensional case, the matrix equation

$$\mathbf{Au} = \mathbf{w}. \tag{5.4.13}$$

The matrix **A** has the form illustrated in Fig. 5.9. The tridiagonal property found for the 1D problem has been lost, but on the other hand the matrix maintains a natural banded structure with only five diagonals with non-zero elements. The matrix **A** is always diagonally dominant, and hence non singular. This is not true in general when a non-uniform mesh is adopted, since off-diagonal terms will arise that may become too large to assure diagonal dominance.

Among the different methods that exist for the solution of Poisson's equation we will deal with the following ones [15]:
a) *Matrix Methods*, based on the solution of the matrix equation (5.4.13) using standard matrix methods such as Gauss elimination;
b) *Rapid Elliptic Solvers* (RES), where special techniques such as Fast Fourier Transform (FFT) and Cyclic Reduction (CR) allow a speed-up of the solution of the system of linear equations (5.4.9).

c) *Iterative Methods*, where an initial guess is given for the unknown in Eq. (5.4.9), that is for the potential at each grid point. An iterative procedure is then used to reach the correct solution by systematically sweeping the mesh and adjusting the mesh values.
Methods a) and b) are exact in the sense that, although errors are introduced in the finite difference formulation of Poisson's equation, an exact solution of the matrix equation (5.4.13) is found.

5.4.2 *Matrix Methods*

The direct solution of the matrix equation (5.4.13) can be obtained in many standard ways, such as Gauss elimination, Choleski decomposition, or Conjugate-gradient methods. Special techniques can be used very efficiently for the case of sparse matrices.
From the point of view of the Monte Carlo simulation of devices, Poisson's equation has to be solved very often (see the next section) in order to follow the evolving carrier distribution. It is therefore very convenient, at least in principle, to invert the matrix **A** at the beginning of the simulation, and to obtain the potential at any time it is necessary by simply multiplying the inverted matrix by the vector **w** containing the updated charge density. Such a method is indeed very fast, and particularly appealing for supercomputers, where vectorized libraries provide optimized routines for matrix inversion and multiplication. The drawback (which is common to all matrix methods) lies in the storage requirements, which limit the applicability of the method to grids with at most a few thousand mesh points [15].
It is important to see where and how the boundary conditions (BC) affect the matrix structure shown in Fig. 5.9. If Dirichlet BC's apply, then the value of the potential is set on the surfaces of the domain. The matrix **A** has then the structure given in Fig. 5.9, that is

$$\mathbf{A} = \begin{bmatrix} [T] & [B] & & & 0 \\ [B] & [T] & [B] & & \\ & [B] & [T] & & \\ & & \cdots & & \\ & & & & [B] \\ 0 & & & [B] & [T] \end{bmatrix}, \tag{5.4.14}$$

where **T** is the square matrix

$$\mathbf{T} = \begin{bmatrix} 4 & -1 & & & 0 \\ -1 & 4 & -1 & & \\ & & \cdots & & \\ & & & & -1 \\ 0 & & & -1 & 4 \end{bmatrix} \tag{5.4.15}$$

and $\mathbf{B} = -\mathbf{I}$, $\mathbf{I}$ being the identity matrix. The known boundary values of the potential enter in the right hand side of the Eq. (5.4.13).
In the case of Neumann BC's the matrix $\mathbf{A}$ is modified in the following way:

$$\mathbf{A} = \begin{bmatrix} [T] & 2[B] & & & 0 \\ [B] & [T] & [B] & & \\ & [B] & [T] & & \\ & & \cdots & & \\ & & & & [B] \\ 0 & & & 2[B] & [T] \end{bmatrix}, \tag{5.4.16}$$

where $\mathbf{B}$ is unchanged and $\mathbf{T}$ is given by

$$\mathbf{T} = \begin{bmatrix} 4 & -2 & & & 0 \\ -1 & 4 & -1 & & \\ & & \cdots & & \\ & & & & -1 \\ 0 & & & -2 & 4 \end{bmatrix}. \tag{5.4.17}$$

The different terms appearing with respect to Eqs. (5.4.14)–(5.4.15) are due to the requirement of zero electric field (i.e. zero derivative of the potential) on the boundaries.

5.4.3 Rapid Elliptic Solvers (RES)

Two approaches developed with the aim of achieving a rapid solution of the matrix equation will be presented. In the simplest method, the complete matrix equation is diagonalized by applying the methods of Fourier analysis along each space dimension [30]. A variation of this method, due to Hockney [31] employs Fourier analysis along only one space dimension; from this, uncoupled tridiagonal matrix equations, which may readily be solved, are obtained along the other space dimension. Extensive reviews on RES methods can be found in [15].

a) Multiple Fourier Analysis

Starting from five-point difference formulations of Poisson's equation (Eq. 5.4.9), the five-point operator on the left-hand side can be represented by P, with

$$P\Phi_{i,j} = (\Phi_{i+1,j} - 2\Phi_{i,j} + \Phi_{i-1,j}) + (\Phi_{i,j+1} - 2\Phi_{i,j} + \Phi_{i,j-1}). \tag{5.4.18}$$

It is possible to show [16] that P satisfies the eigenvalue equation

$$Ps_{ij}(k, l) = \alpha_{kl} s_{ij}(k, l), \tag{5.4.19}$$

where the double Fourier harmonic

$$s_{ij}(k, l) = \hat{\Phi}(k, l) \sin \frac{\pi kj}{J} \sin \frac{\pi li}{I} \tag{5.4.20}$$

are the eigenfunctions of P corresponding to the eigenvalues

$$\alpha_{kl} = 2 \cos \frac{\pi l}{I} + 2 \cos \frac{\pi k}{J} - 4. \tag{5.4.21}$$

Thus, Poisson's equation can be solved for each eigenfunction of P separately.

The method of double Fourier analysis for the solution of Poisson's equation involves five steps. The known source functions (ρ_{ij}) are analyzed, first along one direction, on each 'row' i.

For all k, $0 < k < J$,

$$\hat{\rho}_i(k) = \frac{2}{J} \sum_{j=1}^{J} \rho_{i,j} \sin \frac{\pi kj}{J} \tag{5.4.22}$$

for each Fourier mode of wavelength $2J\,\Delta/k$. The other dimension may be analyzed in the same way.

For all l, $0 < l < I$,

$$\hat{\rho}(k, l) = \frac{2}{J} \sum_{i=1}^{I} \hat{\rho}_i \sin \frac{\pi li}{I}. \tag{5.4.23}$$

The amplitude of the potential for each double Fourier mode may now be determined separately according to the eigenvalues α_{kl},

$$\hat{\Phi}(k, l) = \frac{\hat{\rho}(k, l)\,\Delta^2}{4 - 2\cos(\pi k/J) - 2\cos(\pi l/J)}. \tag{5.4.24}$$

Finally Fourier synthesis is performed along each dimension.

For all i, $0 < i < I$,

$$\hat{\Phi}_i(k) = \sum_{i=1}^{I} \hat{\Phi}(k, l) \sin \frac{\pi li}{I}. \tag{5.4.25}$$

For all j, $0 < j < J$,

$$\Phi_{ij} = \sum_{k=1}^{J} \hat{\Phi}_i(k) \sin \frac{\pi li}{J}. \tag{5.4.26}$$

The boundary conditions determine the particular harmonics we must choose.

In particular, the Fourier half-sine analysis (Eq. 5.4.22) and synthesis (Eq. 5.4.26) is used when the potential is zero or constant at the boundaries. The half-cosine series is used when the derivative of the potential is zero at the boundary:

$$\text{synthesis:} \quad \Phi_j(k) = \frac{1}{2}\hat{\Phi}(0) + \sum_{k=1}^{J-1} \hat{\Phi}(k) \cos\frac{\pi kj}{J}; \tag{5.4.27}$$

$$\text{analysis:} \quad \hat{\Phi}(k) = \frac{2}{J}\sum_{j=1}^{J-1} \Phi_j \cos\frac{\pi kj}{J}. \tag{5.4.28}$$

In the application of the method, the numerical evaluation of sine and cosine functions is avoided by the use of fast Fourier transform techniques, where the harmonic vectors are generated by simple recurrence relations [31].

b) Fourier Analysis and Cyclic Reduction (FACR)

A method that has found wide application because of its speed is to apply Fourier analysis along only one space direction, say along the 'columns' $0 < i < I$. Consequently the problem is reduced to a set of $J - 2$ tridiagonal matrix equations, which couple the variables across columns $0 \leqslant i \leqslant I$ and which may each be solved separately [31].
Taking for example the case where the boundary conditions specify the potential as constant at the end of each column (say $\Phi_{i0} = \Phi_{iJ} = 0$, for all i), we may expand the potential and source functions in terms of Fourier half-sine series

$$\Phi_{ij} = \sum_{k=1}^{J} \hat{\Phi}_i(k) \sin\frac{\pi li}{J}, \qquad \rho_{ij} = \sum_{k=1}^{J} \hat{\rho}_i(k) \sin\frac{\pi li}{J}. \tag{5.4.29}$$

Inserting these expressions for the potential and the source function into Eq. (5.4.9) we obtain:

$$\sum_{k=1}^{J} \{\hat{\Phi}_{i+1}(k) - 2\hat{\Phi}_i(k) + \hat{\Phi}_{i-1}(k)\} \sin\frac{\pi kj}{J} + \left\{2\cos\frac{\pi k}{J} - 2\right\} \hat{\Phi}_i(k) \sin\frac{\pi kj}{J} + \Delta^2 \hat{\rho}_i(k) \sin\frac{\pi kj}{J} = 0, \tag{5.4.30}$$

which must be valid at every point j. By setting the amplitude of each Fourier harmonic separately to zero we obtain that, for each "wave-number" k,

$$\hat{\Phi}_{i+1}(k) - \left(4 - 2\cos\frac{\pi k}{J}\right)\hat{\Phi}_i(k) + \hat{\Phi}_{i-1}(k) = -\Delta^2 \hat{\rho}_i(k). \tag{5.4.31}$$

For each Fourier mode k, a tridiagonal matrix equation is obtained which acts on the unknown vector $\{\hat{\Phi}_i(k)\}$.
The procedure is therefore to Fourier-analyse the source function ρ_{ij} along one dimension only, to obtain the Fourier coefficients for each i,

$$\hat{\rho}_i(k) = \frac{2}{J}\sum_{j=1}^{J} \rho_{ij}(k) \sin\frac{\pi kj}{J}. \tag{5.4.32}$$

The tridiagonal equations (5.4.31) with the known right-hand side are obtained for each wave number. Hockney [31] has solved the tridiagonal equations by a method of cyclic reduction, rather than the inversion method. This way, the system of I tridiagonal equations can be reduced to $\frac{1}{2}I$ tridiagonal equations, of the form

$$\hat{\Phi}_{i+2}(k) - \left[\left(4 - 2\cos\frac{\pi k}{J}\right)^2 - 2\right]\hat{\Phi}_i(k) + \hat{\Phi}_{i-2}(k)$$
$$= -\Delta^2\hat{\rho}_{i+1}(k) - \left(4 - 2\cos\frac{\pi k}{J}\right)\Delta^2\hat{\rho}_i(k) - \Delta^2\hat{\rho}_{i-1}(k). \tag{5.4.33}$$

This procedure may be applied recursively until the central point $\frac{1}{2}I$ is related only to the boundary points $i = 0$ and $i = I$. The reverse procedure is then carried out by determining the variables at the points $\frac{1}{4}I$ and $\frac{3}{4}I$ etc., until the whole mesh is defined.
Consequently, after solving each of the tridiagonal equations, the variables $\hat{\Phi}_i(k)$ are determined for all i and for all wave numbers k. The potential at every mesh point (i, j) is obtained by Fourier synthesis applied to each column i:

$$\Phi_{ij} = \sum_{k=1}^{J} \hat{\Phi}_i(k)\sin\frac{\pi l i}{J}. \tag{5.4.34}$$

This method of Fourier analysis along one direction followed by cyclic reduction (or tridiagonal inversion) along the other direction, is very rapid. A further improvement in the efficiency of the method is obtained by implementing one level of cyclic reduction along the rows and eliminating even lines before Fourier-analysing the source functions along the rows.
Fourier analysis may be avoided entirely, by applying cyclic reduction to the vector of variables $\{\Phi_i\}_j$ for each row j [16].

5.4.4 *Iterative Methods*

The use of a direct method for the solution of a system of equations might at times be too expensive in terms of the amount of storage and computation required, especially when the number of equations become very large. Alternative approaches, based on an iterative procedure, minimize the storage and can achieve considerable speed. They do, however, produce an approximate solution, whose accuracy is determined by a preset tolerance.
An iterative approach starts with an initial guess for the vector $\mathbf{u}$, say $\mathbf{u}^{(0)}$, and proceeds in successive steps in an attempt to improve our *inexact* solution $\mathbf{u}^{(p)}$. In the simplest form, this procedure may be represented by the matrix equation,

$$\mathbf{u}^{(p+1)} = \mathbf{P}\mathbf{u}^{(p)} + \mathbf{c}, \tag{5.4.35}$$

where $\mathbf{P}$ is called iteration matrix, and an improved solution $\mathbf{u}^{(p+1)}$ is derived explicitly from the previous solution $\mathbf{u}^{(p)}$. The matrix $\mathbf{P}$ is related to the given matrix $\mathbf{A}$, and the constant vector $\mathbf{c}$ is related to the known vector $\mathbf{w}$. If, in the limit of many steps (p is very large), convergence is achieved so that

$$\lim_{p\to\infty} \mathbf{u}^{(p)} = \mathbf{P}\mathbf{u}^{(p)} + \mathbf{c}, \tag{5.4.36}$$

then the vector $\mathbf{u}^{(\infty)}$ satisfies the matrix equation

$$(\mathbf{I} - \mathbf{P})\mathbf{u}^{(\infty)} = \mathbf{c}. \tag{5.4.37}$$

Thus $\mathbf{u}^{(\infty)}$ will be the solution we require if this matrix equation (5.4.37) is consistent with the equation of interest (5.4.13). It follows that if the vector $\mathbf{c}$ is related to the known vector $\mathbf{w}$ by the nonsingular matrix $\mathbf{T}$, with $\mathbf{c} = \mathbf{T}\mathbf{w}$, then the iteration matrix must satisfy the condition

$$\mathbf{T}^{-1}(\mathbf{I} - \mathbf{P}) = \mathbf{A}, \tag{5.4.38}$$

or,

$$\mathbf{P} = \mathbf{I} - \mathbf{T}\mathbf{A}. \tag{5.4.39}$$

The advantage of this procedure is that each improved value $\mathbf{u}^{(p+1)}$ is obtained explicitly.

Clearly the technique is simple to apply. In particular, for sparse matrices, it involves only a small number of arithmetic operations at each iteration step. On the other hand, many iteration steps may be required before a reasonable solution is obtained. Two questions are of great relevance: first, is the convergence achieved for any solution guess $\mathbf{u}^{(0)}$ and second, how rapidly the solution vectors $\mathbf{u}^{(p)}$ will converge. Rather than considering the absolute solution vector $\mathbf{u}^{(p)}$, it is useful to examine the error vector $\varepsilon^{(p)}$ which remains at any step: thus if $\mathbf{u}$ is the exact solution of the matrix equation,

$$\varepsilon^{(p)} = \mathbf{u}^{(p)} - \mathbf{u}. \tag{5.4.40}$$

For reasons of consistency, the exact solution $\mathbf{u}$ should satisfy the equation

$$\mathbf{u} = \mathbf{P}\mathbf{u} + \mathbf{c}. \tag{5.4.41}$$

It follows from Eqs. (5.4.35) and (5.4.40) that the error vectors ε are related at successive steps by

$$\varepsilon^{(p+1)} = \mathbf{P}\varepsilon^{(p)}. \tag{5.4.42}$$

In order to have convergence, the magnitude of the error vector must decay, that is

$$|\varepsilon^{(p+1)}| < |\varepsilon^{(p)}|. \tag{5.4.43}$$

This requirement can be analyzed more specifically by looking at the error vectors in terms of the eigenvectors of the iteration matrix $\mathbf{P}$. If $\mathbf{P}$ has a set of n distinct eigenvalues ρ_i, then the starting error $\varepsilon^{(0)}$ in the solution-vector $\mathbf{u}^{(0)}$ can be expanded in terms of the eigenvectors $\mathbf{s}^{(i)}$ of $\mathbf{P}$,

$$\varepsilon^{(0)} = \sum_{i=1}^{n} \alpha_i \mathbf{s}^{(i)}, \tag{5.4.44}$$

where α_i is the amplitude of the initial error vector along the eigenvectors $\mathbf{s}^{(i)}$. By using Eq. (5.4.41), the relation between the error at step p and the initial error is found:

$$\varepsilon^{(p)} = \mathbf{P}\varepsilon^{(p-1)} = \mathbf{P}^p \varepsilon^{(0)}, \tag{5.4.45}$$

which can be written as

$$\varepsilon^{(p)} = \mathbf{P}^p \sum_{i=1}^{n} \alpha_i \mathbf{s}^{(i)} = \sum_{i=1}^{n} \alpha_i \rho_i^p \mathbf{s}^{(i)}. \tag{5.4.46}$$

Consequently, after p iterative steps, the amplitude of the error vector along each eigenvector is proportional to the pth power of each eigenvalue. For large p, the error is clearly related to the eigenvector with eigenvalue of the largest modulus and it follows that, for convergence, the modulus of the largest eigenvalue $\rho = |\rho_m|$ of $\mathbf{P}$ must be smaller than unity. ρ is called the spectral radius of the matrix $\mathbf{P}$. Furthermore, the asymptotic (large p) rate of convergence depends on the spectral radius ρ of the iteration matrix; the smaller the ρ, the more rapid will be the convergence. In the choice of an iterative method and correspondingly of an iterative matrix, the major concern should be in minimizing the spectral radius of the iterative matrix. For the case of the general five-point difference equation (Eq. 5.4.9), the iterative procedure can be set up directly by solving for the value of the potential at the mesh point (i, j) in terms of its surrounding values as follows:

$$\Phi_{i,j}^{*} = \frac{1}{e_{i,j}}(f_{i,j} - a_{i,j}\Phi_{i+1,j} - b_{i,j}\Phi_{i-1,j} - c_{i,j}\Phi_{i,j+1} - d_{i,j}\Phi_{i,j-1}), \tag{5.4.47}$$

$$\Phi_{i,j}^{\text{new}} = \omega\Phi_{i,j}^{*} + (1-\omega)\Phi_{i,j}^{\text{old}}, \tag{5.4.48}$$

where $f_{i,j} = \Delta^2 \rho_{i,j}$ and ω is a constant or variable relaxation factor chosen to optimize the convergence of the iteration.

It might actually be advantageous to express the two previous equations as

$$\Phi_{i,j}^{\text{new}} = \Phi_{i,j}^{\text{old}} - \frac{r_{i,j}}{e_{i,j}}\omega, \tag{5.4.49}$$

where $r_{i,j}$ is called residual and is given by

$$r_{i,j} = a_{i,j}\Phi_{i+1,j} + b_{i,j}\Phi_{i-1,j} + c_{i,j}\Phi_{i,j+1} + d_{i,j}\Phi_{i,j-1} + e_{i,j}\Phi_{i,j} - f_{i,j}. \tag{5.4.50}$$

Different iterative procedures can de devised depending on the selection of ω, the order in which the mesh points are processed, and the time when, during the iteration, the new value replaces the old one.

We will now describe some of the most important relaxation methods.

a) The Jacobi Methods. In the Jacobi method, the solution of the next

iteration is obtained on each mesh point by assuming that all the surrounding values are correct. The new values do not replace the old ones until all the mesh values have been recalculated. Thus, in terms of Eq. 5.4.49, all values of Φ on the right hand side are old values from the previous iteration, and $\omega \equiv 1$.
The rate of convergence of the method is given by

$$\frac{|\Delta\varepsilon^{(p)}|}{|\varepsilon^{(p)}|} = |\rho - 1|, \tag{5.4.51}$$

where, for the case considered here,

$$\rho = \frac{1}{2}\cos\frac{\pi}{J} + \frac{1}{2}\cos\frac{\pi}{I} \tag{5.4.52}$$

and, for large J and I,

$$\rho \sim 1 - \frac{\pi^2}{4J^2} - \frac{\pi^2}{4I^2}\,. \tag{5.4.52}$$

The rate of convergence

$$\frac{|\Delta\varepsilon^{(p)}|}{|\varepsilon^{(p)}|} = \frac{\pi^2}{4J^2} + \frac{\pi^2}{4I^2} \tag{5.4.53}$$

is very small, and the method is of no practical use.
b) Gauss-Seidel Iteration. In each iteration step, each component of the vector $\mathbf{\Phi}^{\text{new}}$ is obtained in sequence. For example, in the calculation of the potential at (i, j), the new iteration values $\Phi^{\text{new}}_{j-1,i}$ and $\Phi^{\text{new}}_{j,i-1}$ have already been determined. Since they are an improvement on the old values, to enhance convergence it is reasonable to use the new values in the iteration procedure. As in the Jacobi method, ω is equal to unity. It can be shown that the rate of convergence of the Gauss-Seidel method is greater than that of the Jacobi method, though only marginally so [32].
c) Successive Over-Relaxation. The convergence of the Gauss-Seidel method can be dramatically improved by taking a linear combination of the old value $\Phi^{\text{old}}_{i,j}$ and the value given by Gauss-Seidel as expressed in Eq. (5.4.49). The relaxation factor ω is taken in the range $1 \leqslant \omega \leqslant 2$ and corresponds to an overcorrection on the value in order to anticipate future corrections (from here the name of over-relaxation)
For the most rapid convergence, the optimized value of ω_b is [31, 33]:

$$\omega_b = \frac{2}{1 + \sqrt{(1 - \rho^2)}}\,. \tag{5.4.54}$$

The asymptotic rate of convergence of the successive over-relaxation method is

$$\frac{|\Delta\varepsilon^{(p)}|}{|\varepsilon^{(p)}|} = \sqrt{1 - \rho^2}\,, \tag{5.4.55}$$

which for our case becomes

$$\omega_b = \frac{2}{1 + \sqrt{\{1 - (\frac{1}{2}\cos \pi/J + \frac{1}{2}\cos \pi/I)^2\}}}, \tag{5.4.56}$$

and for large I and J

$$\omega_b = \frac{2}{1 + \pi\sqrt{\left(\dfrac{1}{2J^2} + \dfrac{1}{2I^2}\right)}}.$$

The asymptotic rate of convergence of the method

$$\frac{|\Delta\varepsilon^{(p)}|}{|\varepsilon^{(p)}|} = \pi\left(\frac{1}{2J^2} + \frac{1}{2I^2}\right)^{1/2} \tag{5.4.57}$$

is considerably enhanced in comparison to the Gauss-Seidel or Jacobi methods, varying inversely as the number of mesh points rather than the square of the mesh points.

A variation on the point successive over-relaxation method can be introduced by noticing that in the matrix equations of interest there is a symmetry over alternate points. Each point (i, j) is coupled to its four neighbouring points, that is, the variables at the even points are directly coupled only to the variables at the odd points and vice versa. It is useful, therefore, to define a double or cyclic iteration where all the variables on the even points are improved first and are then used to improve all the variables on the odd mesh. The optimum value of the relaxation factor $\omega = \omega_b$ is as in Eq. 5.4.54.

d) Cyclic Chebyshev Method. In the method of successive over-relaxation, while the asymptotic rate of convergence is extremely rapid, during the early stages of the iteration the convergence may be slow and, indeed, it is possible for errors to grow temporarily. In the cyclic Chebyshev method, convergence during the early stages of the iteration is considerably improved by varying the relaxation parameter from step to step. The method is the same as that used in cyclic successive over-relaxation, but now a variable relaxation parameter ω is used. Over successive steps, the parameter ω_p is varied to optimize the average rate of convergence,

$$\omega_0 = 1,$$

$$\omega_1 = 1\Big/\left(1 - \frac{1}{2}\rho^2\right),$$

$$\text{for } p \geqslant 1: \quad \omega_{p+1} = 1\Big/\left(1 - \frac{1}{4}\rho^2\omega_p\right). \tag{5.4.58}$$

The first step corresponds to the Gauss-Seidel method ($\omega = 1$) and, thereafter, the relaxation factor ω gradually increases. In the asymptotic limit, ω becomes identical with the optimum value for successive over-relaxation ($\omega_\infty = \omega_b$), so that the asymptotic properties of the Chebyshev method are

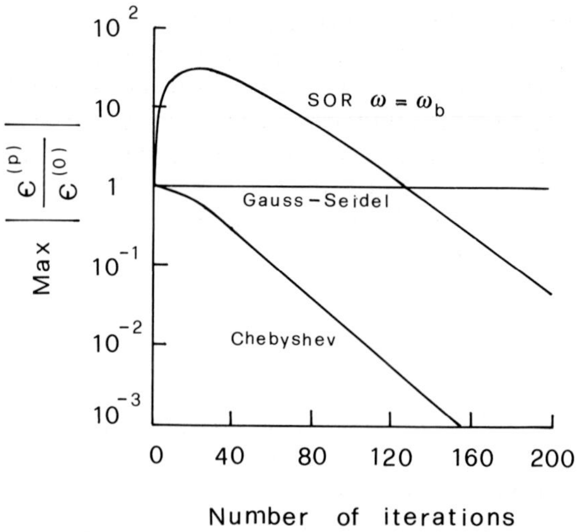

Fig. 5.10. Comparison of the convergence rate for the iterative solution of Poisson's equation using SOR, Gauss-Seidel and Chebyshew methods, respectively [15]

identical with those of successive over-relaxation. However, in the early stages of the iteration, convergence is considerably improved. The rates of convergence for the Gauss-Seidel, Chebyshev and successive over-relaxation methods are illustrated in Fig. 5.10 for the solution of Poisson's equation on a 128 × 128-point mesh [31].

5.4.5 *Calculation of the Electric Field*

Once the potential has been calculated on the grid, the value of the electric field

$$\mathbf{E}(\mathbf{r}) = -\nabla\Phi(\mathbf{r}), \tag{5.4.59}$$

on the grid points is found using centered differences. In the mesh point (i, j) we have

$$E_x = \frac{\Phi_{i+1,j} - \Phi_{i-1,j}}{2\Delta x} \tag{5.4.60 a}$$

and

$$E_y = \frac{\Phi_{i,j+1} - \Phi_{i,j-1}}{2\Delta y}\,. \tag{5.4.60 b}$$

Special care should be taken in determining the value at the boundaries. For example, in the case of Neumann's condition, in order to achieve normal derivatives of the potential equal to zero, an extra fictitious mesh is created

outside the boundary. The values of the potential on the two points adjacent to the boundary (the one just outside and just inside, respectively) are set equal to each other, in order to have zero field on the boundary.
Equations (5.4.60) hold a second order accuracy. To achieve better precision, more complex schemes, such as a nine-point difference formula, should be used, which correspond to higher orders in the Taylor expansion of the differential operator. A drawback in this case would come from the increased complication at the boundaries.

5.4.6 The Collocation Method

The use of finite difference schemes to solve the Poisson's equation may not be satisfactory, particularly when the precision on the electric field obtained is crucial. This is especially true when a reliable solution of the electric field on the boundary is needed [29, 34]. The only way to achieve better precision with finite differences is to increase the number of mesh intervals. But there is a limitation due to the relatively small number of particles which can be used in the Monte Carlo simulation. If the mesh interval is too small, the number of particles included in each mesh may be strongly affected by fluctuations, which would result in an incorrect charge assignment to the mesh points.
Collocation methods for one-dimensional problems provide a dramatic improvement in the precision of the solution for both electric field and potential, and allow more flexibility when a region of the device requires finer detail, since the mesh size can be locally varied without any modification of the algorithm. For two-dimensional problems, collocation methods are available [35, 36]. Even though the computational work increases considerably, they are very appealing because of the large amount of information they can provide simultaneously. Difficulties still remain, when a nonrectangular domain is considered.
We present here the fundamental steps needed to implement a 1D Poisson solver using the collocation method. A complete account of the general theory and of the many variations on the theme may be found in [35, 37].
The one-dimensional domain is divided into mesh intervals, whose size may be in general nonuniform. The potential inside each mesh interval is expanded in a linear combination of spline functions, as

$$\Phi(x) = C_1\phi_1(x) + C_2\phi_2(x) + C_3\phi_3(x) + \cdots + C_n\phi_n(x). \tag{5.4.61}$$

Four Hermite cubics are chosen as a basis on a reference interval [0; 1], as shown in Fig. 5.11. Scaling from the reference to the mesh interval is made using the relation

$$s = [x - x_i]/h_i, \tag{5.4.62}$$

where h_i is the mesh size of the ith interval, whose boundaries are indicated as x_i and x_{i+1}. Since only N_1 and dN_2/ds are different from zero at $s = 0$,

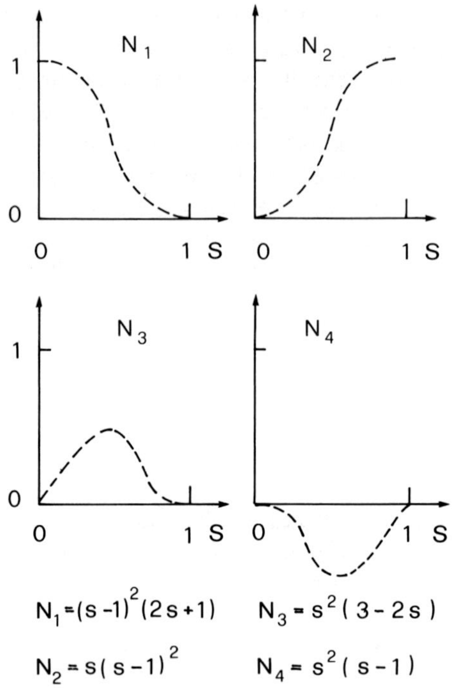

Fig. 5.11. Cubic Hermite polynomials for the collocation method on a reference interval [0; 1] [34]

and so are N_3 and dN_4/ds at $s = 1$ (see Fig. 5.11), letting $E(x)$ be the electric field, we obtain for the coefficients

$$C_1 = \Phi_i, \qquad C_2 = E_i, \qquad C_3 = \Phi_{i+1}, \qquad C_4 = E_{i+1}$$

and we can now write the potential inside the ith interval as

$$\Phi(x) = \Phi_i N_1(s) - h_i E_i N_2(s) + \Phi_{i+1} N_3(s) - h_i E_{i+1} N_4(s).$$

If we now differentiate the previous expression twice, we end up with

$$\begin{aligned} \frac{d^2\Phi(x)}{dx^2} &= \frac{\Phi_i}{h_i^2}\frac{d^2N_1(s)}{ds^2} - \frac{E_i}{h_i}\frac{d^2N_2(s)}{ds^2} \\ &+ \frac{\Phi_{i+1}}{h_i^2}\frac{d^2N_3(s)}{ds^2} - \frac{E_{i+1}}{h_i}\frac{d^2N_4(s)}{ds^2}. \end{aligned} \tag{5.4.63}$$

This represents the left-hand side of Poisson's equation. If the charge density is known in a number of points inside the domain, a system of linear equations can be constructed, whose unknowns are Φ_i and E_i. Two unknowns are left on the boundary, the ones not defined by the boundary conditions. If the potential is assigned on the boundary, then the electric field is obtained, but in the general case the choice is free.

To saturate the unknowns, two equations are obtained for each mesh interval by defining the charge density on two points inside (collocation points). The assignment of charge will impose the interpolation constraints of the solution. When the collocation points are the roots of orthogonal polynomials the procedure is usually called orthogonal collocation. Here, we choose the Gaussian knots of the interval, which are the zeros of the Legendre polynomial of order two. It has been shown that collocation at the Gaussian knots using a basis of piecewise cubic Hermite polynomials gives an order h^4 algorithm to approximate the unique solution to second order boundary value problems [36]. On the reference interval [0; 1], the Gaussian knots are given by

$$s_1 = [1/\sqrt{3})]/2 = 0.211325,$$
$$s_2 = 1 - s_1 = 0.788675.$$

To build the system of equations, we have just to compute the second derivatives of the Hermite cubics at the Gaussian knots of the reference interval. The linear system of equations obtained has a banded matrix, with semibandwidth 3. The matrix is not diagonally dominant, so we need to use an efficient direct solver with pivoting.
We have seen previously that the solution of the 1D problem using the 3-point finite difference formula are h^2 approximates and only for the potential, while a further discrete differentiation is needed for the electric field. To get results comparable to the ones provided by collocation, one has to reduce the mesh size but this is not always possible in Monte Carlo methods, particularly when we are in a region of the device with very few carriers. Also from a purely mathematical point of view, one cannot generally assert that decreasing the mesh size always increases the accuracy, because if it is true that the discretization error decreases, the round-off error is generally increased [37].
The method presented may be readily extended to a two-dimensional rectangular domain, on which a rectangular grid is defined [22]. Again, the mesh size does not have to be uniform along the two sides. In each subrectangle of the grid 16 Hermite bicubics are taken, obtained performing the possible multiplications between the 4 Hermite cubics, defined on each side, as the ones given in the one-dimensional case. Following analogous steps, one ends up with an expression for the potential inside the subrectangles, in which the unknowns are the potential, the electric field along the x and along the y direction, and the second mixed derivative of the potential, at the grid points. The interpolation constraints are imposed assigning the charge to the 4 Gaussian knots inside each subrectangle, obtained from the ones along the sides. On each boundary point two boundary conditions have to be assigned (three on the corners). One is given by setting to zero the second mixed derivative on every point, and the second will be a fixed potential where there are no contacts, or otherwise the normal component of the field set to zero. The third condition on the corners will be given by setting to

zero one of the components of the field, following physical considerations. The matrix to be solved is much more complex in this case. Again, it is banded and asymmetrical and requires a very efficient direct solver with pivoting. The number of equations tend to be very large, as there are 4 of them for each grid point. However the method is again very appealing, since field and potential are solved simultaneously and the boundary conditions of zero field normal to a boundary are naturally included. In addition, since it is possible to modulate the mesh size without any formal change in the equations (while it turns out to be very cumbersome in the case of finite differences) the application of this algorithm is very interesting for devices with two-dimensional effects (HEMT, MOSFET) in which most of the conduction carriers are confined within a narrow region. Using finite differences, accuracy of the same order (for the potential only) is obtained only with the 9-point formula.

5.5 The Monte Carlo Simulation of Semiconductor Devices

We are in the position to examine the modification to the standard Monte Carlo program (presented in chapter 2) so that the non-uniformities of a device can be accounted for. In the previous sections it has been shown how the geometry of the device enters the simulation, and what the role of the boundary conditions is. The notion of superparticles has also been introduced. Several additional points need to be deal with:

— the use of initial conditions and the discretization scheme for the time evolution;
— the presence of a non-homogeneous field and the possibilities available to optimize the M.C. algorithm using space-dependent scattering rates (variable Γ);
— the presence of non-uniform doping (variable Γ for different regions of the device, degeneracy due to high doping densities)
— the possibility of studying noise.
— the optimization of the overall simulation;

The present section will deal with each of those issues.

5.5.1 Initial Conditions

At the beginning of the simulation, after defining the geometry of the system and the doping profile, positions and velocities of the mobile carriers must be assigned. Such initial conditions for the simulation are somewhat arbitrary. One possible choice is to ensure the initial neutrality of the device, by randomly distributing an equal number (per unit area) of electrons as there are dopants. An example is given in Fig. 5.12 [38]. Carrier velocities

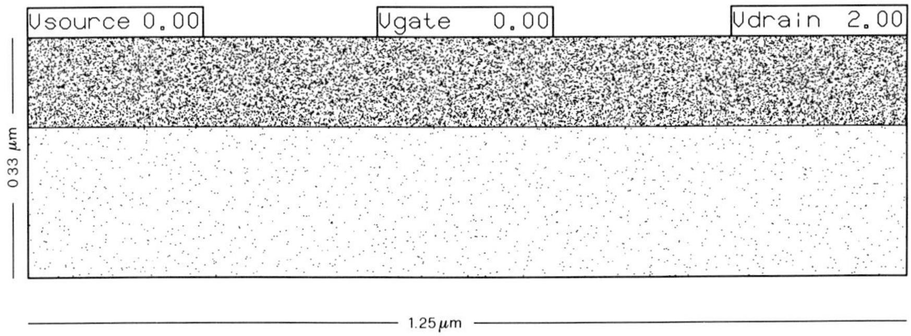

Fig. 5.12. Initial carrier distribution for the prototype MESFET shown in Fig. 5.3, corresponding to charge neutral condition. The simulated electrons are distributed according to a thermal distribution at the lattice temperatute (300 K). No bias is applied. The epilayer and substrate doping are, respectively, 5×10^{17} cm^{-3} and 10^{16} cm^{-3}

can be generated randomly following a Maxwellian distribution. The simulation will in general evolve naturally towards a steady state situation not necessarily characterized by the overall charge neutrality of the device. In a MESFET, for example, a depletion will be formed under the gate, while in a MOSFET the channel will accumulate carriers at the interface Si-SiO$_2$. The simulated transient towards the steady state will reproduce the real transient of the device as long as the boundary conditions are treated correctly. This point will be dealt with more extensively in Ch. 6.
During the transient regime, when the local fields vary very rapidly, it is necessary to account for the displacement current in the device. The equivalent total charge $Q(t)$ as a function of time for each electrode is given by

$$Q(t) = q(N_a - N_i) + \kappa \int E_y(x, t)\, dx, \tag{5.5.1}$$

where the integral extends over the global area of the electrode. In Eq. (5.5.1), q is the superparticle charge, N_a and N_i are respectively the total number of particles absorbed and injected at the electrode up to the time t, and $E_y(x, t)$ is the y-component of the electric field on the electrode surface at time t and position x. The time behavior of $Q(t)$ starting from an initial condition of charge neutrality is shown in Fig. 5.13. When a voltage is applied from source to drain at $t = 0$, charges start flowing in and out of the electrodes until stationarity is reached. Initially, carriers are repelled by the potential barrier at the gate contact, and a depletion region is formed under the gate. The difference in the equivalent charge flowing from the drain and source contacts corresponds to such depletion charge. During the transient, the field that builds up at the gate induces a displacement current at the gate. A steady-state condition (that is a constant current flow from source to drain) is reached in a few pisoseconds. Figure 5.14 shows the steady-state distribution for zero bias on the gate and 2 Volts on the drain [38].

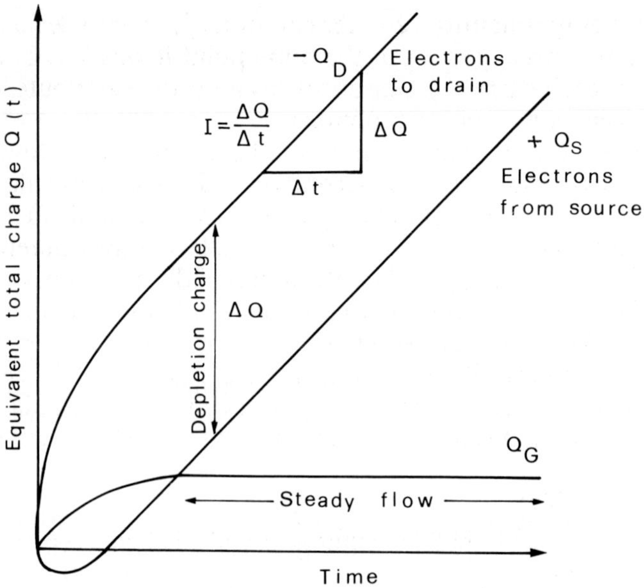

Fig. 5.13. Qualitative plot of the time evolution of the equivalent total charge for source, drain and gate contacts, after a step potential has been applied to the drain [15]

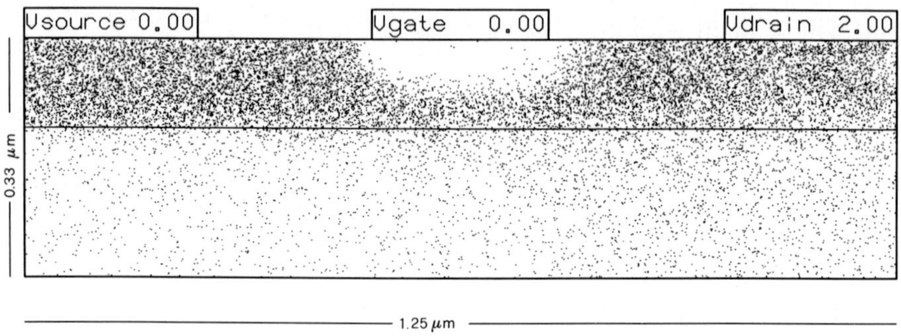

Fig. 5.14. Steady-state distribution with 2 V on the drain and −0.8 V on the gate (corresponding to the Shottky barrier between metal and semiconductor)

5.5.2 Time Cycles

There are two characteristic times for the simulation: the time between successive scattering events δt, determined stochastically starting from the microscopic scattering rates (see chapter 3.5), and the field-adjusting time, that determines the equally spaced time intervals ΔT after which potential and fields are updated. The ensemble of carriers is simulated up to the field-adjusting time. At this point, the simulation is frozen and the carrier variables (such as position, velocity, energy, etc.,...) are stored. After the calculation of average quantities and the solution of the particle field equations, the simulation starts again from the same point it was left. The remaining free flight for each carrier can be stored as a particle attribute, or it can be calculated using a new random number.

The equations of motion (5.1.1) and (5.1.2) are solved via a finite difference scheme on the time interval δt. A second order Runge-Kutta algorithm [39] can be used as a good compromise between accuracy and efficiency.

An alternative method [5] is based on a three level time discretization, that uses a fixed time step Δt_{ball} for the calculation of the accelerative part, a fixed time step Δt_{sc} for the scattering events, and the field adjusting step ΔT for the solution of Poisson's equation. Although such a method is very appealing (it does not require all the controls that are necessary when the free flights between scattering are determined stochastically), care should be taken in the choice of each time step, which are subject to the following constraints:

(*i*) $\Delta t_{sc} \ll 1/\tau_{\max}$,

where $1/\tau_{\max}$ is the maximum scattering rate for the simulation;

(*ii*) $\Delta T_{\text{ball}} \leqslant \Delta t_{sc} \leqslant \Delta T$.

In this method, all carriers are accelerated ballistically up to a time Δt_{sc}. At this point, the final k-vector and energy ε are calculated, and a random number (for each particle) is used to decide whether a scattering event will occur according to the ratio $\Delta t_{sc}/\tau_0(\varepsilon)$, where $1/\tau_0(\mathbf{k})$ is the total scattering rate of the particle. The type of scattering is chosen in the usual way with a weighted selection from all scattering mechanisms.

An upper limit for the field-adjusting times, at which the particle-mesh calculation of sections 5.3–5.4 is performed, can be taken from plasma simulations [15]. It should be pointed out, though, that the frequent scatterings present in a semiconductor device act to disrupt collective (and coherent) behaviour typical of a plasma, thus relaxing the requirement of stability of the numerical algorithm.

The stability requirement of the particle-mesh calculation is

$$\omega_p \, \Delta T < 2,$$

where ω_p is the plasma frequency, with $\omega_p = (4\pi n e^2/m)^{1/2}$, and ΔT is the field-adjusting time step.

The above condition can be determined from the rates of heating and collision time in a coulombic plasma [15]. For typical values encountered in a MESFET, ($n \sim 10^{17}\ \text{cm}^{-3}$, $m = 0.07\, m_0$), the plasma frequency is $\sim 2 \times 10^{13}\ \text{sec}^{-1}$, and an appropriate ΔT is found to be equal to 5×10^{-14} sec. In a typical situation of a M.C. run for GaAs a high energy particle would have on average about 2 scatterings during one interval ΔT (since its total scattering rate is $\sim 0.5 \times 10^{14}\ \text{sec}^{-1}$), while a low energy particle will scatter once every 10 "field adjusting" time steps.

A very different situation is found for example in a MOSFET, where the high densities ($\sim 10^{20}\ \text{cm}^{-3}$) in the drain and source regions lead to values

of ω_p as high as 2×10^{14} sec^{-1}. Consequently, the field calculation has to be performed on a femtosecond time scale, in order to sample correctly the plasma oscillations. The failure to do so produces catastrophic instabilities, as explained very nicely by Fischetti and Laux [5]. The instabilities are caused by the fluctuations in the charge density owing to the long range of the coulomb interaction. In the presence, for example, of a negative fluctuation of electrons, the large positive charge corresponding to the electron "vacancy" causes a large field, that attracts nearby electrons. Unless the field is updated rapidly in comparison with the charge rearrangement, an uncontrollable oscillation builds up.
The carrier instabilities can be avoided by using a smoothed charge density in the solution of Poisson's equation [40]. There, the electron concentration $n(\mathbf{r})$ is expressed in terms of the quasi-fermi level $\phi_n(\mathbf{r})$ defined by

$$n(\mathbf{r}) = n_i \exp\left[\frac{\phi(\mathbf{r}) - \phi_n(\mathbf{r})}{K_B T_e(\mathbf{r})}\right], \tag{5.5.2}$$

where n_i is the intrinsic electron concentration, and T_e the position dependent electron temperature. Even in the presence of large fluctuations in the particle density, the quasi-fermi level does not vary significantly. The Poisson's equation becomes non-linear in ϕ, but the instabilities are suppressed. The use of the electron temperature is very helpful in achieving a fast convergence in the Poisson solver [41].

5.5.3 *Free Flight*

The free flight between successive scattering events is calculated using Eq. (3.4.5). Electrons are accelerated according to Eqs. (5.1.1) and (5.1.2). Since we have a position-dependent field, it is in principle necessary to check whether or not the carriers move from one cell to another during the free flight. In fact, while the field is constant within each cell, it changes abruptly at the interface with the neighbouring ones. Such a variation can be very high in special regions, such as the drain-end of the gate in a FET.
The algorithm can be devised as follows:

i) The free flight time is split into two components $\delta t = \delta t_i + \delta t_g$, respectively the time necessary to reach the border of the adjacent cell and the remainder of the flight up to the scattering;
ii) The time step δt_i is calculated knowing the initial position and velocity of the electron
iii) The electrons, initially in the cell, are accelerated for a time δt_i with a field E_i, and for the remaining δt_g with a field E_g. If several cells are crossed within a free flight, the procedure can be repeated for each cell.

In the regions where the variations of the electric field are small (typically the source and drain regions), the change of cell by the accelerating electron is always a fairly smooth process. Problems might arise where the electric field changes over very small distances. However, in such a situation, electrons are usually very hot, and the rate of scattering very high. Thus, electrons are likely to scatter very often, and the number of free flights out of a cell should be strongly reduced.

The previous considerations justify the choice that is often made to accelerate electrons with a constant value of the electric field (the one corresponding to the starting cell), updating the value of the field only at the beginning of the successive flight (in case the electron has moved to a new cell).

5.5.4 *Scattering*

The use of the self-scattering algorithm (with constant Γ) certainly simplifies the calculation of the free flight. It is, though, very time consuming when the scattering rate varies greatly with energy, as many self scatterings are generated. Several procedures can be adopted to improve such a situation. In general, it is advisable to tabulate the scattering probabilities $\lambda(k)$ and $\sum_i \lambda_i(k)$ for the various processes (including self scattering) at the beginning of the simulation. At the end of a free flight, rather than recalculating the scattering rates corresponding to the final electron energy, those rates can be read from the table (or calculated with a linear interpolation). In the selection of the scattering mechanism, the self-scattering event should be checked first. As was discussed in section 3.4, a piecewise-constant Γ technique can be used. An example of this was shown for the Γ valley of GaAs above or below the threshold for intervalley scattering.

An iterative technique was suggested by Warriner [15, 42] in order to obtain a further reduction in the number of self-scatterings. Although very attractive and fast, this procedure does not yield the correct distribution of free flight [43] since the Γ it uses depends on the random number.

In a device it might be useful to use a position dependent Γ. For example, undoped and doped regions will have very different scattering rates for ionized impurity scattering. It is then useful to have a Γ for each region. A multivalue Γ selection rule has been suggested, where Γ is a position dependent quantity [8]. Details of the procedure can be found in the reference.

All of the methods just examined are based on a stochastic generation of the free flights. As was illustrated earlier in section 5.5.3, it is possible to use a fixed time step for the free flight. In this case, no self scattering is needed.

5.5.5 *Carrier-Carrier Scattering*

When a device presents highly doped regions, the interaction among carriers can be very important. The careful treatment of this interaction leads to the identification of new channels for the exchange of energy and momentum, as shown in section 2.4.4. The procedures that were introduced in section

3.9.4 [44] can be applied to device simulation [45]. In the calculation of the scattering rates for carrier-carrier scattering, local distribution functions (defined over limited portions of the device) should be used. This would account for the spatial inhomogeneities (and correspondingly for the different carrier densities and temperatures) in the device. Specific examples will be given in the next chapter. A very good discussion of the complications arising when both short range (screened electron-electron interaction) and long range (electron-plasmon interaction) have to be included in a self consistent simulation can be found in [5].

5.5.6 *Degenerate Statistics*

A further problem found in devices that present highly doped regions is that degenerate statistics have to be accounted for (this is the case for example of the source and drain regions of a MOSFET or the base of a PDB device). The algorithm that was presented in section 3.9.3 can be used in the case of a device simulation with minor modifications. Since only limited regions of a device are generally heavily doped, it might be convenient to divide the device into different areas. Each one will be characterized by its own volume, doping and free carrier concentration, which will determine the normalization term for the k-space occupancy factor (see the discussion in 3.9.3). The rejection algorithm to account for Pauli exclusion principle is then applied to each individual area, each characterized by a defined electron distribution, calculated as an average over all the electrons found in that area. This procedure has been applied to the simulation of the THETA device (see section 6.8) [45]. The drawback of the algorithm lies in the tabulation of the distribution function if too many areas have to be defined for the degeneracy calculation.

An alternative approach [5] is to approximate the electron distribution function at a position $\mathbf{r}$ and time t by a Fermi-Dirac function calculated using the position-dependent electron temperature and Fermi level determined self consistently during the simulation. A rejection technique is used (exactly like in the previous algorithm) in the calculation of the final state, to account for the possible occupancy of such a state.

5.5.7 *Statistics*

Once the steady state is reached, average values can be taken as ensemble averages over single cells. If, for example, the local velocity distribution inside the device has to be calculated, then each cell will be assigned a value

$$v^{(j)} = \frac{1}{N} \sum_{i=1}^{N_p} v_i^{(j)}$$

over the N_p electrons found at the instant in the p-th cell. Time averages can also be used to reduce the statistical uncertainty.

In areas where the electron population is very small, it is possible to extend a technique originally proposed by Phillips and Price [46] (see chapter 3.12.3), that allows good statistics to be obtained in rarely visited energy regions. Two situations are of particular interest. If a device presents regions with a high doping density N_D^+ connected to regions with low densities N_D, the carrier concentration will reflect (except at the interface between the different regions) the doping distribution. Therefore, most of the simulated carriers (roughly in a ratio N_D^+/N_D) will populate the high doping, low field regions, requiring an extensive amount of computation for the simulation of a quasi-thermal distribution. This is the case for example of the MOSFET or a MESFET with ion implanted source and drain contacts.
A similar case is found in **k**-space, when we are interested in the population of high energy states, which are rarely touched by the carriers, but might lead to very important physical phenomena (a typical example is the carrier injection into SiO_2 for the channel of a MOSFET).
The latter situation is the one examined by Phillips and Price. The population of the high energy states can be enhanced by generating a fixed number N of carrier histories each time one of the simulated particles enters the rarely-populated region. Each of the N generated particles will have the same initial conditions (equal to the state of the "parent" particle at the moment of the multiplication) and a weight of $1/N$ for the calculation of average quantities. The multiplication algorithm can be repeated several times at higher energies, resulting in an "avalanche" of carriers which fill up the tail of the distribution function at higher and higher energies [8].
In real space a similar multiplication technique can be used [5, 8, 23]. An example is shown in Fig. 5.15 [23]. In the self-aligned MESFET structure shown there, only a fraction N_D/N_D^+ of the total carriers would be found in the channel (this simple estimate neglects the spilling of charge into the channel). If such a number is too small, high fluctuations and unreliable

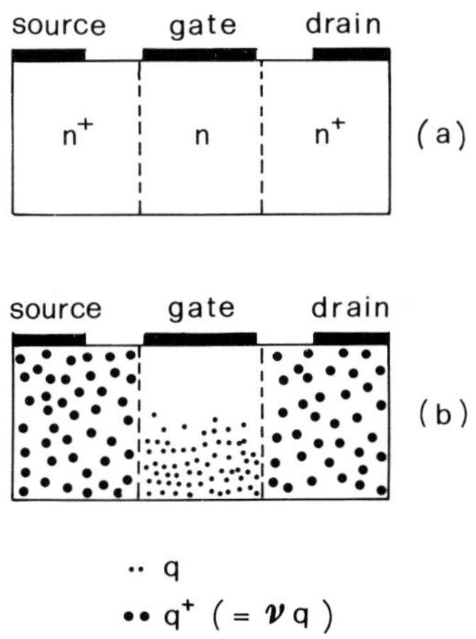

Fig. 5.15. Pictorial description of the multiplication technique (*b*) described in the text, applied to the MESFET structure shown in (*a*) [23]

statistics are found. The low density region is then assigned an enhancement factor v. Every particle that enters that region is replicated v times. The charge of the generated superparticles is reduced by the same factor v. This concept is shown schematically in Fig. 5.15b. With respect to the traditional simulation, the multiplication method guarantees great reductions of computer time. To ensure current continuity at the two interfaces, the probability for a particle leaving the enhancement region to be kept in the simulation is taken proportional to $1/v$. It is important, in the application of such a multiplication scheme, to make sure that the algorithm does not produce excessive noise or instability on the results. It is only "a posteriori" that an optimal value for v can be found. Application of the real-space multiplication algorithm to MOSFET simulation can be found in Refs. [5, 8].

5.5.8 Static Characteristics

As seen previously, the steady-state current at each electrode is given by the net number of superparticles crossing the electrode per unit time. By performing several computer runs, it is possible to construct the I-V characteristics of the device. A typical example is given in Fig. 5.16 [42]. In performing different runs at successive values of V (or other external parameters), the initial conditions of each run are taken equal to the final conditions of the previous one, in order to minimize the time involved in the transient. The values of the gate voltage reported in the figure are those used as the boundary condition in the simulation. To compare those values with experimental ones, the height of the Schottky barrier (0.8 V for aluminum gates) should be added. The values of source and drain voltages are directly comparable with the experimental ones, since those are ohmic contacts.

Important device parameters can be extracted from the simulation. For

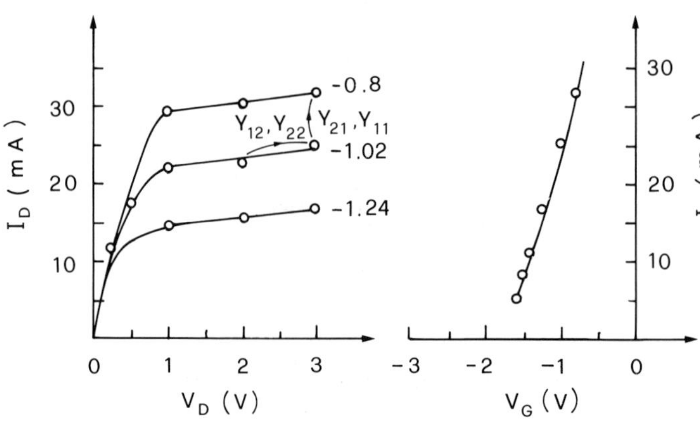

Fig. 5.16. Typical static I-V characteristics for a simulated MESFET [15]

example the output resistance

$$R_{\text{out}} = \frac{\Delta V_D}{\Delta I_D} \quad \text{for constant } V_G, \tag{5.5.3}$$

and the transconductance

$$g_m = \frac{\Delta I_D}{\Delta V_G} \quad \text{for constant } V_D, \tag{5.5.4}$$

can be obtained from a series of computer runs, starting from normal operating conditions ($V_D = 0.3$ V and $V_G = -1.02$ V in Fig. 5.16) and varying the drain and gate bias, respectively.

It was stressed in Sec. 4.3.5 that the definition of an equivalent circuit (as the one shown in Fig. 4.24) is very helpful in understanding the behavior of a device. Modeling is usually done on the basis of a limited set of lumped parameters, associated with different regions of the device, that describe locally distributed resistances and capacitances. Warriner [42] has devised the following procedure to extract the lumped parameters (for the equivalent circuit of Fig. 5.17) from the simulation.

The total gate capacitance C_G is distributed between source C_{sg} and drain C_{gd}. If a step change is applied respectively to the gate (ΔV_G) and drain voltage (ΔV_D), the source and drain capacitances are obtained from

$$C_{sg} = \frac{\Delta Q_G}{\Delta V_G}, \qquad C_{gd} = \frac{\Delta Q_G}{\Delta V_D}, \tag{5.5.5}$$

where ΔQ_G is the total equivalent charge flowing from the gate in response

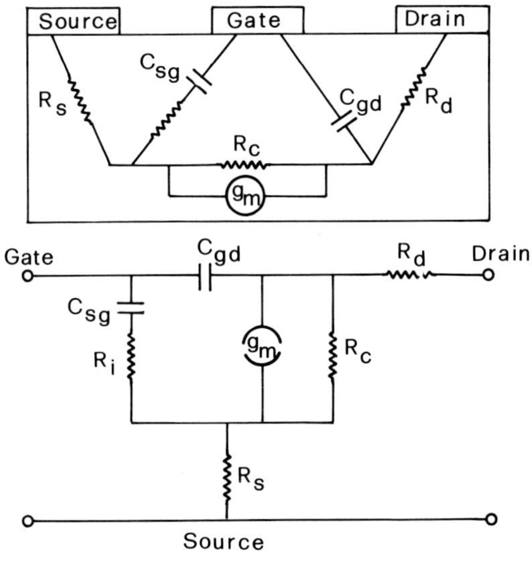

Fig. 5.17. Equivalent circuits for a MESFET [15]

to the step potential. Such flow will vary in time according to the law

$$Q_G(t) = \Delta Q_G(1 - e^{-t/\tau}), \tag{5.5.6}$$

where the time constant τ is given by

$$\tau = \tfrac{1}{2}(R_s + R_i)C_{sg} \tag{5.5.7}$$

and

$$\tau = R_d C_{gd} \tag{5.5.8}$$

for source-to-gate and gate-to-drain path, respectively.
Equations (5.5.6) and (5.5.7) allow the calculation of the corresponding parasitic resistances.
The channel resistance R_C is generally much larger than source and drain resistances. It is given by the slope of the static characteristic near the operating point

$$R_C = \frac{\Delta V_D}{\Delta I_D}. \tag{5.5.9}$$

5.5.9 *A.C. Characteristics*

The parameters described in the previous section characterize the behavior of a device under d.c. conditions.
A convenient dynamic description of the device is given by the Y parameters defined (for a two-port device) by the matrix relation [15]

$$\begin{bmatrix} \Delta I_1(\omega) \\ \Delta I_2(\omega) \end{bmatrix} = \begin{bmatrix} Y_{11}(\omega) & Y_{12}(\omega) \\ Y_{21}(\omega) & Y_{22}(\omega) \end{bmatrix} \begin{bmatrix} \Delta V_1(\omega) \\ \Delta V_2(\omega) \end{bmatrix}, \tag{5.5.10}$$

where ΔI and ΔV denote small sinusoidal variation of current and voltage, respectively, of frequency ω about the normal operating point of the device.
The zero-frequency values of some of the parameters have already been seen in the previous section, namely $g_m = Y_{21}(0)$ and $R_{\text{out}} = 1/Y_{22}(0)$.
The a.c. characteristics could be studied directly, by following the device response to different frequencies. Such a method is though prohibitive in terms of computer time.
Instead, the Y parameters can be obtained by applying a step change in voltage ΔV_l at a given port l, and calculating the Fourier transform of the response current I_m at port m. We have then

$$Y_{lm}(\omega) = i\omega \frac{I_m(\omega)}{\Delta V_l} \tag{5.5.11}$$

with $\{m, l = 1, 2\}$ [15].
The frequency dependence of the output impedance $Z_{\text{out}} = [1/Y_{22}(\omega)]$ of a MESFET calculated from the above procedure is shown in Fig. 5.18 [42].

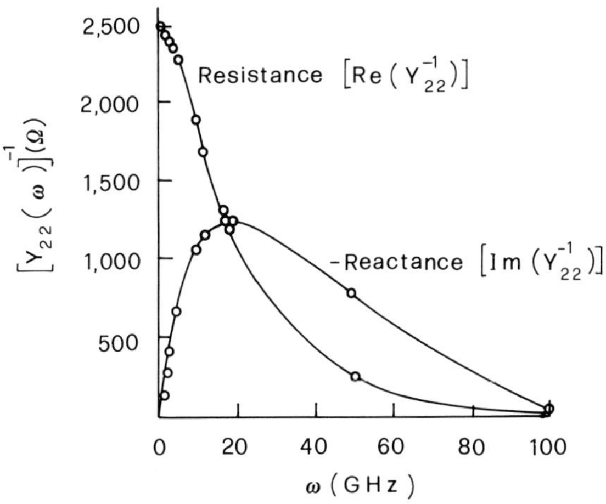

Fig. 5.18. Frequency dependence of the calculate output impedance $[Y_{22}(\omega)]^{-1}$ [15]

The four Y parameters of Eq. (5.5.10) can be obtained (as a function of frequency) in two computer runs.
In the first run the step bias is applied on the drain (denoted by the sub-index 2), in the second on the gate. The current response for gate and drain respectively is calculated in both cases, and substituted in Eq. (5.5.11) to obtain the Y-parameters. An example can be found in Ref. [47]. It should be noted that the Y-parameters are intrinsic to the device itself, and do not account for the parasitic effects due to the external environment (such as contacts, metalizations, etc., ...).

5.5.10 *Noise*

Noise in a device is related to the fluctuations of the output current around its mean value and is due to the stochastic nature of the microscopic processes that govern carrier transport in the device. As was already pointed out in chapter 3, the Monte Carlo simulation is particularly well-suited for the study of noise, since it naturally incorporates all the microscopic noise sources.
The most important figure of merit for noise in a transistor is the minimum noise figure NF, given by the ratio of the signal to noise power at the gate, $P_1/\delta P_1$, divided by the corresponding ratio at the drain, $P_2/\delta P_2$, or [15, 48]

$$NF = \frac{P_1/\delta P_1}{P_2/\delta P_2} = \frac{\delta P_2}{\delta P_1} \times \frac{P_1}{P_2}, \tag{5.5.12}$$

which can be expressed as

$$NF = 1 + \left(\frac{Y_{11}}{Y_{22}}\right)^2 \cdot \frac{\langle \delta I_2^2 \rangle}{\langle \delta I_1^2 \rangle}. \tag{5.5.13}$$

The second factor on the right-hand side of Eq. (5.5.13) is the ratio of the mean-square current fluctuations at the drain to those at the gate. Such quantities are measured directly during the MC simulation.

In general, current fluctuations as measured in a simulation are magnified by several orders of magnitude due to the use of superparticles. However, such an error will be cancelled in Eq. (5.5.13) since it effects both numerator and denominator.

The minimum and maximum values of the frequencies that can be measured by a M.C. simulation are connected to the parameters of the simulation. In particular, for a simulation of length T, subdivided in N time steps ΔT, the minimum and maximum measurable frequencies are [15]

$$f_{\max} = \frac{1}{2\Delta T}$$

and

$$f_{\min} = \frac{1}{T}.$$

For typical values of $\Delta T = 10^{-15}$ sec and $N = 10^4$, we have $f_{\max} = 50$ THz and $f_{\min} = 10$ GHz.

We have focused in this chapter on the Monte Carlo simulation of a prototype device, identified as a GaAs MESFET. The various steps involved in the simulation have been discussed in some detail.

As pointed out in the previous chapter, a variety of semiconductor devices exist, each presenting very characteristic physical features. Therefore, al though quite general, the algorithms and techniques introduced so far might require major changes when they have to be applied to specific devices. The next chapter will deal with applications of the MC simulation to various devices.

References

[1] Lugli, P., Jacoboni, C.: In: ESSDERC 87, Proc., 17th European Solid State Device Research Conference (Calzolari, P. U., Soncini, G., eds.), p. 97 (1987), and references therein.

[2] Awano, Y., Tomizawa, K., Hashizume, N.: IEEE Trans. Electron Dev. *ED31*, 448 (1984).

[3] Tomizawa, M., Yoshii, A., Yokoyama, K.: IEEE Electron Dev. Lett. *EDL6*, 332 (1985).

[4] Ravaioli, U., Ferry, D. K.: IEEE Trans. Electr. Dev. *ED33*, 677 (1986).

[5] Fischetti, M. V., Laux, S. E.: Phys. Rev. *B38*, 9721 (1988).

[6] Moglestue, C.: COMPEL *1*, 7 (1982).

[7] Wang, T., Hess, K.: J. Appl. Phys. *57*, 5336 (1985).

[8] Sangiorgi, E., Riccò, B., Venturi, F.: IEEE Trans., Computer-Aided Design *CAD-7*, 259 (1988).

[9] Baccarani, G., Jacoboni, C., Mazzone, A. M.: Solid State Electr. *20*, 5 (1977).

[10] Littlejohn, M. A., True, R. J., Hauser, J. R., Golio, J. M.: J. Vac. Sci. Technolog. *B1*, 449 (1983).

[11] Hwang, C., Navon, D. H., Tang, T.: IEEE Trans. Electron Dev. *ED-34*, 154 (1987).

[12] Bandyopadhyay, S., Brown, M. E., Maziar, C. M., Datta, S., Lundstrom, M. S.: IEEE Trans. Electron Dev. *ED-34*, 392 (1987).

[13] Park, Y., Navon, D. H., Tang, T.: IEEE Trans. Electr. Dev. *ED31*, 1724 (1984).
[14] Hockney, R. W., Warriner, R. A., Reiser, M.: Electron Lett. *10*, 484 (1974).
[15] Hockney, R. W., Eastwood, J. W.: Computer Simulation Using Particles. New York: McGraw-Hill. 1981.
[16] Potter, D.: Computational Physics. London: Wiley. 1973.
[17] Arnold, D., Hess, K.: Solid-State Electron. *31*, 593 (1988).
[18] Brandt, A.: Math. Comp. *31*, 333 (1977).
[19] Dwyer, H. A., Smoke, M. D., Kee, R. J.: In: Numerical Grid Generation (Thompson, J. F., ed.), p. 339. New York: Elsevier. 1982.
[20] Winskow, A. M.: J. Comp. Phys. *2*, 149 (1967).
[21] Selberherr, S.: Analysis and Simulation of Semiconductor Devices. Wien-New York: Springer. 1984.
[22] Ravaioli, U.: Ph.D. Dissertation, Arizona State University, unpublished (1988).
[23] Yamada, Y., Ikeda, S., Shimojok, N.: In: Proc., XVII European Solid State Device Research Conference ESSDERC87 (Calzolari, P., Soncini, G., eds.), p. 111. Bologna: Tecnoprint. 1987.
[24] Hockney, R. W.: Phys. Fluids *9*, 1826 (1966).
[25] Birdsall, C. K., Fuss, D.: J. Comp. Phys. *3*, 494 (1969).
[26] Eastwood, J. W., Hockney, R. W.: J. Comp. Phys. *16*, 342 (1974).
[27] Thurner, M., Selberherr, S.: In: Proc., NASECODE IV Conf. (Miller, J. J. H., ed.), p. 327. Dublin: Boole. 1987.
[28] The Mathematics of Finite Elements and Applications (Whiteman, J. R., ed.). London: Academic Press. 1973.
[29] Ravaioli, U., Lugli, P., Osman, M., Ferry, D. K.: IEEE Trans. Electr. Dev. *ED32*, 2097 (1985).
[30] Boris, J., Roberts, K. V.: J. Comp. Phys. *4*, 552 (1969).
[31] Hockney, R. W.: In: Methods in Computational Physics, Vol. 9 (Alder, B., Fernbach, S., Rotenbers, M., eds.), p. 135. New York: Academic Press. 1970.
[32] Varga, R. S.: Matrix Iterative Analysis. Englewood Cliffs, N.J.: Prentice-Hall. 1962.
[33] Young, D.: In: Survey of Numerical Analysis (Todd, J., ed.), p. 380. New York: McGraw-Hill. 1962.
[34] Lugli, P., Ravaioli, U., Ferry, D. K.: Monte Carlo Simulation of Contacts in Submicron Devices. In AIP Conf. Proc., No. 122, The Physics of VLSI, p. 162. Palo Alto: AIP. 1984.
[35] Prenter, P. M.: Splines and Variational Methods. New York: J. Wiley, 1975.
[36] Trabka, E. A., Burkey, B. C., Lavine, J. P.: Calculation of electrostatic potentials in charge-coupled devices using Hermite orthogonal collocation. In: Proc., Custom Integrated Circuits Conference, p. 50. Rochester, N.Y.: CIC. 1979.
[37] Ames, W. F.: Numerical Methods for Partial Differential Equations. New York: Academic Press. 1977.
[38] Saraniti, M., Lugli, P.: unpublished.
[39] Handbook of Tables for Mathematics, p. 640. Cleveland: CRC Press. 1975.
[40] Venturi, F., Smith, R. K., Sangiorgi, E., Pinto, M. R., Riccò, B.: In: Simulation of Semiconductor Devices and Processes (Baccarani, G., Rudan, M., eds.), p. 383. Bologna: Tecnoprint. 1988.
[41] Sangiorgi, E., Venturi, F., Riccò, B.: IEEE Trans. Comp. Aided Des. (to be published).
[42] Warriner, R. A.: Ph.D. Dissertation, Reading University, unpublished (1976).
[43] Yorston, R. M.: J. Comp. Phys. *64*, 117 (1986).
[44] Lugli, P., Ferry, D. K.: Physica *B129*, 532 (1985); Lugli, P., Ferry, D. K.: IEEE El. Dev. Lett. *EDL6*, 25 (1985).
[45] Antonelli, F., Lugli, P.: In: Proc., XVII European Solid State Device Research Conference ESSDERC87 (Calzolari, P., Soncini, G., eds.), p. 177. Bologna: Tecnoprint. 1987.
[46] Phillips, A., Price, P. J.: Appl. Phys. Lett. *30*, 528 (1977).
[47] Moglestue, C.: IEEE Trans. Computer-Aided Design *CAD-5*, 326 (1986).
[48] Moglestue, C.: IEEE Tran. Computer-Aided Design *CAD-4*, 536 (1985).

6 Applications

6.1 Introduction

In the present chapter, we will present a series of applications of the Monte Carlo method to the simulation of specific devices. It will become obvious that the Monte Carlo simulation of devices has followed a very different path than the one that has governed the actual evolution of the real devices. It was shown in chapter 4 that while the first steps towards the development of semiconductor devices for microelectronics were based on covalent semiconductors (Si and Ge), GaAs and other polar materials entered the scene only at a later stage. Furthermore, a large gap existed (and still exists) between the state of the art of fabrication, which is still dominated by Si C-MOS and bipolar structures, and the realm of "research and development", where GaAs fied-effect transistors with very sophisticated geometries and extremely reduced dimensions are currently produced. As pointed out in chapter 4, one exception is certainly found in the microwave area, where GaAs single devices are currently used.
On the other hand, the application of Monte Carlo algorithms to the study of devices starts much later, as an exercise of the computer scientists, directly on a GaAs MESFET [1].
Only several years later, other simulations started appearing in literature, and other devices were taken into consideration. Several reasons contribute to this delay. First of all, MC techniques were first used to set up realistic models for semiconductor materials and to obtain a detailed knowledge of the microscopic processing controlling carrier transport in such materials. Furthermore, the common belief (partially true) was that computers were, at that point, too slow for a full-device simulation.
As a general comment, it can certainly be stated that the Monte Carlo simulation of devices has taken gigantic steps forward within the last few years. In this, it has been helped by the advances in technology that have allowed the fabrication of structures more and more transparent from the physical point of view. The possibility of getting rid of spurius effects, and of being able to characterize the device performances directly from its physical processes, has opened the way for an extensive use of the Monte

Carlo as an instrument of modelling. In the near future, the availability of very fast and dedicated minicomputers with vectorized capabilities will provide a further incentive for the use of this technique. Furthermore, the possibility of coupling Monte Carlo and traditional simulators will allow the study of devices that are today out of the range of the Monte Carlo simulators.

A complete list of the major work done on the field is given in [2–78]. As it can be seen, several academic and industrial groups are currently performing Monte Carlo simulations of devices. The following sections will concentrate on specific aspects of each simulators.

6.2 Diodes

We start the presentation of specific examples of MC simulation of semiconductor devices from two terminal structures (diodes). The simulation of such structures is simple in the sense that it can be reduced to a one-dimension description. Nevertheless, the simulation can provide much useful information on the transport properties of more complex structures.

In the following pages we will examine two basic devices, namely an n^+-n-n^+ diode and a Metal-n-n^+ Schottky diode. Monte Carlo simulations of GaAs avalanche photodiodes [20, 21] and GaAs photodetectors [22] have also been performed.

6.2.1 n^+-n-n^+ Diodes

The general structure of an n^+-n-n^+ diode is shown in Fig. 6.1 [2]. The highly doped n^+ regions act as cathode and anode. Abrupt homojunctions are assumed for simplicity. The 1-D simulation domain is discretized according

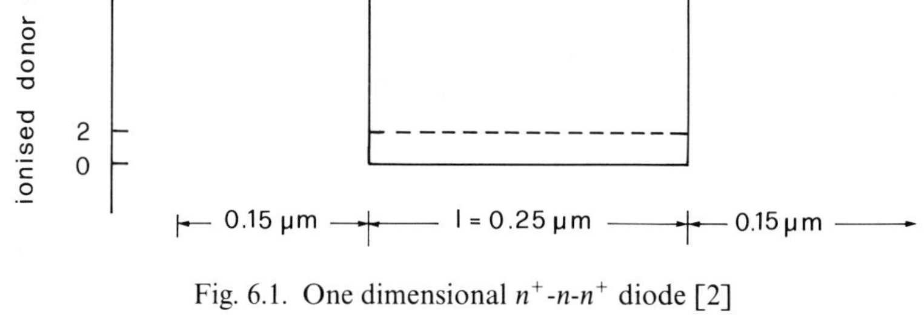

Fig. 6.1. One dimensional n^+-n-n^+ diode [2]

to the guidelines presented in the previous chapter. For the simple geometry examined here, the electric field can be obtained directly through the solution of the equation

$$E(x) = E(0) + \frac{4\pi q}{\kappa} \int_0^x [n(x) - N_D(x)]\, dx, \tag{6.1}$$

which, on the grid, assumes the discrete form

$$E_i = E_0 + \frac{4\pi q}{\kappa} \Delta x \sum_{j=1}^{i} [n(j\,\Delta x) - N_D(j\,\Delta x)], \tag{6.2}$$

where n and N_D denote, respectively, the electron and impurity concentration at a given position. The direct calculation of the electric field avoids the numerical errors involved in the finite difference derivation of the field from the potential obtained as a solution of Poisson's equation. The n^+-n-n^+ structure has been studied by several groups [2–6] in connection with ballistic transport.

A series of typical MC results for a GaAs submicron diode, with an intrinsic ($N_D = 0$) active layer of 0.25 μm surrounded by regions doped at $N_D = 2 \times 10^{17}$ cm^{-3}, are shown in Fig. 6.2 [2]. The lattice temperature is 77 K and the anode voltage V_a is 0.75 V. Several points are worth noticing. The free-electron distribution (Fig. 6.2.a) shows that the electrons diffuse from the doped regions into the intrinsic layer. The dipole of charge at the two interfaces induces a field that counteracts this tendency. When a voltage is applied to the structure, the potential drops mostly inside the intrinsic layer (c). A very high electric field is found (b), that reaches a maximum of 50 kV/cm near the anode. Carriers that enter the active layer overcoming the small potential barrier at the cathode are accelerated quasi-ballistically for about half of the intrinsic region. Correspondingly, the average electron velocity (in the x direction) increases up to a value of about 10^8 cm/sec. The following decrease is due to the electron transfer into the satellite valleys, and to backscattering from the anode n^+-layer (caused by impurity scattering). Several other issues were investigated in [2]. Raising the temperature up to 300 K, produces a reduction of the maximum velocity in the active layer (down to a value of 5×10^7 cm/sec), due to enhanced scattering. Increasing the length of the active layer also reduces the electron velocities, as the carrier motion moves from quasi-ballistic to collision-dominated conditions. As a final result, it was shown that a Si diode, with 0.25 μm active layer operating at 77 K, does not display, under the same bias conditions, the quasi-ballistic transport of the GaAs analogue.

One important point in the MC simulation of submicron structures involves the role of boundary conditions (BC's). In the situation presented in Fig. 6.2, the electric field drops rapidly (within a length of few hundred Å) as we move into the doped region. Carriers are then in thermal equilibrium at both ends of the simulation domain. Different schemes can be used to handle the BC's: for example, carriers reaching one boundary can be reinjected at the opposite end with the same energy and **k**-value (periodic

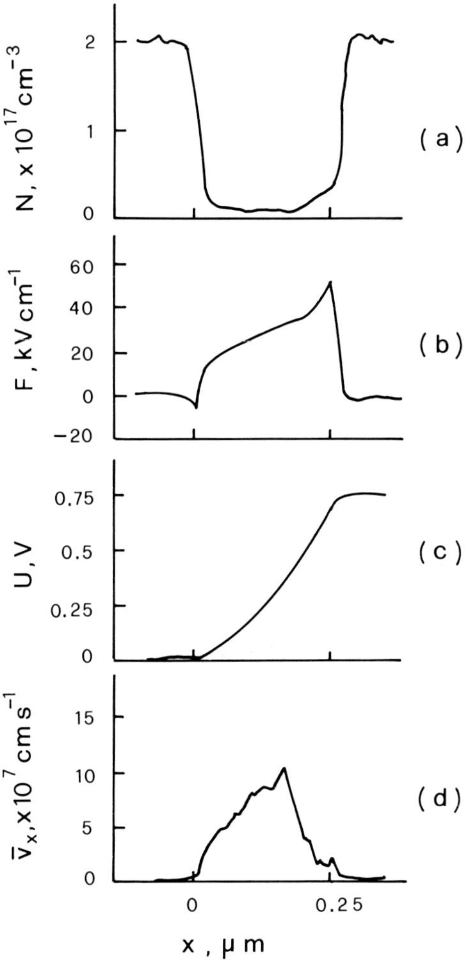

Fig. 6.2. Free-carrier concentration (*a*), electric field (*b*), potential (*c*), and velocity (*d*) profiles for a submicron GaAs diode with intrinsic layer of 0.25 μm [2]

BC's). In an alternative scheme, the existing carrier can be still reinjected at the other end, but with a **k**-vector taken randomly from a thermal distribution. Both strategies impose equal field at each contact and conservation of total number of carriers (and therefore of charge neutrality) during the simulation. They are certainly adequate for the situation examined in Fig. 6.2, where the contacts (i.e. the boundaries of the simulated structure) are far removed from the active layer of the device. Problems start when the contacts directly influence the device performance (an example will be seen in the next section). In such a case, alternative schemes have to be adoped, trying to control the current through the contacts and letting the number of simulated electrons fluctuate during the simulation [3]. An analysis of the effect of contacts on transport in submicron structures is given in [4].

Simulations of n^+-n-n^+ diodes have been performed to study noise properties of GaAs structures [5, 6] and the quasi-ballistic transport in GaAs [3, 7–10], InP [11] and AlGaAs-GaAs-AlGaAs [9–12] structures. The case of a p-type active layer in an n^+-p-n^+ diode has also been studied [13, 14].

6.2.2 Schottky Diode

Metal-semiconductor contacts are of great importance in a number of semiconductor devices, whose applications range from high-speed logic to microwaves. As the dimensions of these devices reach the submicron limit, contacts become the limiting factor in the performance in the ballistic or quasi-ballistic mode of operation. The standard theory of Schottky barriers becomes less accurate when the devices have submicron dimensions, since the typical relaxation times in the system are comparable to the transit time through the device. For this reason, the Monte Carlo simulation can give more reliable answers than other simulators.

A one-dimensional (1-D) model of a metal-n-n^+ structure, which is a realistic simplification of a Schottky diode, was presented in [15]. The simulation of such a structure is very interesting in that the device is never charge neutral, except under flat-band conditions. This is due to the presence of a depletion or an accumulation region near the interface. Since the value of the electric field at the two boundaries $x = 0$ and $x = w$ (see Fig. 6.3) is related through Gauss' law to the net charge inside the device, it is necessary to allow the

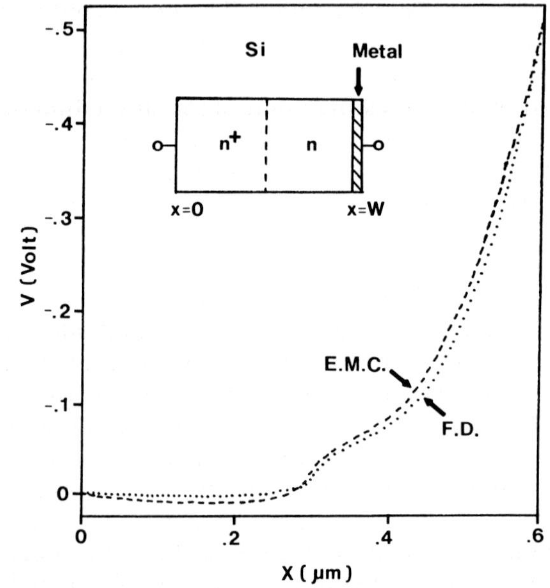

Fig. 6.3. Equilibrium potential distribution for a 0.6 μm W/Si Shottky diode, as obtained from the MC simulation (dashed line) and from Fermi-Dirac statistics (dotted lines) [15]

number of the electrons simulated to vary during the simulation, as a constant number of the electrons would give incorrect results. In this respect, the use of a cubic Hermite collocation method presented in section 5.4 to solve the 1-D Poisson's equation is crucial [16]. As explained there, in this method the potential is represented by an interpolation consisting of a linear combination of cubic spline functions, whose coefficients are the values of potential and electric field at each mesh point. The choice of uneven mesh spacing is allowed by the method without further complications. Boundary conditions are imposed by setting the potential on the two boundaries (at 0 and w), and interpolation constraints are set by assigning the charge density in two collocation points (Gaussian knots) inside each mesh interval. In the collocation method the values of potential and electric field on the mesh points are obtained simultaneously.

The knowledge of the electric field E_0 at the $x = 0$ boundary, gives us directly the value of the current J through that interface, through the relation $J = n^+ e\mu E_0$, where n^+ is the impurity concentration in the highly doped region, e is the electron charge and μ the low field mobility at that concentration. This relation assumes that the electrons are in thermal equilibrium with the lattice near the $x = 0$ boundary which is valid as long as the n^+ region width amounts to a few Debye lengths. The flux of electrons out of the device through the $x = 0$ boundary is controlled by the MC process. Thus, at every time step, the flux of the electrons in the opposite direction is given directly by the value of E (i.e. of J) at the interface. This procedure allows us to update constantly the numbers of electrons simulated, in accordance to the field distribution inside the device. In this way, the interface at $x = 0$ is modelled as a perfectly injecting contact. The metal contact at $x = w$ acts as an absorbing boundary: electrons with energy sufficient to overcome the barrier are injected into the metal (no image-force lowering is included in the simulation). Another interesting point of the simulation of the Schottky diode is the inclusion of tunnelling through the barrier [16].

The probability that an electron with energy ε at a distance x from the barrier will tunnel is given by [80]

$$T(\varepsilon) \cong \exp\left\{-2/\hbar \int_x^W dx\; 2m^*[qV(x) - \varepsilon]^{1/2}\right\} \tag{6.2.1}$$

where m^* is the effective mass, V the potential seen by the electron, and $\hbar$ is Planck's constant. As a MC electron reaches the barrier, the tunnelling probability is calculated from Eq. (6.2.1) using a parabolic least square interpolation of the potential $V(x)$ obtained from the solution of Poisson's equation. A random number is then used to decide whether the electron will tunnel or not. It is important to notice that no assumptions on the electron distribution function near the contact or on the shape of the potential barrier are needed. Quantum mechanical reflection and field emission from the metal into the semiconductor could be easily included along the same line.

The validity of the self-consistent algorithm was checked by comparing the equilibrium potential distribution (reflecting the shape of the conduction band) obtained from the MC simulation with the result of an iterative process which combines the use of Fermi-Dirac statistics and Poisson's equation. The charge distribution across the device is obtained from the Fermi probability and then integrated to give the corresponding potential distribution. The fact that the two methods agree so well (see Fig. 6.3) confirms the correctness of the MC (and of the boundary conditions used) simulating a Schottky diode. Fig. 6.3 corresponds to a 0.6 μ-long Si/W structure at room temperature, with an impurity concentration of 10^{16} and 10^{17} cm^{-3} for the n and n^+ regions, respectively, and electrics fields along the $\langle 111 \rangle$ direction.

Under high injection conditions, when a strong forward bias is applied, the barrier on the semiconductor side is greatly reduced and the electrons are easily injected into the metal. The potential and the charge distributions corresponding to a band bending of $3k_B T$, where T is the lattice temperature, are shown in Fig. 6.4 and the forward bias J-V characteristic is given in Fig. 6.5. The carrier concentration in the vicinity of the barrier is much higher than the one obtained by Baccarani and Mazzone [17] assuming a parabolic fixed shape for the potential distribution. This is probably due to the tailing of the electron distribution from the highly doped region and to the consequent deviation of the potential distribution from the parabolic form characteristic of the depletion approximation. The present method is suitable for studying forward bias conditions. Reverse bias would require the

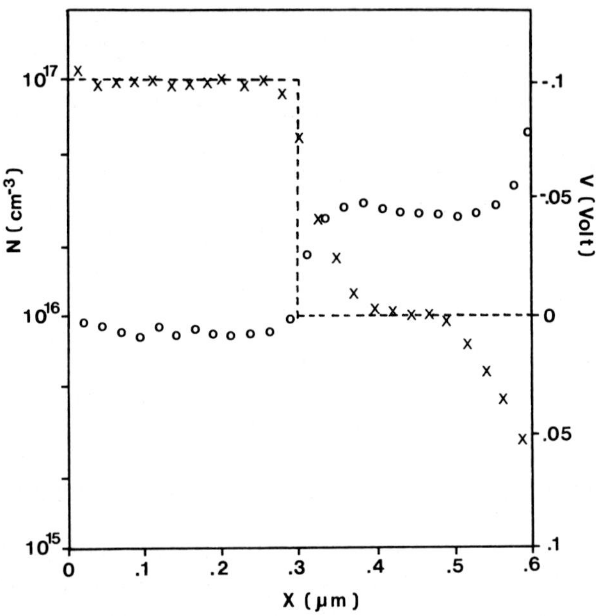

Fig. 6.4. Carrier (x) and potential (o) distributions under forward bias conditions for the W/Si Shottky diode. The dashed line shows the doping profile [15]

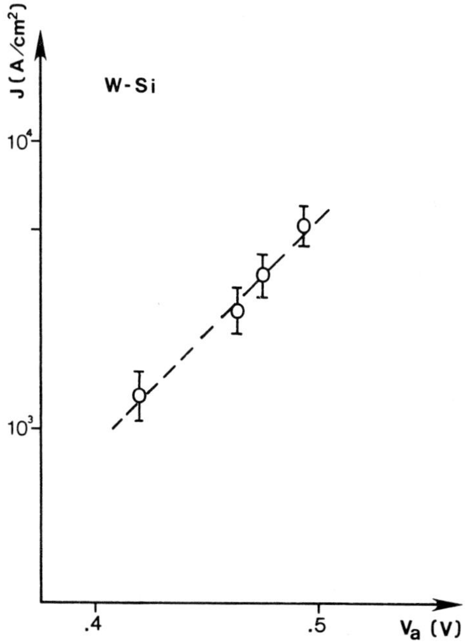

Fig. 6.5. Forward-bias J-V curve for the W/Si Shottky diode [16]

simulation of an exceedingly large number of particles, in order to avoid large fluctuations that would otherwise affect too much the reverse current. The same kind of MC simulation has been also implemented for a 0.7-μm Au/GaAs diode, with the same doping concentrations as above. The results for potential and electron distribution in equilibrium are presented in Fig. 6.6.

At room temperature, for the electron densities considered here, current is provided solely by thermoionic processes, while tunnelling is insignificant. Higher densities cause a narrowing of the barrier, thus increasing the tunnelling probability. Significant tunnel currents start to appear for electron concentrations above $5 \times 10^{18}\ \mathrm{cm}^{-3}$, in agreement with previous theoretical results [81]. This indicates that the method proposed here might be suitable also for the simulation of ohmic contacts.

A MC analysis of carrier transport in the space-charge region of a GaAs Schottky barrier has been presented in [17, 18]. These studies (performed on a fixed field configuration) showed that the electron distribution function deviates significantly from a Maxwellian near the metal-semiconductor interface.

Schottky barriers diodes are used as mixers up to frequencies of several hundred GHz. To study the frequency response of the diode, a Monte Carlo simulation has been performed [19] applying a voltage that varies in a sinusoidal way. Significant electron heating was detected at high frequencies (above 400 GHz).

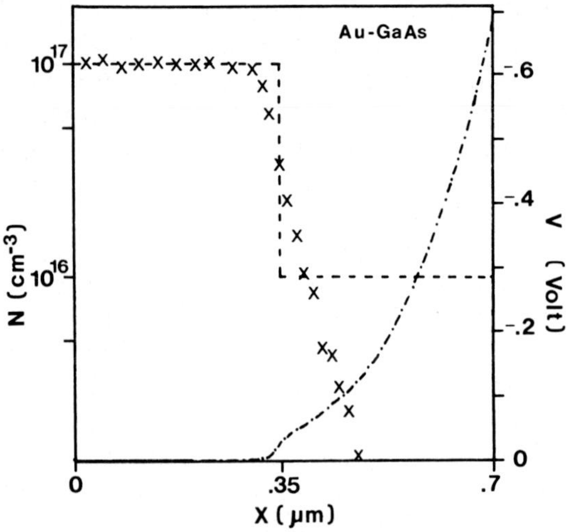

Fig. 6.6. Carrier (x) and potential (o) distributions under forward bias conditions for a 0.7 μm Au/GaAs Shottky diode. The dashed line shows the doping profile [16]

6.3 MESFET

The GaAs MESFET was taken in chapter 5 as the prototype for the illustration of M.C. simulators of devices. The choice was suggested by the fact that the physical model is a fairly simple one, at least as long as the transport properties of the carriers are concerned.

The complications arise if one tries to properly account for the presence of contacts. It should be already clear to the reader that the simulation of ohmic and injecting contacts is one of the major problems of MC simulations. In the pioneering work of the Reading' group [1], the condition of charge neutrality (number of free carriers always equal to the fixed charge) on the layer of cell adjacent to the contacts was assumed to accurately reproduce the ideal ohmic contact at the source and drain electrodes. The Schottky barrier at the gate was simulated simply by applying on the gate a potential equal to the barrier height of that characteristic metal-semiconductor interface. Although not completely satisfactory, such prescription has still been widely used in MC device simulations.

Several simulations of MESFET's have been presented after the work of Hockney's et al. [1]. From the physical point of view, the various simulations can be divided into two groups, depending on the GaAs model used (two [82] or three [82] valley model, or the full band diagram [84]). The scattering mechanisms are also taken from the two references just listed, and include non-equivalent intervalley ($\Gamma \leftrightarrow X$ or L for the two valley model, $\Gamma \leftrightarrow L$, $L \leftrightarrow X$, $\Gamma \leftrightarrow X$ for the three valley model), equivalent intervalley ($L \leftrightarrow L$ in the first case, $L \leftrightarrow L$ and $X \leftrightarrow X$ in the second), polar optical and acoustic phonon scatterings. Furthermore the programs include interaction with ionized impurities. It is hard to evaluate the results of a Monte Carlo simulation of a device in terms of the simple physical parameters of the program. The overall device acts in a sense as a filter, and the final output hardly reflects the individual contributions. The outcome of a simulation also reproduces such uncertainty, where all the single ingredients are intermixed to give a certain result (those with experience in modelling will feel very sensitive towards this issue).

Furthermore, it is currently not possible in most cases to give a direct comparison of theoretical and experimental I-V characteristics of a device. Important information is still missing (for example the parasitic effects of contacts and substrate), and a detailed model of the underlining physics would be necessary to account for such effects in a simulation.

Nevertheless, it is possible to check the general validity of a given adopted model, for example comparing the results of special simulations with experimental data. For instance, experimental values of drift velocity, diffusivity, luminiscence etc. in GaAs can be used to determine the effectiveness of the coupling between central and satellite valleys [83, 85], while a direct estimate of the energy separation between different valleys provides an important input for the simulation.

Once the physical model is carefully checked, other features of the simulation can be used as modelling tools. In the following we will give a brief review of most of the work that has been performed on GaAs MESFET. The specific issues dealt with by the different programs will be stressed and discussed.

In conclusion of the previous discussion, we would say that simulators using 3 valley models are certainly to be preferred, even if the differences between the two models are only important in the presence of very high electric fields ($\geqslant$20 KV/cm). Probably more important are the values of the coupling constants and valley separation, which strongly influence the transport properties of GaAs and the results of the simulation.

6.3.1 Short Channel Effects

It was shown in the previous chapter that the MC method is particularly suitable for the simulation of ultrasmall devices. On the one hand, all of the features of the non-stationary transport of such devices are naturally accounted for. On the other, the small dimensions allow the simulation of larger ensembles of carriers, which leads in turn to a reduction in the statistical uncertainty of the results. Submicron devices are usually characterized by the phenomenon of velocity overshoot: due to the very high fields present in the device, carriers can accelerate quasi-ballistically and reach a velocity well above the saturation velocity characteristics of that given material (see section 2.9). The effect of gate length on the average electron velocity on GaAs MESFET has been studied by Yoshii et al. [23]. The simulated device is shown in Fig. 6.7. The Monte Carlo simulation uses a 2

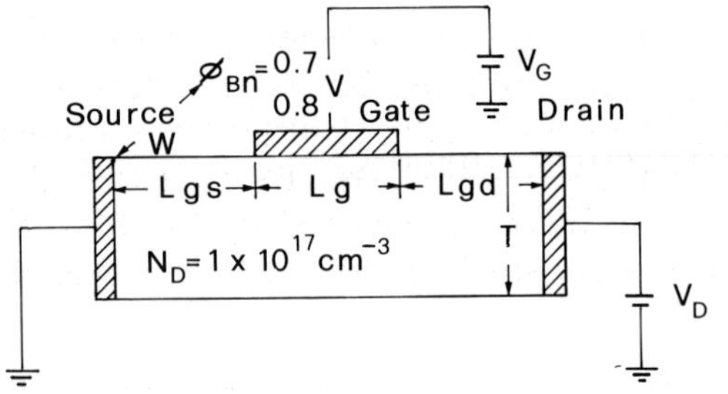

Fig. 6.7. Cross section of the simulated GaAs MESFET [23]

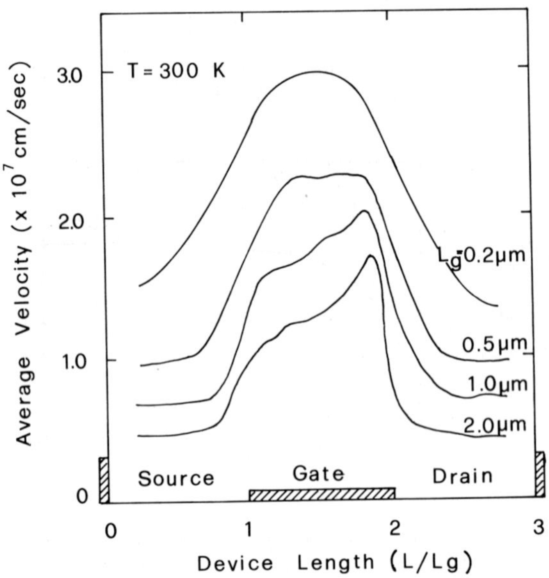

Fig. 6.8. Average electron velocity in the channel of a MESFET with different gate length at $V_D = 1.0$ V [23]

valley model for GaAs. Figure 6.8 illustrates the average electron velocity in the channel of the MESFET for various gate lengths. In this case $V_D = 1$ V and $T = 300$ K. The long devices ($Lg \geqslant 1$ μm) are characterized by stationary transport, with the carriers that are in quasi-equilibrium with the local fields in every point of the device. The maximum velocity ($\sim 1.8 \times 10^7$ cm/sec) in this case corresponds to the peak velocity of the static v-E curve of GaAs, and denotes the absence of intervalley transfer. As the device size is reduced, higher and higher velocities are reached, well above the peak value. As pointed out by Awano et al. [24], a remarkable feature of submicron devices is that very high electric fields can already be found near the source region (see for example Fig. 6.11). The high value of the field at the source-end of the gate is responsible for the almost ballistic acceleration of

the electrons as soon as they are injected into the channel region. The simplified geometry of the simulated device (with source and drain contacts on the lateral surfaces) does not alter the physical considerations, although they might be of importance in the comparison with real devices.
Unfortunately, the field increase achieved by the reduction of device size activates the transfer to the satellite valleys of the channel electrons. This is

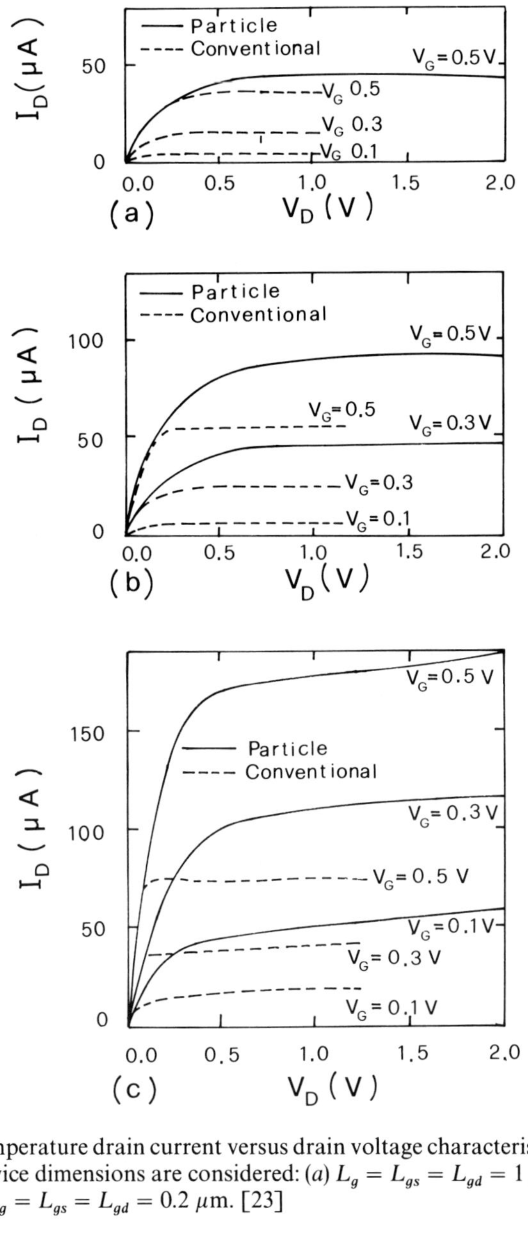

Fig. 6.9. Room temperature drain current versus drain voltage characteristics at various gate voltages. Three device dimensions are considered: (*a*) $L_g = L_{gs} = L_{gd} = 1\ \mu m$, (*b*) $L_g = L_{gs} = L_{gd} = 0.5\ \mu m$, (*c*) $L_g = L_{gs} = L_{gd} = 0.2\ \mu m$. [23]

clearly shown in Fig. 6.8 by the velocity curve corresponding to the shortest gate length (equal to 0.2 μm). The velocity reaches its maximum at the center of the channel, but drops before the end of the gate, because of the occurrance of intervalley transfer. It is then necessary to choose the device dimension carefully in order to obtain high velocities without suffering the performance degradation deriving from the intervalley transfer.

The drain current versus drain voltage characteristics obtained in [23] are shown in Fig. 6.9 for various device dimensions. It is clear that the non-stationary carrier dynamics have a very important effect on the current. With the gate length reduction, the transconductance is also greatly improved.

Based on their Monte Carlo procedure, Yokoyama et al. [25] have also suggested a new scaling scheme for GaAs submicron MESFET's, showing that ultrashort switching times (down to 2 ps for gate length of 0.2 μm) can be achieved by exploiting the overshoot phenomenon.

The properties of short-channel MESFET's were also studied by Awano et al. [24] using the geometry shown in Fig. 6.10. There the source and drain regions are heavily doped, and the gate is self-aligned (channel length equal to gate length 0.25 μm in this case). The model used in the simulation is the same as the previous one. The profiles of electric field E_x, average lateral velocity v_x and concentration N as a function of distance x from the source are presented in Fig. 6.11. The electric field on the source side of the channel is already much higher than the threshold field ($F_{th} \simeq 3$ kV/cm) for negative differential resistivity of GaAs. Such a high field can never be observed near the source of a long channel GaAs FET [86]. The average velocity reaches

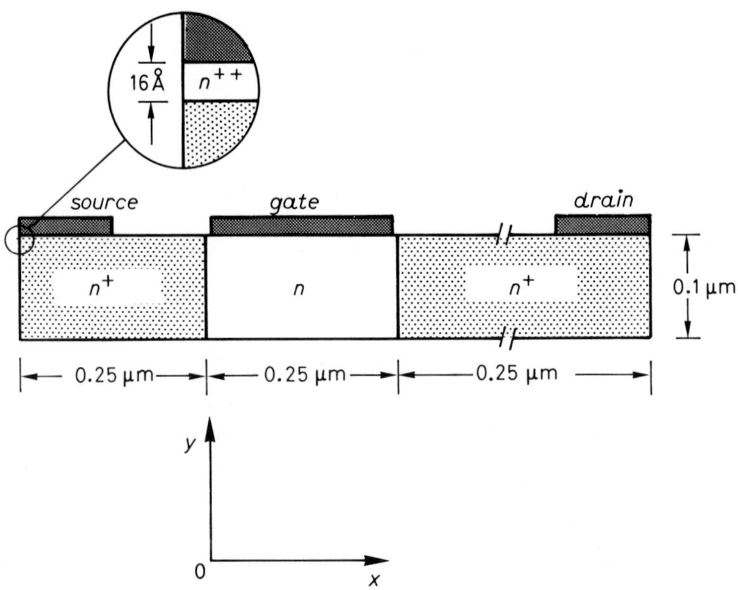

Fig. 6.10. Two dimensional structure of the simulated self-aligned MESFET [24]

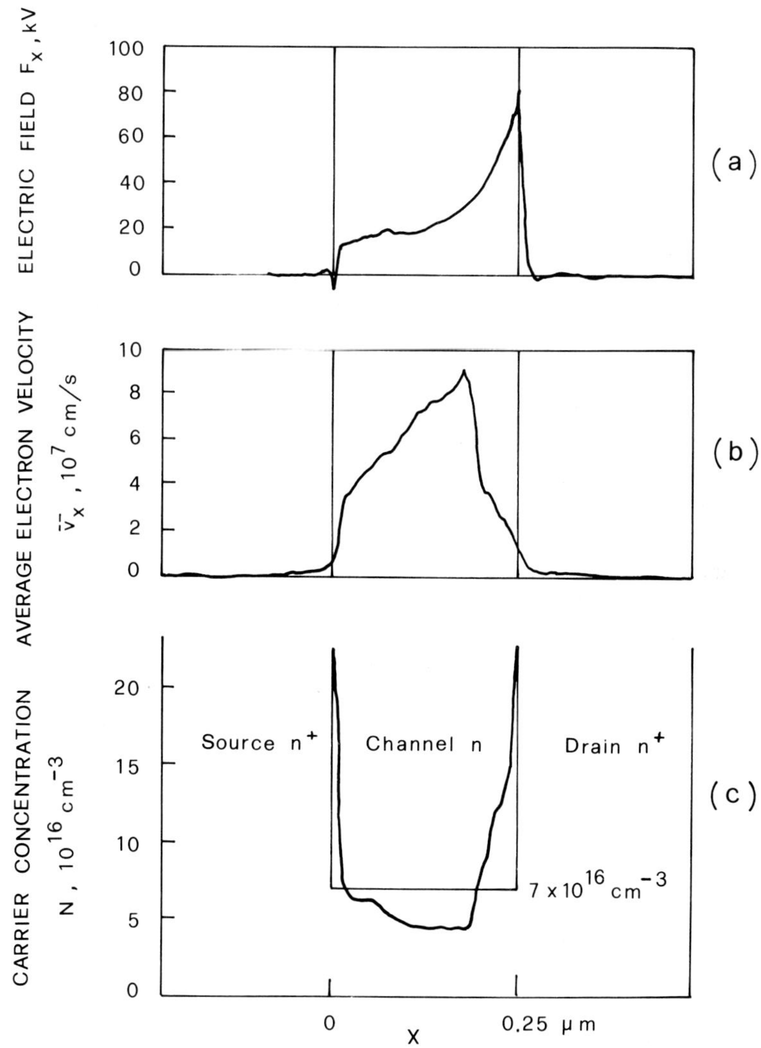

Fig. 6.11. (*a*) Electric field, (*b*) average lateral velocity, and (*c*) carrier concentration of a quartermicron MESFET [24]

about 3.5×10^7 cm/s at $x = 0.2$ μm, as a result of the quasi-ballistic acceleration caused by the steep potential variation. This in turn leads to extremely large drain currents. Near the drain the velocity exhibits a sharp decrease, due to the accumulation of *L*-valley slow electrons in that region. A large electron accumulation is also present near the source, caused by the electron spillover from the doped region. Such high concentration balances the rather small electron velocity to sustain the large drain current of the device.

Figure 6.12 shows the kinetic energy and lateral velocity distribution of the

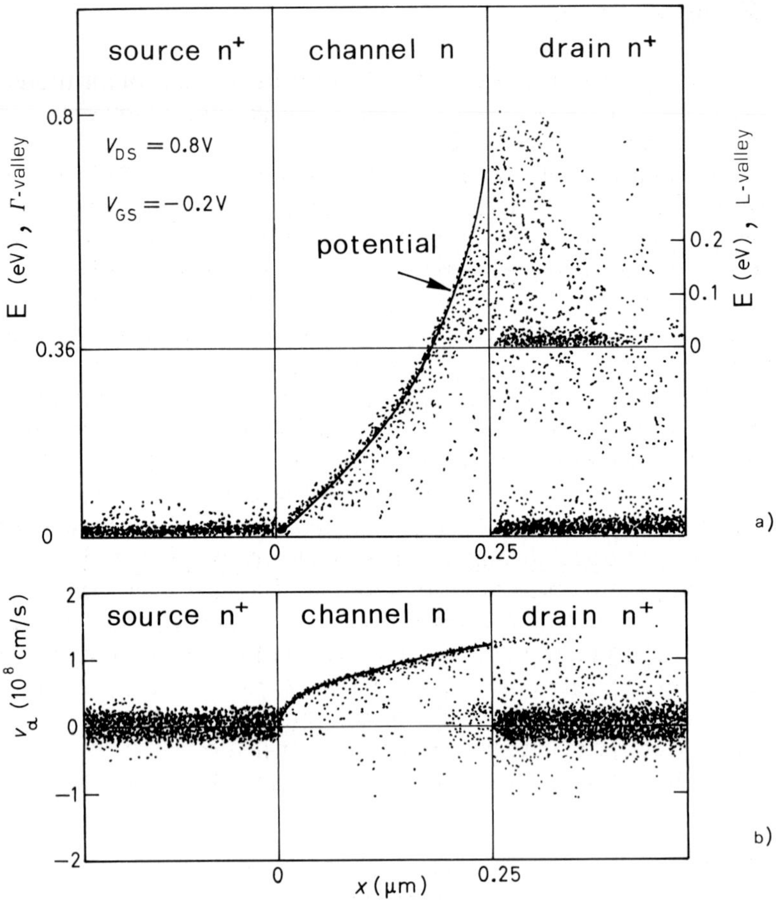

Fig. 6.12. Distribution of electron (*a*) energy and (*b*) lateral velocities in the simulated MESFET. The dots reported here are only representative of the original figure in [24]

channel electrons. In part (a) we can see that the carriers can be found mostly around the potential profile of the device (solid line). This indicates that only small fractions of the kinetic energy are lost in the scattering processes. A detailed observation of the distribution indicates the presence of L-valley electrons near the drain. Some of the low-mobility electrons are back-scattered by the impurities of the n^+ drain region and pile up at the end of the channel.

It has also been possible to analyze the current control mechanisms. In particular, by switching-off the intervalley transfer in the simulation, it has been shown that, contrary to long devices, such mechanism has only a limited influence on the saturation characteristics of the drain current of submicron structures. In addition, it has been found that the gate voltage controls the current through the modulation of the depletion layer depth as well as of the electron velocity.

6.3.2 Geometry Effects

The great versatility of the Monte Carlo programs as a tool for modeling is probably best demonstrated by the study of different geometries of GaAs FET on the device performance.

Using the simulator developed in Reading, Sanghera et al. [29] have studied the effect of the semi-insulating substrate on the noise properties of FET's. Warriner [30] had previously pointed out that the absence of the substrate provides better performances (better saturation of the drain current at increasing gate voltages, and higher transconductance). It was shown that, in order to obtain high values of transconductance, it is necessary to prevent the current from flowing through the substrate. The ideal structure, therefore, would display a high potential barrier at the lower boundary of the epilayer in order to contain the current flow within the epilayer. By using the structure shown in Fig. 6.13, Sanghera et al. presented a careful analysis of the effect of an additional *p*-type layer introduced between epilayer and substrate. Furthermore, they looked at the variation of the noise figures as a function of width and doping density of the various layers. It was found that the transconductance is extremely sensitive to the thickness of the epilayer and to its doping concentration. Although the *p*-type buffer leads in general to an improvement in the performance of the device over the structure with the substrate, only one particular geometry (with $d_e = 0.18\ \mu\text{m}$, $d_b = 0.20\ \mu\text{m}$, $d_s = 0.26\ \mu\text{m}$ and $n_b = 10^{16}\ \text{cm}^{-3}$) was found to compare favorably with the device with no substrate.

An additional important remark involves the type of simulated interface. In

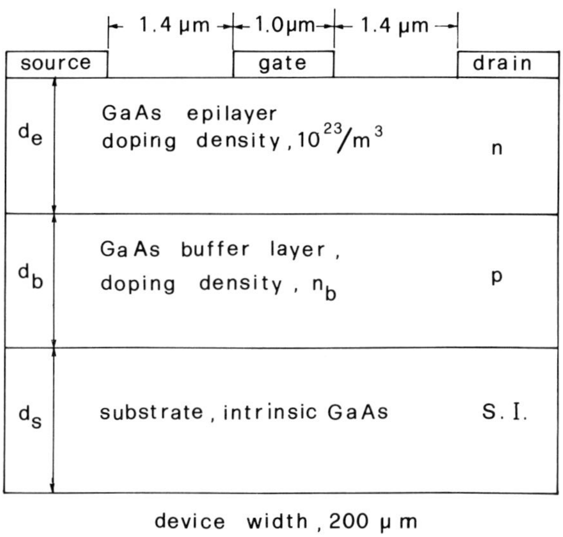

Fig. 6.13. Two dimensional structure of the simulated MESFET with buffer layer of width d_b [29]

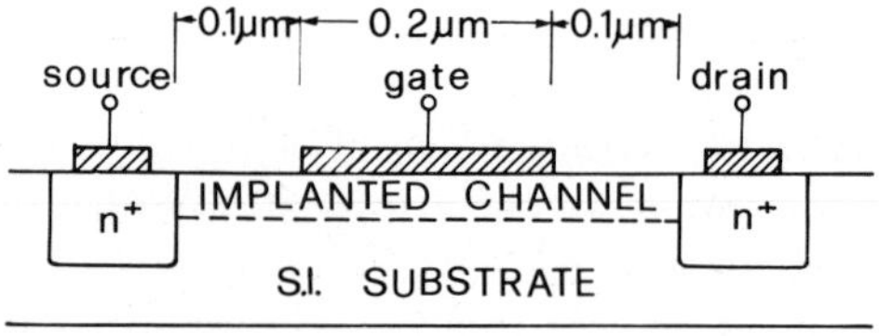

Fig. 6.14. Two dimensional structure of the simulated MESFET ion-implanted source, drain, and channel [32]

the original program of Hockney and co-workers, there is a sharply-defined step in the doping density going from the epilayer into the substrate, which reproduces an abrupt interface. Real devices, on the other hand, are characterized by interfaces that extend for at least a few manometers. Due to the strong sensitivity of the transconductance and of other parameters on the channel width, it is important to study the general effects of the doping profile. Moglestue [31] analyzed the performance of several FET's with various smooth doping profiles and an ion-implanted one, comparing them with the ideal abrupt-interface one. The Monte Carlo results (obtained with the original Reading program) indicate large variations of the transistor figures at different bias levels. An important conclusion was that ion-implanted devices exhibit the lowest noise figure and the highest maximum power gain.

Another example of MC simulation of FET with an implanted channel is provided by the work of Williams et al. [32]. In this case, a three-valley model is used to simulate the submicron structure of Fig. 6.14. The doping profile of the channel region, as well as of the heavily doped n^+ contact regions, are shown in Fig. 6.15. The treatment of the boundary conditions and the Poisson's solver for the actual simulated device follows the prescriptions set by Hockney et al. as described in the previous sections.

The equipotential contours of the MESFET for $V_{gs} = 0$ and $V_{Ds} = 2$ V are presented in Fig. 6.16. Most of the voltage drop occurs between gate and drain, however high fields can be found along the whole channel. As pointed out earlier, such high fields can substain high velocities in the gate region. The present simulation indicates that an average carrier in the channel exhibits velocity overshoot for distances of the order of 0.2 μm. An estimate of the cut-off frequency derived from the transit time gives a value of f_T equal to 230 GHz for $V_{DS} = 1$ V and $V_{gs} = -0.5$ V. As a reference, it should be remembered that the device simulated by Awano et al. [24] gave a value of f_T of 160 GHz.

To conclude the brief discussion of the geometry effect, it is worth mentioning that the influence of particular gate profiles can be studied with the M.C. procedure. For instance, Fig. 6.17 shows some of the structures studied by Moglestue [35, 36]. In the simulation, the stray fields from the metal gate have been introduced to study the effect of gate trench shape and metalization on the interval fields and on carrier dynamics. Details of this particular work and of other interesting papers can be found in the References.

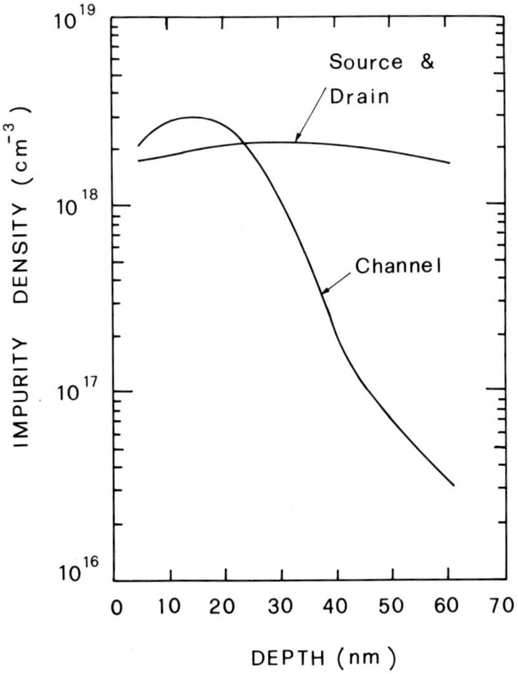

Fig. 6.15. Impurity density profile of the ion-implanted MESFET [32]

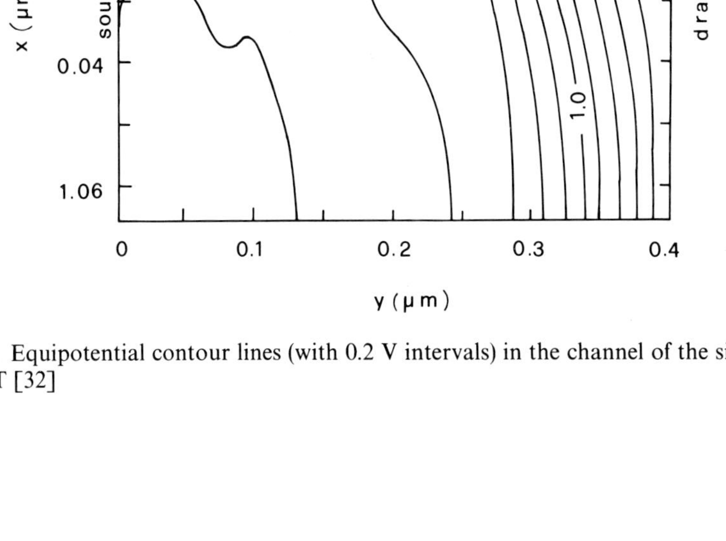

Fig. 6.16. Equipotential contour lines (with 0.2 V intervals) in the channel of the simulated MESFET [32]

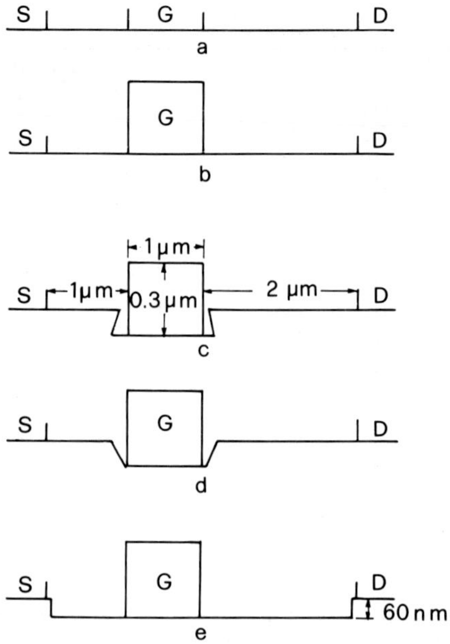

Fig. 6.17. Various gate profiles for MESFETs [35]

6.3.3 *Space-Charge Injection FET*

A symmetrical n^+-i-n^+ FET structure has been simulated with a 3-valley Monte Carlo program by Fauquemberg et al. [37]. Such a transistor structure was originally proposed by Eastman [87]. The main feature is a submicron undoped active layer in which carriers are injected by neighbouring n^+ layers. As shown in Fig. 6.18, the source and drain contacts are deposited on the n^+ layers (here at 10^{17} cm^{-3}), while the two opposed gates are located on the active layer. The electron distribution indicates a large depletion region under the gates. Thus, the electron flow is confined to a narrow channel in the middle of the undoped region of the device. Very high velocities can be obtained (up to 7×10^7 cm/s for 0.3 μm active layer), which lead in turn to a predicted cut-off frequency of 400 GHz.

6.3.4 *Conclusions*

As shown by the previous examples, GaAs MESFET's have been widely studied. Several specific features of the devices have been investigated by Monte Carlo methods, achieving important information on the principles of operation of the device and on possible optimizations.
There seem to be three main problems still outstanding:

a) The simulation of the contacts: as already seen, all of the simulators presented in this section make use of the boundary conditions for source

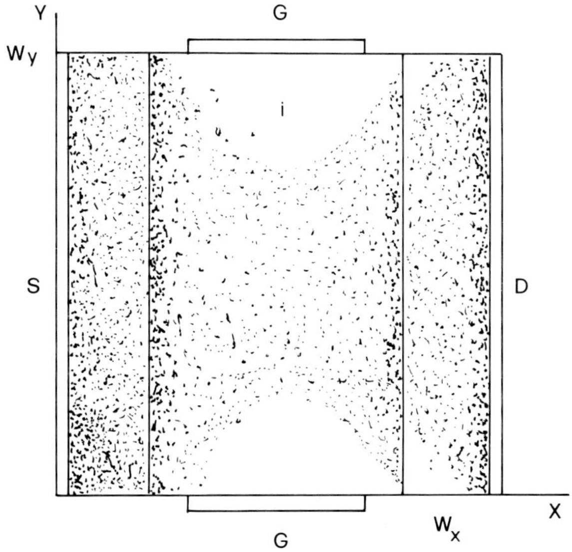

Fig. 6.18. Electron distribution in a n^+-i-n^+ FET structure, with $W_x = W_y = 0.75\,\mu$m, $V_{DS} =$ 1 V, $V_{GS} = -0.1$ V. The dots reported here are only representative of the original figure in [37]

and drain proposed by Hockney and co-workers. A detailed study of the influence of the proposed conditions on the overall performance of the devices should be performed. It would also be an advantage to be able to model the contacts directly, for instance extending the simulation presented in section 6.2.

b) The simulation of heavily-doped regions: the simulation of devices is usually restricted to doping densities well below 10^{18} cm^{-3}. In this case it is possible to use the semi-classical Monte Carlo scheme, where no correlation exists among carriers. Higher densities and lower temperatures require the use of degenerate statistics, that is the inclusion of Pauli exclusion principle in the simulation via the algorithm described in section 5.5. Furthermore the simulation of high electron densities, when carrier densities exceed 10^{17} cm^{-3}, the interaction among electrons can become important. Then, the electron-electron interaction should be explicitly accounted for in the simulation.

6.4 HEMT and Heterojunction Real Space Transfer Devices

As we have seen in chapter 4, the High Electron Mobility Transistor (HEMT) can be taken as a prototype of planar heterostructure devices. The main feature is given by the AlGaAs/GaAs single heterojunction, that creates the potential well that confines the electrons in the so-called two-dimensional electron gas. The contacts are usually laid as in the traditional MESFET, with the additional problem in that it is necessary to create a direct channel from the source and drain contacts to the conducting channel in the GaAs, in order to avoid large currents through the doped alloy and high contact resistance.

From the simulation point of view, many additional complications arise with respect to the MESFET case:

a) The device presents regions where quantization of the electron motion is important (channel) as well as regions where the three-dimensional characteristics are prevalent. It is necessary in principle to use an algorithm capable of handling both situations, that is a two-dimension quantized description of the electronic motion (as described in section 2.10) in the channel, a three dimension description for the overall device, and a possibility of transfer between the two.
b) When current flows in the device, the width of the channel is going to widen going from source to drain. The 2D properties of the electron gas would also change (subband spacing, wave-function). Thus the scattering rates should always be calculated self-consistently with the electron distribution in the quantum well at different points in the channel.
c) As the electrons confined in the well are heated up by the field, they might achieve enough energy to surmount the barrier between GaAs and AlGaAs. An accurate model is then required for the real space transfer.

We will see in the following section that different groups have recently started tackling some of these problems, some of which are still unresolved due to the lack of physical models to describe the microscopic processes (this is certainly true for the points a and c listed above.)

6.4.1 HEMT

One approach that has been taken is to ignore the two-dimensional nature of transport along the channel, and run a complete 3D k-space simulation in the same self-consistent manner as for MESFET's. Figure 6.19 shows the device simulated in this manner by Wang et al. [41]. The presence of the potential discontinuity at the heterojunction is handled in the Poisson solver by simply adding the appropriate band gap discontinuity (here 0.34 eV) on the interface mesh point. For electrons crossing the heterojunction, the conservation of energy and momentum parallel to the interface are required. Charge neutrality is assumed in the regions adjacent to the source and drain contacts in the way proposed by the Reading group (see [88]) as a boundary condition for the electron concentration. As for the MESFET simulations examined in the previous chapter, Dirichlet and Von Neuman boundary conditions are chosen respectively for the metal-semiconductor and for the free surfaces of the device.

The ensemble Monte Carlo model incorporates a complete Γ-L-X structure for both GaAs and AlGaAs layers. The scattering mechanisms used in the

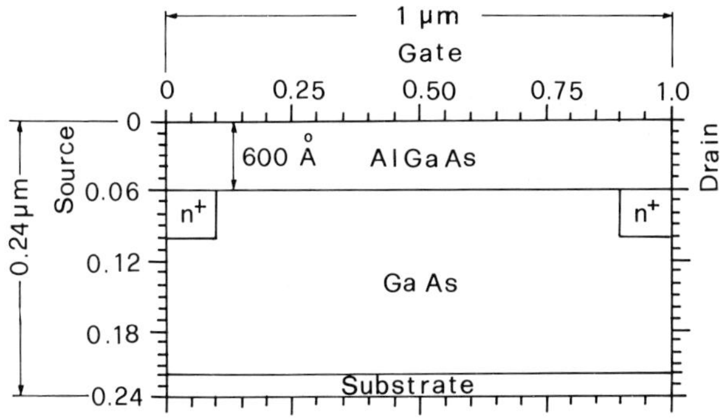

Fig. 6.19. Two dimensional structure of the simulated HEMT. A doping concentration of 2×10^{17} cm^{-3} is used for the n-AlGaAs and the n^+ GaAs contacts [41]

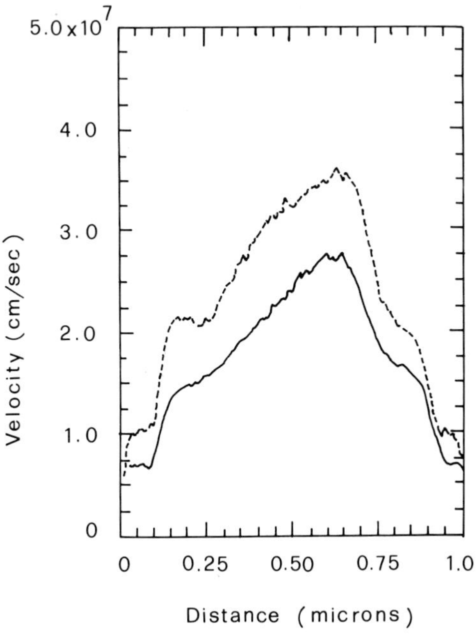

Fig. 6.20. Profile of the electron velocity at 300 K (solid line) and 77 K (dashed line). The effective gate bias is -0.35 V (including the Shottky barrier height) and the drain bias is 0.5 V [41]

simulation include polar, optical, acoustic, intervalley and ionized impurity scatterings. Impact ionization and electron-electron interactions are also accounted for. Figure 6.20 shows the electron velocity profile in the GaAs channel. A significant overshoot is detected, which increases with decreasing lattice temperature (the maximum velocity under the gate is 2.8×10^7 cm/sec at 300 K and 3.7×10^7 cm/sec at 77 K). The higher velocity reached by the carriers at lower temperatures are a consequence of the reduced impurity scattering achieved by modulation doping. In fact, when the impurities are removed from the channel, a reduction in temperature causes a direct increase in mobility as a result of the reduced efficiency of the polar optical scattering. Both real space transfer and intervalley transfer act to restrict the velocity overshoot. By varying the band-gap discontinuity, Wang et al. were able to show that the k-space transfer to upper valleys tend to dominate if the interface barrier potential is comparable to the intervalley separation.

The influence of technological parameters (GaAs and AlGaAs layer thickness, doping level, aluminium composition and gate length) on the HEMT performance have been studied by Fauquemberg et al. [42]. The approach is analogous to the previous one, since the simulation couples a 3 valley-bulk model for GaAs and AlGaAs with a traditional Poisson's solver. By looking at an inverted HEMT structure with a recessed gate, the authors found that the maximum transconductance value g_m increases when the GaAs layer thickness is reduced (in Fig. 6.21 g_m, at room temperature, goes from 750 mS/mm to 1350 mS/mm moving from a 750 Å down to a 150 Å layer), and when the temperature is lowered (in Fig. 6.22, $g_m = 2200$ mS/mm at 77 K for the 125 Å layer). The gate length in both cases is 0.3 μm. From Fig. 6.21 it can be seen that the transconductance is low (around 100mS/mm) for large negative bias voltages V_{gs}, and varies slowly with V_{gs}. In that condition, the two-dimensional electron gas is fully depleted, and conduction only occurs in the low mobility AlGaAs layer. On the contrary, for a small negative bias, the main conducting channel is through the GaAs quantum well, and g_m displays a very steep variation with V_{gs}. Such a high slope can lead to very interesting developments in high speed logic, leading to very

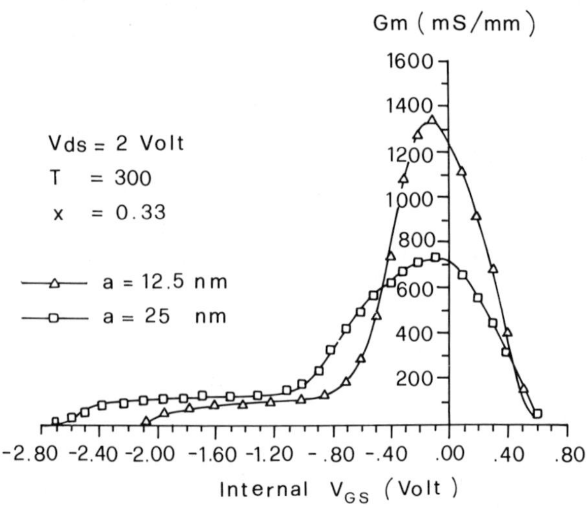

Fig. 6.21. Transconductance versus gate-to-source voltage at room temperature for two HEMTs with different layer thickness at 0.3 μm gate length [42]

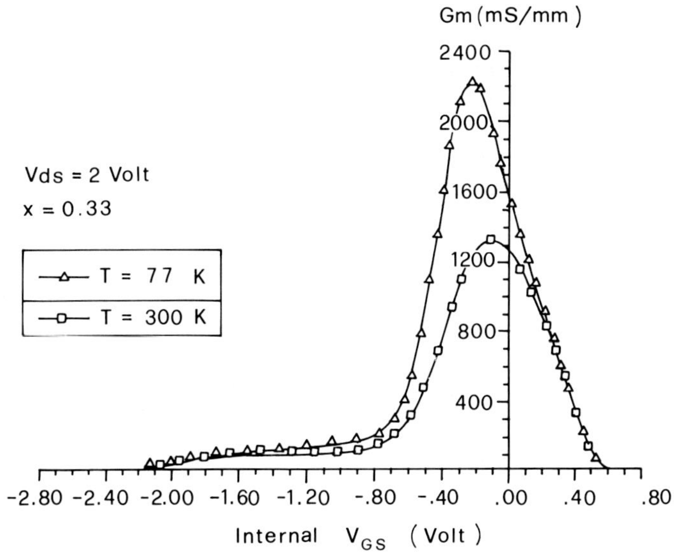

Fig. 6.22. Transconductance versus gate-to-source voltage at two different temperatures for a 0.3 μm-gate HEMT, with 12.5 nm active layer thickness [42]

short switching times. At room temperature, the maximum transconductance decreases at decreasing gate lengths Lg ($g_m = 2000$ mS/mm for $L_g = 0.77$ μm and $g_m = 1350$ mS/mm, $L_g = 0.3$ μm), evidencing a more efficient gate control for the longer devices. The degradation of performance in the ultrashort device might also be an indication of increased intervalley (or real space) transfer. Since the gate to source capacitance C_{gs} increases at longer gate length, the cut-off frequency (proportional to the ratio g_m/C_{gs}) increases in smaller devices, with a maximum value of 100 GHz at room temperature for $L_g = 0.3$ μm.

A simulation of HEMT's including size quantization was performed by Tomizawa et al. [43, 44]. In their simulation, the width of the channel for the device shown in Fig. 6.23 was taken as a parameter. No self-consistency was introduced, and the electrons moved in a given fixed potential distribution. Two-dimensional scattering rates for the channel electrons were used, calculated by relating the density of the two-dimensional electron gas to the constant transverse electric field. A three-dimensional scattering rate was used for the other regions of the device. Figure 6.24 represents typical drain current I_d versus drain voltage V_D characteristics, at various gate voltages V_g for two different temperatures. It is again clear that much higher currents, and higher transconductances, are obtained at low temperatures. The gate length considered here was 1.2 μm.

In agreement with Fauquemberg et al. [42], it was also found that the maximum transconductance increases when the AlGaAs layer width is reduced. This improvement is related to the increased gate control obtained when the gate electrode and the conduction electrons are closer together.

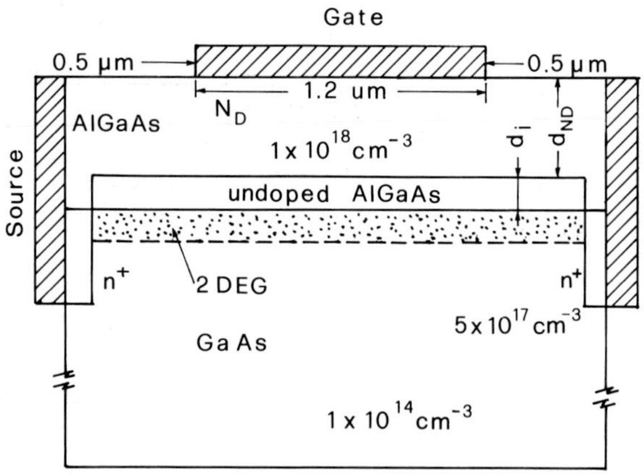

Fig. 6.23. Two dimensional structure of the simulated HEMT [43]

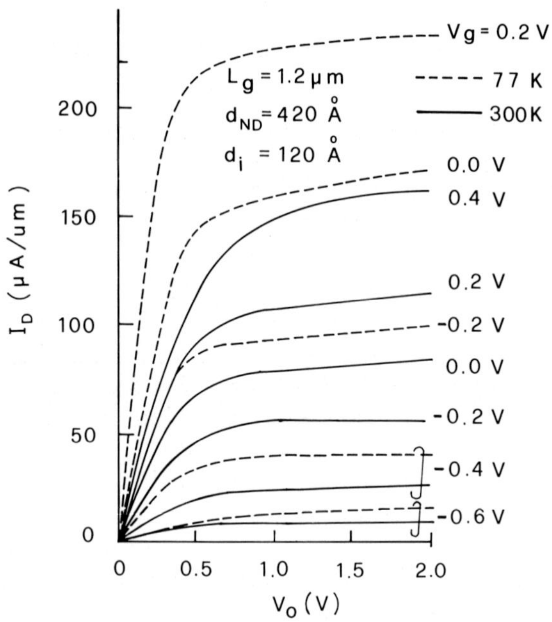

Fig. 6.24. Drain current versus drain voltage characteristics at various gate voltages for nitrogen (dashed lines) and room temperature (solid lines) operation [43]

The first self-consistent simulation of HEMT structures including size quantization was performed by Ravaioli et al. [45–48]. The simulated structure is presented in Fig. 6.25. The gate length is 0.3 μm, and the total thickness of the AlGaAs layer is 600 Å, including a 100 Å undoped spacer layer. The doping concentration of the AlGaAs and of the n^+ contact regions in GaAs is taken to be 2×10^{17} cm^{-3}. The remaining GaAs regions are considered undoped. The information about the difference in the equilibrium Fermi level of the

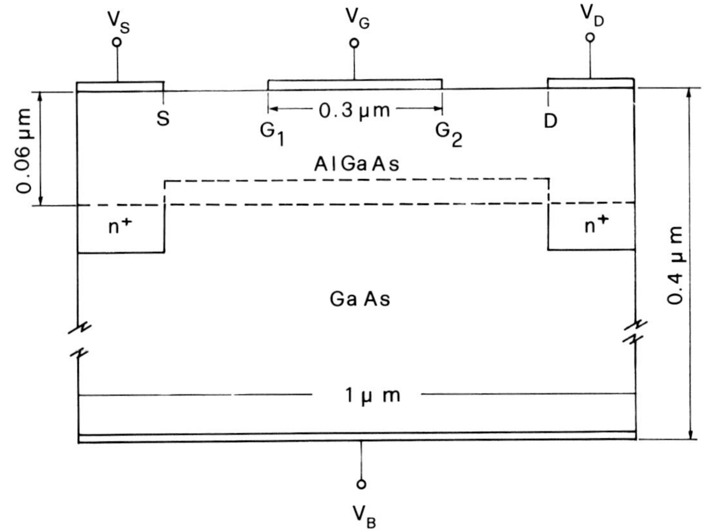

Fig. 6.25. Two dimensional structure of the simulated HEMT [48]

two materials is introduced in the model including a back contact at the bottom of the device. The surface voltage applied is used as a boundary condition for Poisson's equation in order to properly reproduce the shape of the conduction band in the GaAs regions.

Both three- and two-dimensional electrons are considered. The 3D electrons in GaAlAs and GaAs are treated using a complete three-valley non-parabolic model which includes, in addition to intervalley and intravalley scatterings, polar optical, acoustic and ionized impurity scatterings. The quantum well is treated in the triangular approximation, in which only the two lowest subbands are considered for electron conduction. Inter and intra-subband transitions via polar optical and acoustic scattering are considered. The two dimensional scattering rates are related to the local subband levels, which are in turn determined by the local carrier concentration in the channel. In order to account for the effect of the channel non-uniformity the following procedure has been devised: at the end of each field-adjusting time step, the local charge density in the channel is determined, and the subband level spacing is calculated in the triangular well approximation. The scattering rates are then obtained using the wave functions proposed by Ando et al. [89], and the scattering event is randomly chosen. Use is made of the self-scattering algorithm, with a maximum total probability determined at the beginning of the simulation using a reference surface charge density in the channel never to be exceeded during the simulation. Such maximum probability is used to calculate the self-scattering rate at every stage and to normalize all scattering probabilities. Transitions between two- and three-dimensional states in the GaAs are also allowed, under the requirement of conservation of energy and momentum parallel to the interface.

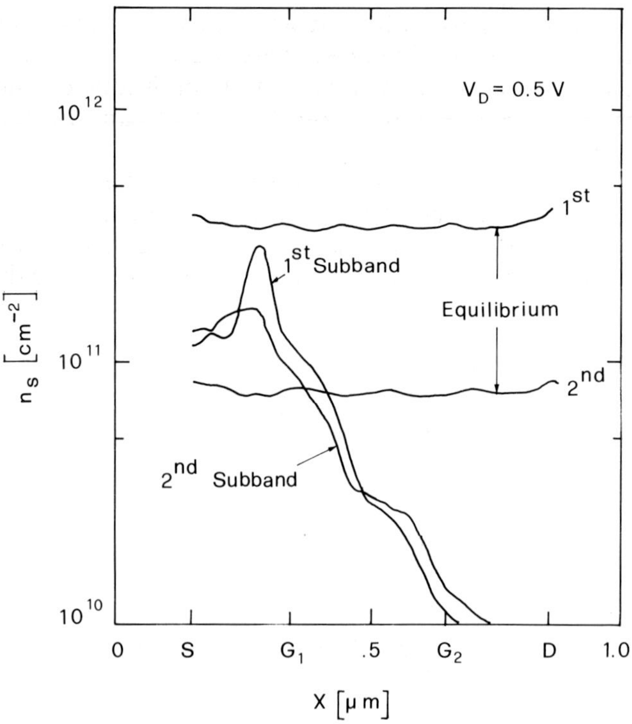

Fig. 6.26. Surface electron densities in the two lowest subbands for $V_D = 0.5$ V [46]

In the solution of Poisson's equation, a finite difference three-level multigrid algorithm is used. The method is based on a series of partial iterative relaxations on grid with increasing mesh size. The intermediate solution is passed between levels using an interpolation scheme. Such an algorithm is ideal for an extension to local refinement schemes, since finer grid levels can be introduced in any region of the device.
The electron distribution along the channel in the two subbands is presented in Fig. 6.26 for a drain voltage of 0.5 V. The subband populations are calculated with the M.C. simulation starting from an analytical initial condition given by Delagebeandeuf et al. [91]. In the portion of the channel to the left of the gate there is a substantial accumulation of electrons in the higher subband, due to intersubband transfer. As we move towards the drain, the occupation of the subbands drops rapidly, since the carriers are heated up by the high field along the channel and tend to escape the quantum well (here the barrier from GaAs to AlGaAs is taken equal to 0.25 eV). At $V_0 = 0.5$ V, the total subband population is found to be less than 15 percent of the equilibrium one. At increasing drain voltages, a high carrier density is found in the GaAs and in the AlGaAs. In these conditions, a significant part of the drain current tends to flow through a stray path in the doped AlGaAs layer, thus leading to a degradation in the device performance. The

parallel channel that opens in the AlGaAs can lead to negative differential resistance, as seen in the drain current versus drain to source voltage (Fig. 6.27). The curve corresponding to flat-band conditions on the gate is included there for reference. In this case the current can flow freely through the AlGaAs layer.

A similar simulation based on a self-consistent Monte Carlo program has been presented by the group at the University of Paris, Orsay [49–52]. It uses a 3-valley GaAs model, with bulk scattering rates for both GaAs and AlGaAs. The simulated structure is shown in Fig. 6.28. The drain charac-

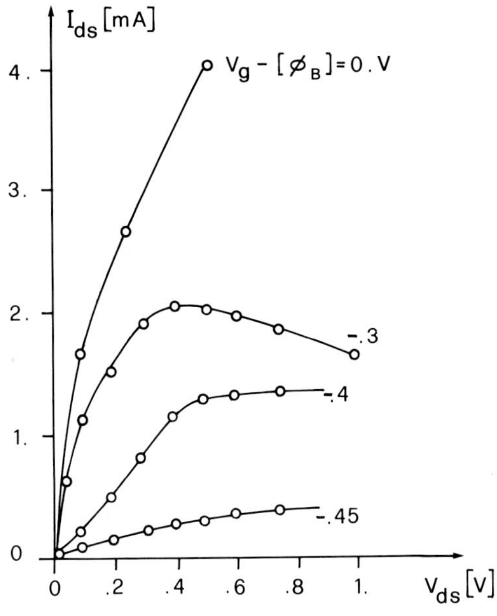

Fig. 6.27. Drain current versus drain voltage characteristics for various gate voltages [46]

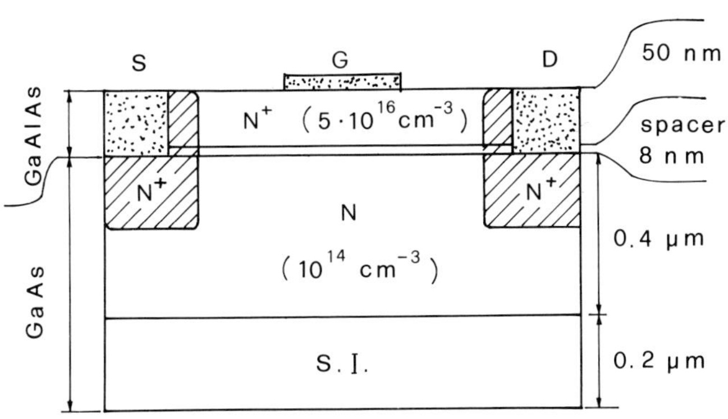

Fig. 6.28. Two dimensional structure of the simulated HEMT. The gate length is 0.1 μm or 0.5 μm [49]

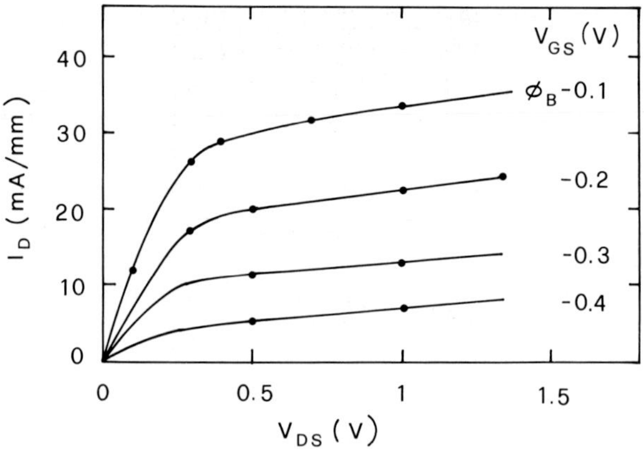

Fig. 6.29. Drain current versus drain voltage characteristics at various gate voltages for the 1 μm HEMT [51]

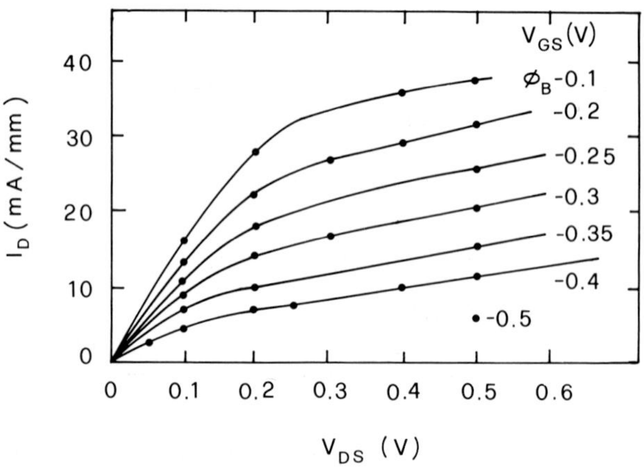

Fig. 6.30. Drain current versus drain voltage characteristics at various gate voltages for the 0.5 μm HEMT [51]

teristics are given in Fig. 6.29 and 6.30, respectively, for a 1 μm and a 0.5 μm channel devices. It was found that the transverse field vanishes near the end of the gate, implying that in the longer device channel pinch-off is the reason for current saturation. A different situation characterized the shorter device, due to the occurrence of both *k*-space and real-space transfer near the drain, as evidenced by the spatial electron distributions plotted in Fig. 6.31. The calculated transconductance are 115 mS/mm and 130 mS/mm, respectively, in good agreement with measured values for similar devices.

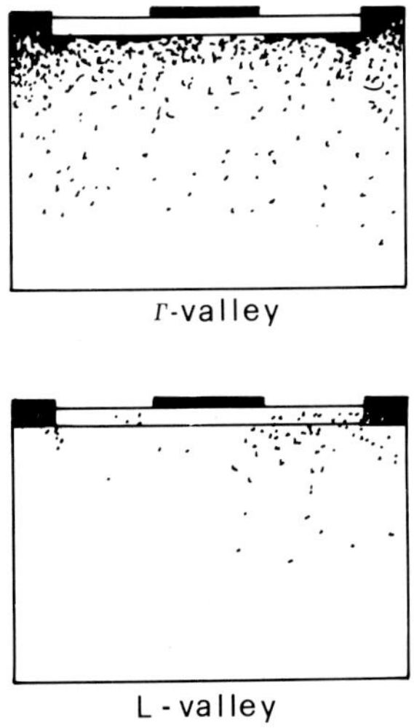

Fig. 6.31. Electron spatial distribution in the Γ and L valleys for the 0.5 μm HEMT. The dots reported here are only representative of the original figure in [52]

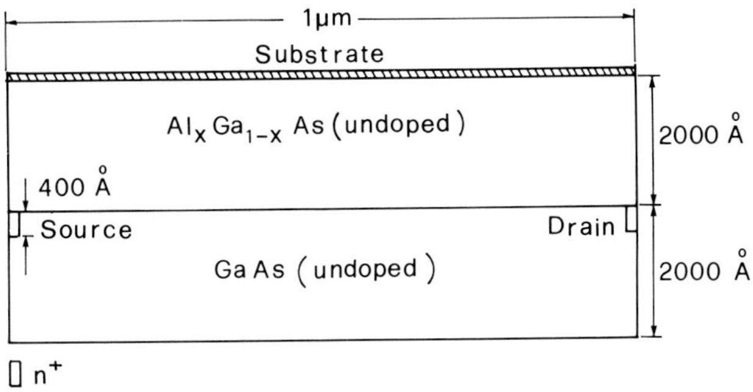

Fig. 6.32. Two dimensional structure of the simulated real-space transfer device [53]

6.4.2 *Real-Space Transfer Devices*

The possibility of building devices that exploit the real-space transfer phenomenon has been recently investigated by the group at the University of Illinois, Urbana [53, 54]. The self consistent simulation described previously has been applied to the structure of Fig. 6.32, which is a prototype of the negative differential resistance field effect transistor (NERFET) and charge injection transistor (CHINT) proposed by Luryi [92].

The current-voltage characteristics of the NERFET at 77 K is shown in Fig. 6.33. At small drain biases, the device is in its linear regime, with a negligible substrate current. The electrons are confined in the GaAs layer and the

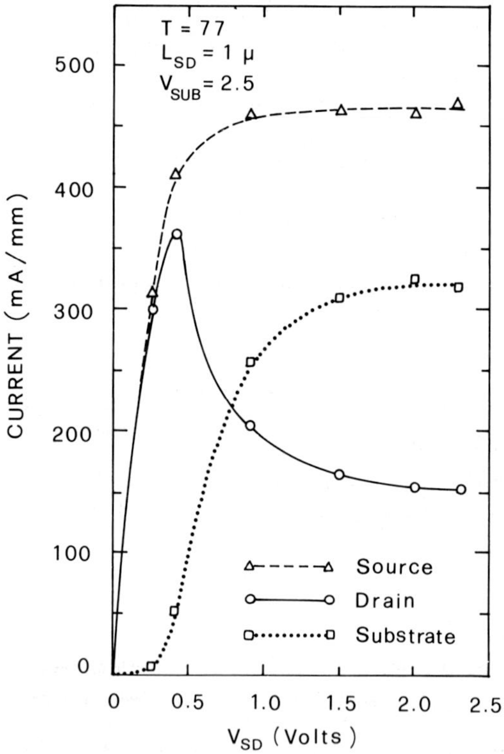

Fig. 6.33. Current-voltage characteristics of the real-space transfer device at 77 K with a bias of 2.5 V [53]

AlGaAs is depleted. When the drain voltage is increased to values above 0.4 V, real-space transfer starts occurring. The drain current exhibits a sharp decrease (negative differential resistance) as only a fraction of the electrons emitted at the source electrode are collected by the drain electrode. A peak-to-valley ratio of 2.5 is obtained for a substrate voltage of 2.5 V. Current saturation is reached at higher drain voltages.

6.4.3 *Velocity-Modulation Field Effect Transistor*

The same Monte Carlo program as section 6.4.2 has been used to simulate the velocity-modulation transistor (VMT) proposed by Sakaki [93] as a possible candidate for subpicosecond switching times. The simulated VMT (Fig. 6.34) [54–56] consists of a double *n*-AlGaAs/GaAs/*n*-AlGaAs heterostructure. Two parallel channels are found at the two interfaces in the GaAs. One of the channels (*A*) is heavily doped, and therefore exhibits a low mobility, while the other channel (*B*) is undoped. The off-state of the device is shown in Fig. 6.35 a. The gate voltages of the devices are, respectively, 0.6

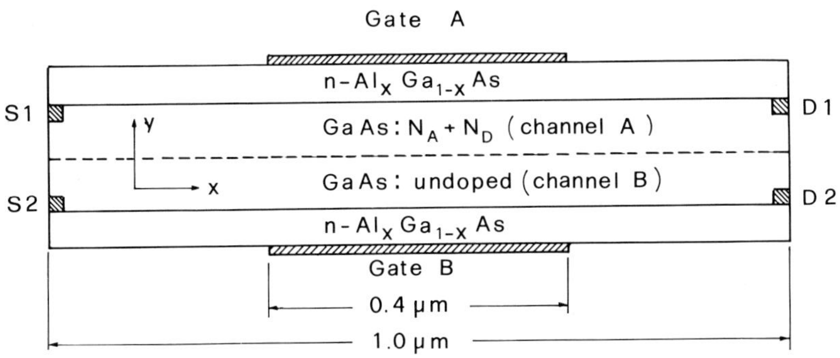

Fig. 6.34. Two dimensional structure of the simulated velocity-modulation transistor [54]

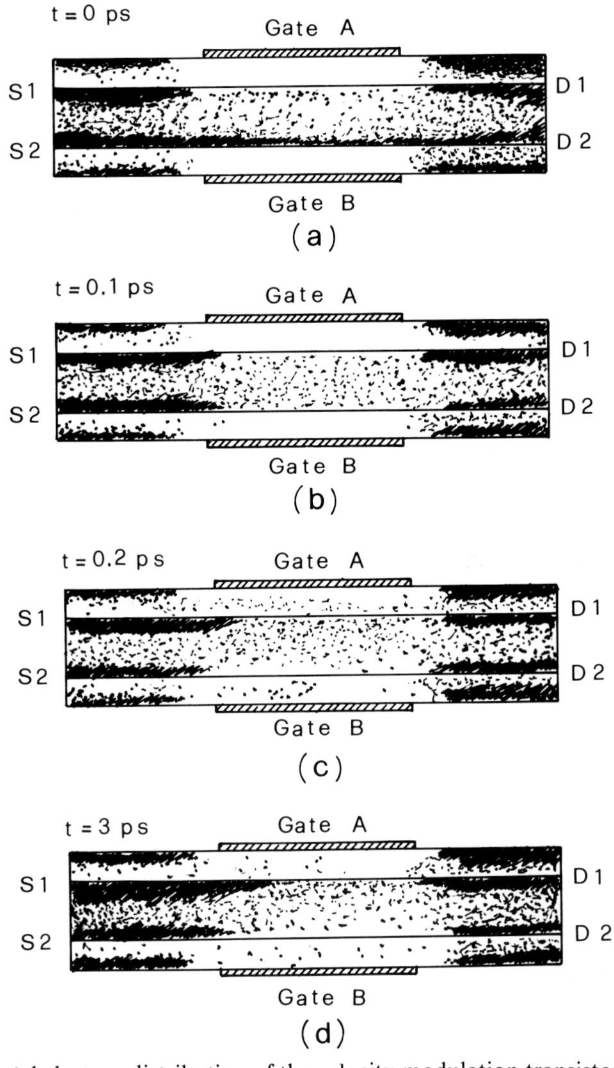

Fig. 6.35. Spatial electron distribution of the velocity-modulation transistor during switching. The dots reported here are only representative of the original figure in [54]

and 0.0 V for channel A and B. A source to drain voltage of 0.8 V is applied. In these conditions, the current is carried by electrons in the low mobility channel. The device is switched on by instantaneously reversing the gate biases at $t = 0$. As seen in Figs. 6.35 b and 6.35 c, channel A is completely depleted in about 0.2 ps, corresponding to the average transit time for perpendicular transport through the 1000 Å GaAs layer. Moreover, a stationary current is achieved after only about 3 ps. (Fig. 6.35 d), which is the time necessary for the redistribution of electrons in the given channel. The channel current modulation is higher at 77 K (where the drain current goes from an off-state value of 32 mA/mm to a on-state value of 98 mA/mm) than at 300 K ($I_{OFF} = 49$ mA/mm and $I_{ON} = 80$ mA/mm). Finally it was estimated from the MC results that the switching speed of VMT would be superior to that of conventional GaAs field effect transistors.

6.5 Bipolar Transistor

Bipolar transistors are devices in which both electrons and holes contribute to the transport process. The simultaneous presence of both types of carriers poses the first problem for the simulation, since it requires in principle a Monte Carlo procedure for each type. Furthermore, electrons and holes interact, especially when the minority carriers move in a region which is heavily doped. The doping of emitter and base is, in fact, an additional difficulty, that poses all the problems already faced in section 5.5 (carrier-carrier scattering, degeneracy, low field regions etc.).

A Monte Carlo study of the transport of electrons across the base and the collector space-charge region of a silicon bipolar transistor was presented by Baccarani et al. [57]. The potential profile was kept fixed during the simulation. Holes were not considered. Velocity overshoot was observed at the beginning of the space-charge collector region, although it was found of limited influence on the collector transit time.

In the present sections we will deal in some detail with a simulation of a Si bipolar transistor based on the hybrid MC (see section 5.5).

The doping profile of the one-dimensional Si n-p-n device considered by Park et al. [58] is shown in Fig. 6.36. The drift-diffusion equation is solved for the entire device structure. When its applicability is questionable (for example in the presence of high field gradients), a one particle MC simulation is performed and the average velocity thus calculated is fed back into drift-diffusion equation. The equation is solved self-consistently with Poisson's equation until convergence is achieved. For the simulated device, very steep gradients in the electric field are found in the base; as seen in Fig. 6.37, for a base-collector bias V_{CB} of 0.8 V the field reaches a value of 100 kV/cm over a distance of less than 0.1 μm. The dynamics of the minority electrons through the base and the collector is followed with the Monte Carlo assuming that the carriers are injected by the emitter with a hemi-Maxwellian velocity distribution. Absorbing boundary conditions are assumed when the

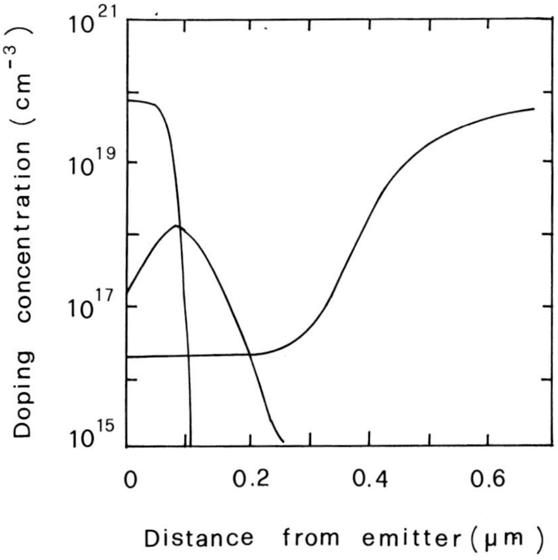

Fig. 6.36. One dimensional doping profile of the simulated *n-p-n* bipolar device [58]

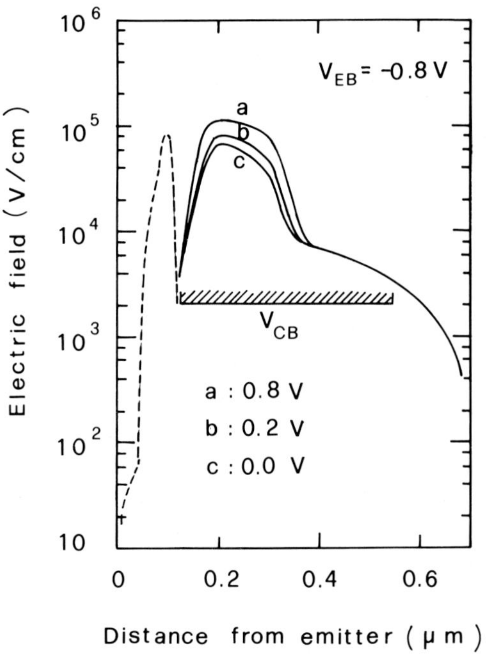

Fig. 6.37. Electric field distribution as a function of position for three different values of the collector voltage [58]

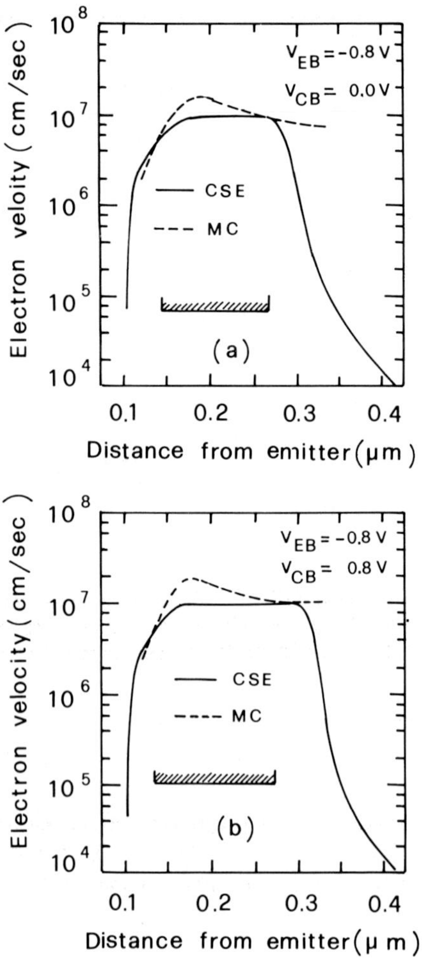

Fig. 6.38. Average electron velocity as a function of distance for $V_{BE} = -0.8$ V, and (*a*) $V_{CB} = 0$ V; (*b*) $V_{CB} = 0.8$ V. The solid line shows the result of a static model, the dashed line shows the MC results [58]

MC carriers cross either the left or the right boundary. Acoustic and optical phonons as well as impurity scattering are considered in a simple spherical model for Si. The spatial distribution of electron velocity in the base and base-collector space-charge regions is displayed in Fig. 6.38 for two values of V_{CB}. The Monte Carlo velocity (solid curve) is higher than the one calculated with the static model, and exhibits a slight overshoot in correspondence with the high field region.

The simulation does not include back diffusion from the highly doped collector region. A perfectly absorbing metal contact is assumed 0.7 mm from the emitter contact. For this reason, no decrease of the electron velocity in the collector region is found in the MC results. The collector current

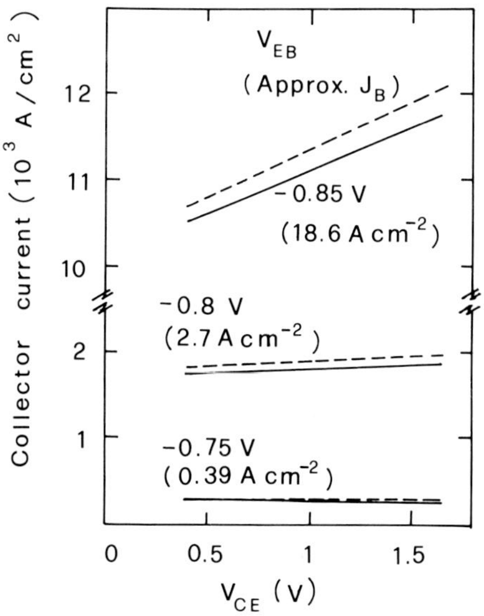

Fig. 6.39. Collector current versus emitter-to-collector voltage characteristics at different values of V_{BE} obtained from the static model with (dotted lines) and without (solid lines) MC correction. The values in the brackets indicate approximate values of the base current [58]

versus emitter-collector voltage characteristics are shown in Fig. 6.39 with the emitter-base voltage as a parameter. The increase in current connected with the velocity overshoot (solid curve) in the base region is very small (less than 5 percent) under the present operating conditions. At constant V_{EB}, the collector current increases with V_{CE}, due to the penetration of the base-collector space charge region in the base region, and consequent reduction of the effective base width.

6.6 HBT

A self-consistent ensemble MC simulation of a AlGaAs/GaAs heterojunction bipolar transistor (HBT) was presented by Tomizawa et al. [59]. Two types of *n-p-n* transistor were simulated with a linearly-graded region extending from the emitter region up to the base-collector junction (type A), and to the emitter-base junction (type B). A schematic representation of the two structures is given in Fig. 6.40. The same two valley model for GaAs and AlGaAs presented previously [43] was used. The scattering processes taken into account include intravalley and intervalley scatterings, impurity scattering, electron-hole scattering in the p^+-base region (no details are given in the reference about the way such scattering is treated), and electron-plasmon scattering in the n^+ collector region [12]. The recombination process of electron-hole pairs is also considered using the phenomenological recombination times for electron and holes. While the electron dynamics is

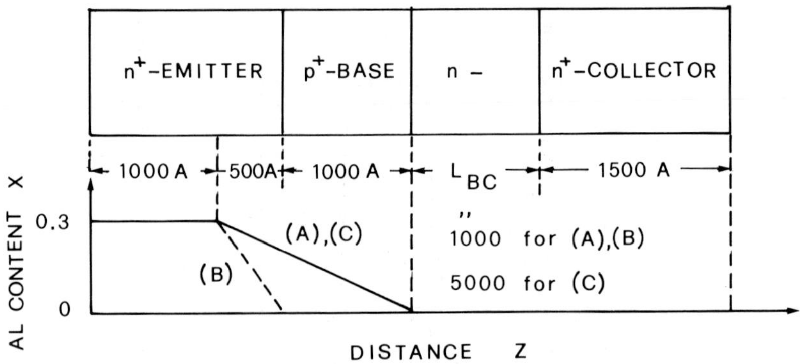

Fig. 6.40. One dimensional structure of three different *n-p-n* heterojunction transistors [59]

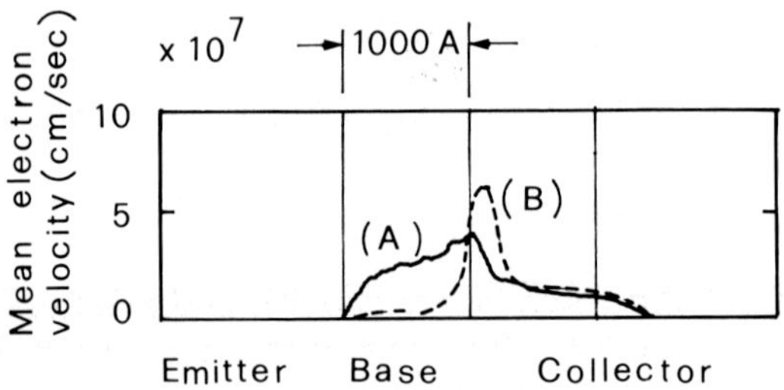

Fig. 6.41. Profile of the average electron velocity in type A (solid line) and type B (dotted line) devices. The bias conditions are $V_{BE} = 1.6$ V; $V_{CE} = 1.5$ V for device A, and $V_{BE} = 1.4$ V; $V_{CE} = 1.5$ V for device B [59]

followed by the MC algorithm, the hole motion is solved using a hydrodynamical model (continuity equation in a finite-difference scheme). One-dimensional Poisson's equation is solved self-consistently with the carrier dynamics. In order to minimize the computation time, the simulation consists of two separate stages. In the first stage, the spatial steady-state for both electrons and holes are determined by the hydrodynamical model. This preliminary stage is needed because it would be extremely time-consuming to establish the steady-state hole distribution in a bipolar device by solving the electronic motion using the Monte Carlo particle model. At the end of the first stage, the initial conditions for all the electrons are determined by distributing them according to the distribution just determined. At this point, the MC procedure for electrons is started.

The profile of the mean electron velocity for the two structures is shown in Fig. 6.41. The electrons in the graded base region (solid curve) move in a quasi-ballistic fashion as a result of the accelerating force of the quasi-electric field (the grading of type A transistor produces a quasi-electric field of about 20 kV/cm in the base). This field sustains the high electron velocity (up to a value of 4.5×10^7 cm/s near the base-collector junction) by constantly supplying the carriers with energy and forward momentum, thus balancing

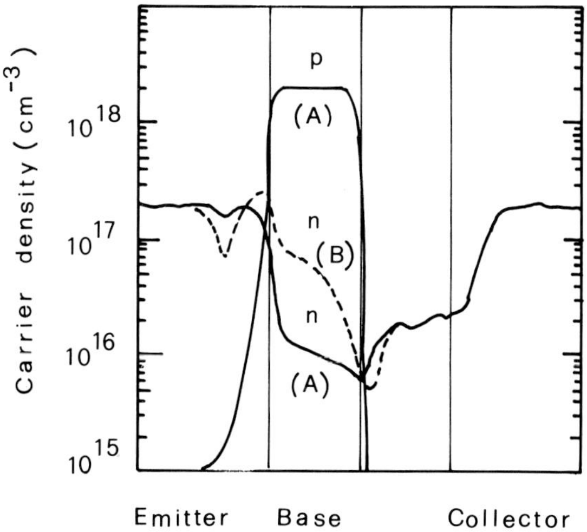

Fig. 6.42. Electron (n) and hole (p) density profiles for type A (solid lines) and type B (dotted lines) devices [59]

the obstructive effect of ionized impurity and hole scattering. A very different velocity profile is found in type B transistor (broken curve). There, the mean velocity is as low as 5×10^6 cm/s over a substantial part of the base region, suggesting that the electron motion in the non-graded base is diffusion dominated. The bias voltages are adjusted for the two transistors to give a collector current density of about 0.45 mA/mm^2.

The electron and hole spatial distributions for the two devices under the same bias condition as before are plotted in Fig. 6.42. Because of the near ballistic transport of electrons in the graded base, the electron density in the base of type (A) devices (solid curve) is smaller by a factor of seven or eight than that of type (B) transistor. The amount of holes which neutralizes the space charge of electrons in the base region should be much smaller for type (A) transistor. As a consequence, lower base capacitance and higher cutoff frequency is expected for this device. This is confirmed by Fig. 6.43, where the current-gain cutoff frequency f_T is plotted as a function of collector current for a constant emitter-collector voltage. It is seen that f_T for the graded device (A) reaches a very high value of 150 GHz at $I_c = 1.4$ mA/mm^2. Virtually the same values of cutoff frequency were obtained for a 0.2 mm base transistor with a smaller Al content. The steep decrease of the cutoff frequency above is due to the accumulation of holes in the emitter region for a large forward bias voltage. Due to the large base capacitance caused by the diffusion dominated electron transport in the base, the frequency f_T of device (B) only reaches a value of 58 GHz for $I_c = 0.6$ mA/mm^2.

The simulation of a structure similar to type (A) with a much larger n-collector region ($L_{BC} = 5000$ Å) showed a substantial degradation of the cutoff frequency.

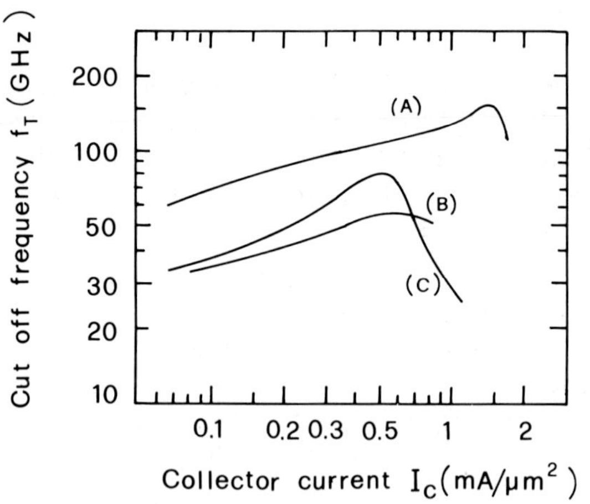

Fig. 6.43. Current-gain cutoff frequency as a function of collector current for the three simulated HBTs with $V_{CE} = 1.5$ V [59]

A full self-consistent MC simulation of a double HBT (DHBT) based on InGaAs/InGaAlAs has been performed by the group of Orsay [60–62]. The MC results were used to interpret the experimental characteristic of the actual devices. A significant amount of electrons were found to move quasi-ballistically through the base. An analytical model for the HBT was also built starting from the MC results.

6.7 MOSFET and MISFET

The MOSFET certainly constitutes one of the most difficult structures to deal with using a Monte Carlo simulation. The difficulties arise from the following features that characterize the device:

a) The conduction of carriers takes place in the inversion (or accumulation) layer at the interface between Si and SiO_2. The electron motion in the well is quantized in the direction perpendicular to the interface.
b) When a drain voltage is applied, the width of the channel changes moving from source to drain. Thus subbands levels, wave-function and scattering probabilities should be always calculated self-consistently with the charge distribution.
c) The Si-SiO_2 interface is very dirty, and the scattering of the electrons with the surface roughness is very important.
d) Source and drain regions are heavily doped (about 10^{20} cm^{-3}).

While considerable work has been done on these particular issues (see, for example, sections 2.9 and 2.10), it is only very recently that Monte Carlo simulations of the overall device have been performed. Traditional simulators have found a much wider application on MOSFET's.

6.7.1 *MOSFET*

A very interesting simulation has been performed by a group at the University of Bologna (Italy) [64, 65]. Several of its features have already been presented (section 5.5), since they contain algorithms that could easily be applied to any Monte Carlo simulator. In particular we would like to recall some of the essential points:

1. Optimization of the self-scattering scheme to space-varying field distributions (Sangiorgi et al., [65]);
2. Multiple repetition scheme, to handle high density regions and areas of the energy distribution function rarely populated;
3. Careful use of the boundary conditions, to correctly simulate nonstationary transport in submicron structures.

The simulator avoids the complications of the 2D nature of the electron transport at the interface by using three-dimensional scattering rates. Scattering mechanisms with acoustic and optical phonons are considered in a simplified spherical model of Si. Ionized impurity, surface roughness scattering, and impact ionization mechanisms are also taken into account. A one-particle Monte Carlo is used, where the field distribution for the device is taken from a drift-diffusion 2D simulator and kept fixed during the simulation. A schematic representation of a MOSFET is given in Fig. 6.44. The simulated region is shown by the dashed line. The main goal of the simulation was the study of very high energy effects, in particular of the injection of electrons into the gate. Fig. 6.45 compares the measured and calculated gate injection efficiency (i.e. the ratio between gate and drain currents) as a function of gate voltage for two types of devices.
Despite the simplicity of the model, good agreement is obtained which indicates the importance of the high-energy tails of the electron distribution in controlling gate injection. It should be remembered that only the novel algorithms for the magnification of rare events introduced here allowed the detection of the electron population at very high energy. Figure 6.46 illus-

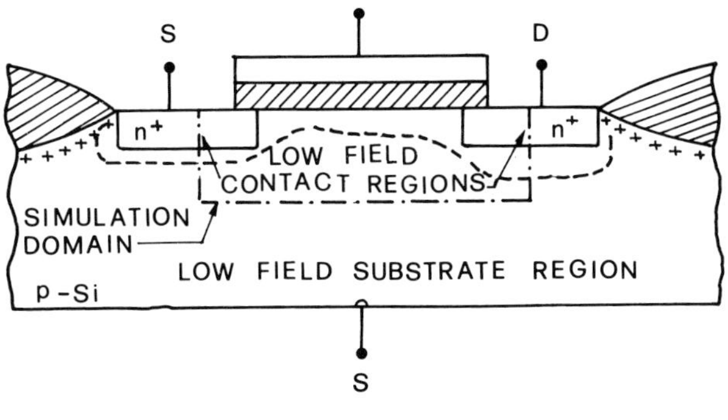

Fig. 6.44. Two dimensional structure of the simulated MOSFET [64]

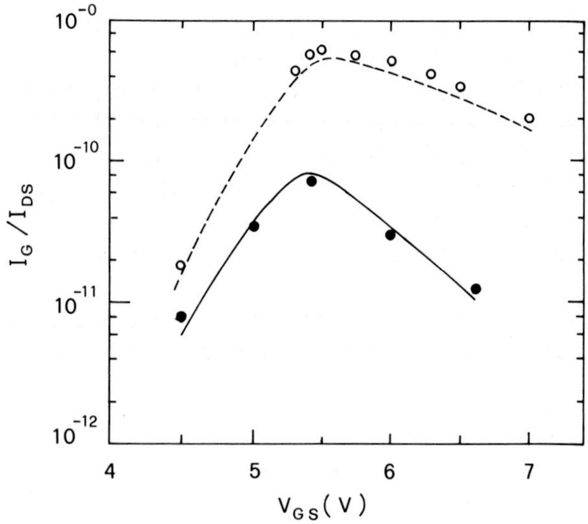

Fig. 6.45. Gate injection efficiency versus gate voltage for $V_{DS} = 5$ V. The open circles and the dotted line refer to experimental and MC results respectively for device A (effective gate length of 0.8 μm and layer oxide of 25 nm). The full circles and the solid line refer to experimental and MC results respectively for device B (effective gate length of 0.8 μm and layer oxide of 15 nm) [64]

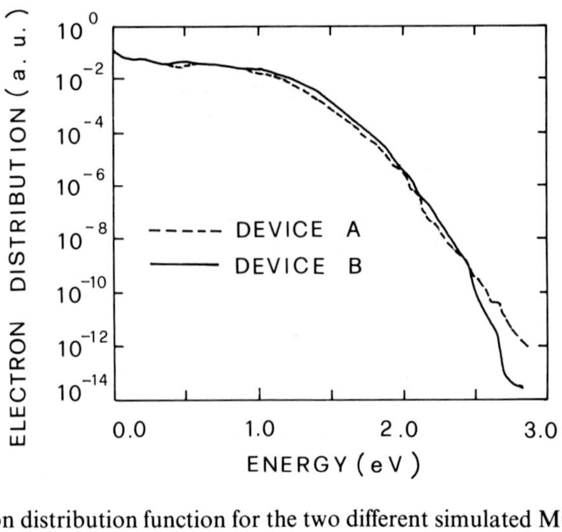

Fig. 6.46. Electron distribution function for the two different simulated MOSFETs taken at the drain end of the channel for $V_{GS} = 6$ V and $V_{DS} = 5$ V [64]

trates the energy distribution function up to 3 eV at the maximum energy point along the channel. The two distributions (respectively for device *A* and *B*) are markedly different from the Maxwellian function which is found at low fields. The field distribution along the channel responsible for the considerable electron heating detected by the simulation is presented in Fig. 6.47 for the two devices. As can be seen, very high values (up to 300 kV/cm) of the electric field can be found near the drain end of the gate. A self-consistent version of the previous program has recently been presented [66, 67].

A slightly different structure has been studied by Throngnumchai et al. [68] (see Fig. 6.48). In this case, the MOSFET is made in an ultrathin nearly

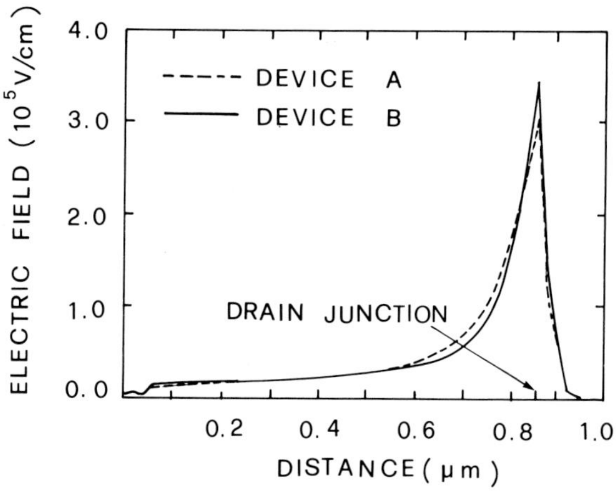

Fig. 6.47. Parallel electric field profile for the two simulated MOSFETs for $V_{GS} = 6$ V and $V_{DS} = 5$ V [64]

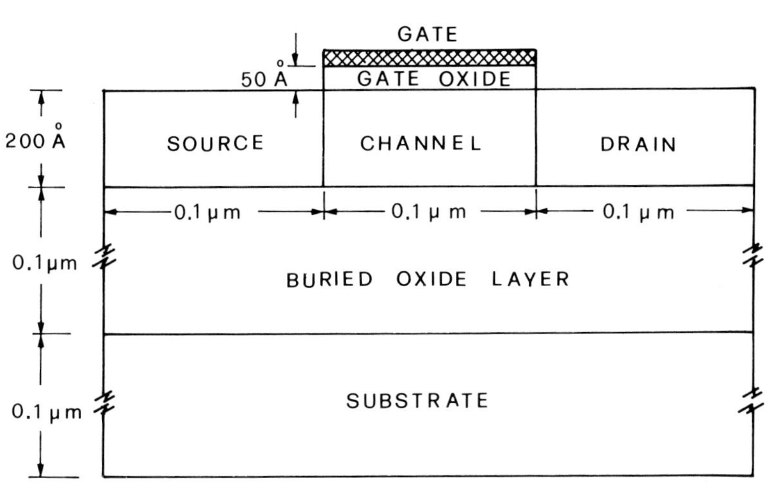

Fig. 6.48. Two dimensional structure of the simulated MOSFET [68]

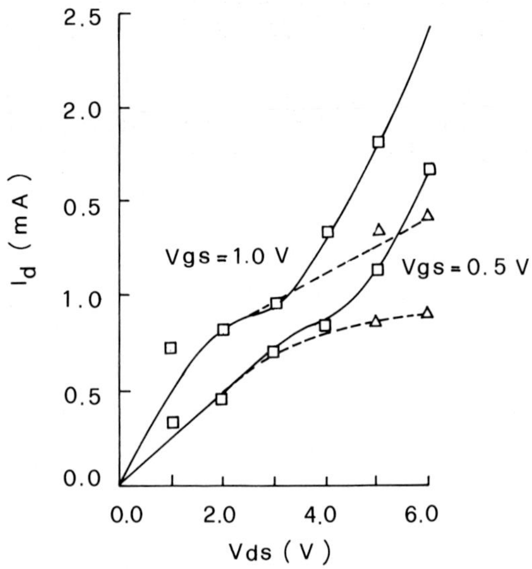

Fig. 6.49. Drain current versus drain voltage characteristics for the simulated 0.1 μm MOSFET at two different gate voltages, with (solid lines) and without (dashed lines) impact ionization [68]

intrinsic silicon on insulator (MOSFET's SOI) in order to reduce punch through phenomena. A self-consistent many particle MC simulation is performed, where both electrons and holes are taken into account. The same scattering mechanisms (acoustic and optical phonons, surface roughness and impact ionization) are considered for both types of carriers with the modification of deformation potential and effective mass. A simplified spherical non-parabolic band structure is considered, with no size quantization. Poisson's equation is solved using a SOR method, with charge distribution assigned according to the cloud-in-cell method (cfr. section 5.3).
The drain current I_d versus drain-source voltage V_{DS} characteristics are illustrated in Fig. 6.49, for two different gate-source voltages. When impact ionization phenomena are taken into account (solid curves), a steep increase in the characteristics is observed. The avalanche breakdown voltage is found to be around 3 V for the 0.1 mm-channel device, while it moves up to 6 V for a longer device (0.5 mm channel length). The contour lines for the potential distribution and for the average electron energy (expressed in K degrees) are shown, respectively, in Figs. 6.50 and 6.51. The electron average energy reaches its maximum value (8000 K $\approx$ 0.7 eV) at the drain end of the channel, in the region where the electric field also peaks. A closer look at the electron distribution function reveals strong deviation from a Maxwellian behaviour in the high field regions.
Tomizawa et al. [69] have presented a comparison between a simulation based on the relaxation time approximation and the self-consistent MC

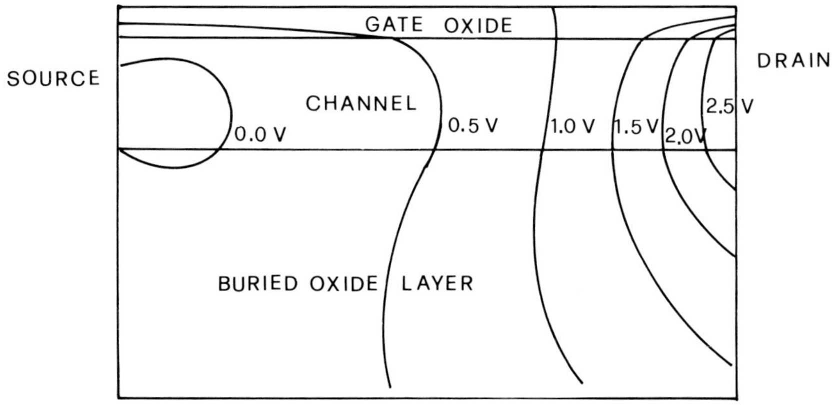

Fig. 6.50. Equipotential contourns (with 0.5 V spacing) for the simulated MOSFET with $V_{ds} = 2.5$ V and $V_{gs} = 0.5$ V [68]

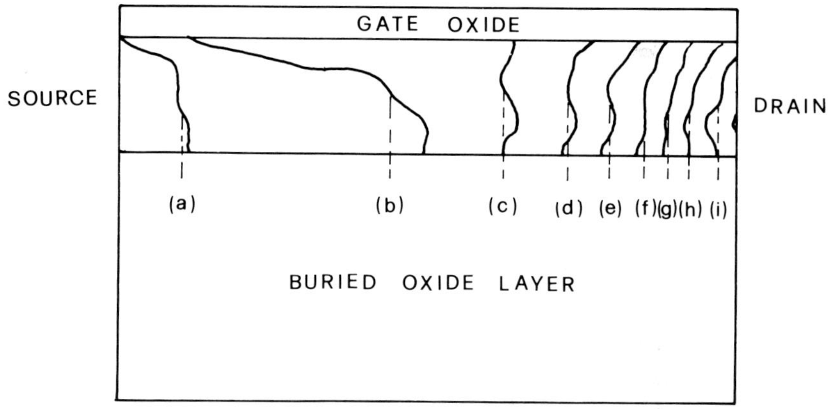

Fig. 6.51. Profile of constant electron temperatures in the simulated MOSFET (in the bias conditions of Fig. 6.50) with spacing of 1000 K from (*a*) 1000 K up to (*i*) 9000 K [68]

procedure for a simplified quarter-micron Si MOSFET structure. They found considerable overshoot and carrier heating in the channel, which can be properly analyzed from a full MC particle simulator.

Another self-consistent simulation of a MOSFET has also been recently performed by Fischetti and Laux [70]. Some of the outstanding features of this simulation have already been outlined in chapter 5. It uses a full band description for Si, where the energy band diagram and the related density of states are calculated from an empirical pseudopotential model [94] and tabulated at the beginning of the simulation. Velocity and energy of an electron with a given **k**-vector are obtained interpolating over the tabulated values. The scattering rates (obtained with a numerical integration of Fermi

Golden rule expression) are stored in a look-up table that is read during the simulation. The model includes electron scattering with optical phonons (LA, TA, LO, and TO), impurities (calculated in two steps, first obtaining a rate from the Brook-Herring formula corrected for the band structure effects, which is then used in Ridley's statistical screening model [95]), other electrons (including both short range binary collisions as well as long range plasmon effects). A very detailed discussion is given of the difficulties of properly including the latter scattering mechanism into a two dimensional self-consistent simulation. Impact ionization is included using Keldysh formula [96]. Degeneracy effects are also accounted for. All scattering rates are three-dimensional. Quantization in the channel has not been considered. After each collision process, the final state is chosen from those that satisfy the energy-conservation requirement with an interpolation scheme similar to the one adopted in the calculation of the scattering rates.
The simulation uses a fixed free-flight time, chosen to be very small (typically $\sim 2 \times 10^{-16}$ s) in order to correctly reproduce the particle dynamics. Poisson's equation is solved by a direct method using a conjugate gradient technique. The field-adjusting time step is also taken very small (around 4×10^{-16} s) to be able to resolve in time plasma oscillations in the drain region (see section 5.5). The simulation is applied to a submicron Si-MOSFET with effective channel length ranging from 0.25 μm down to 60 nm. The MC results are compared with experimental measurements on Si-MOSFET fabricated using direct-writing electron-beam lithography for the gate, and ion implantation for source and drain contacts region (where the dopant concentration reaches a value of 1.5×10^{20} cm^{-3}) [97].
Fig. 6.52 shows the average electron energy (a) and velocity (b), along a 60 nm channel and the electron energy distribution (c) at the drain-end of the channel, calculated at nitrogen and room temperature (continuous and dashed lines, respectively). The profiles are taken about 1 nm away from the Si-SiO$_2$ interface. The bias conditions are $V_{DS} = 0.6$ V and $V_{GS} = 0.7$ V. Significant velocity overshoot occurs near the drain-end of the channel, even at room temperature. Figure 6.52a shows that electrons can reach high averages energies, up to 0.35 meV at low temperatures, suggesting that very few collisions occur in the high-field region of the channel. This is confirmed by the high population of high energy states just inside the drain contact (Fig. 6.52c). The structure of the energy distribution function shows, for the 77 K result, a typically far-from-equilibrium structure, with a peak at 0.5 eV and a depletion at lower energies (the very-high population at the lowest energy comes from the large number of thermal carriers in the drain). The room temperature results are smoother, since the enhanced scattering in the channel washes out the "quasi-ballistic" structure observed at 77 K. The calculated transconductance g_m (for the same bias conduction of Fig. 6.52) is compared with the experimental value in Fig. 6.53 at different effective-gate length. The outstanding agreement proves that the sophisticated theoretical model used in the simulation can predict the macroscopic behaviour

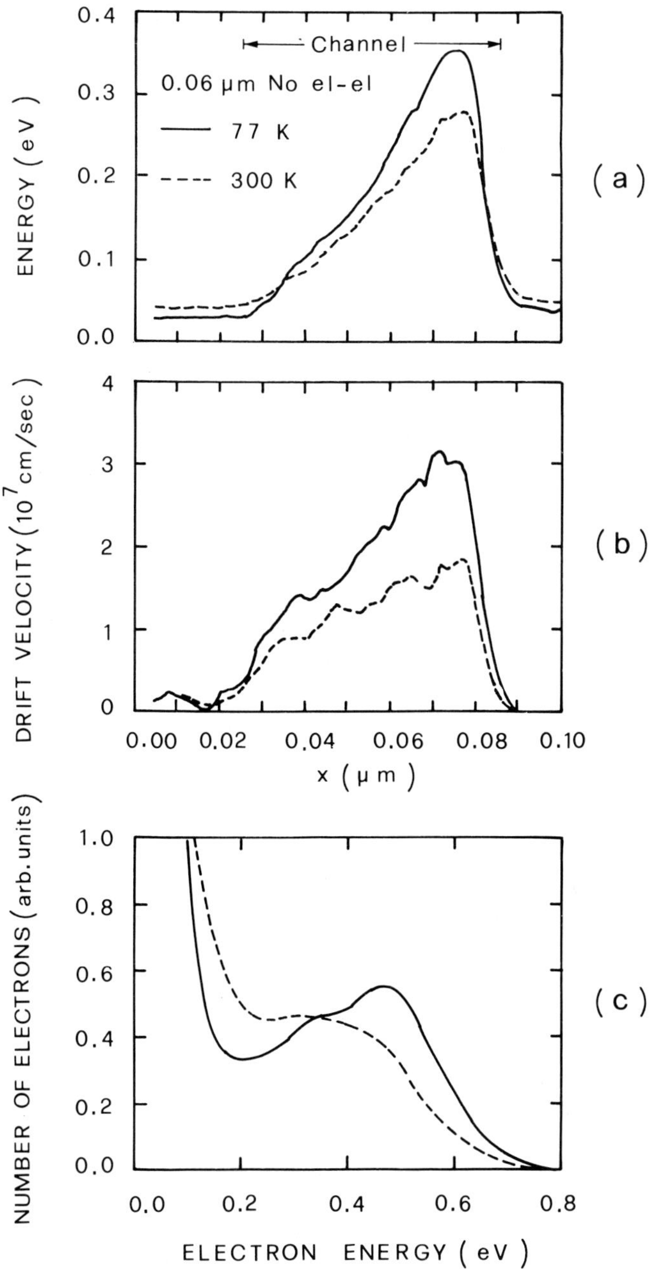

Fig. 6.52. Profile of the average electron (*a*) energy and (*b*) velocity for the 60 nm-channel simulated MOSFET. The electron distribution at the drain and of the channel is also shown (*c*). Solid and dotted lined refer respectively to nitrogen and room temperature operation. The bias condition was $V_{GS} = 0.7$ V and $V_{DS} = 0.6$ V [70]

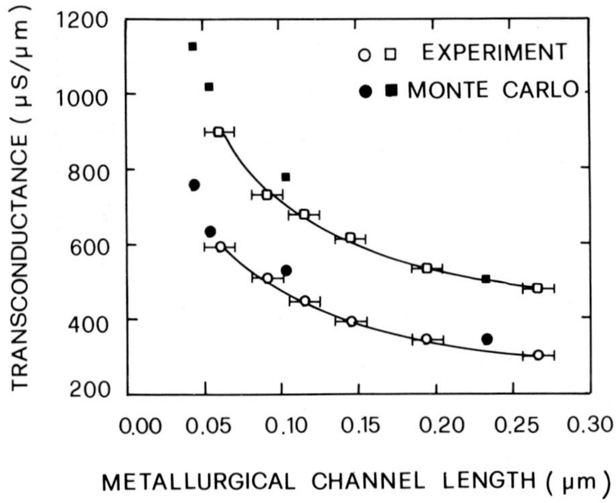

Fig. 6.53. Comparison of the experimental (open symbols) and MC (full symbols) transconductance as a functon of gate length at 77 K and 300 K [70]

of the real device. High values of the transconductance are usually taken as an indication of velocity overshoot. The microscopic M.C. analysis confirms that, especially at low temperatures, this is indeed the case. A simpler drift-diffusion model with a reasonable value for the mobility and the saturated velocity could not reproduce the low-temperature situation and would give underestimated values of the transconductance [98].

The effect of the band-structure on the simulation results are presented in Fig. 6.54, where three situations corresponding to a full band-structure model (continuous lines), a non-parabolic-band model (dashed lines) and a parabolic-band model (dotted lines) are compared. Room temperature operation with $V_{GS} = V_{DS} = 2.5$ V is presented, for an effective gate length of 0.25 μm. The parabolic model greatly overestimates the average velocity and energy in the channel. A large difference is also found in the energy distribution function. In the parabolic model, electrons are allowed to accelerate continuously along the band under the effect of the electric field. On the contrary, using the full band structure, the highest energy regions are characterized by low values of group velocity, which do slow-down hot carriers even in the absence of scattering mechanisms. Considerable improvement with respect to the parabolic model can be obtained including a non-parabolicity correction (with $\alpha = 0.5$ for the results shown in the figure). The difference between the various models are even higher at lower temperatures. It should be mentioned that, despite such large differences, very similar values of the terminal currents are found.

The results shown here did not include electron-electron scattering. A significant effect of this interaction was found, which favours the redistribution of energy and momentum among the channel electrons, enhancing, for example, the high-energy tail of the energy distribution function.

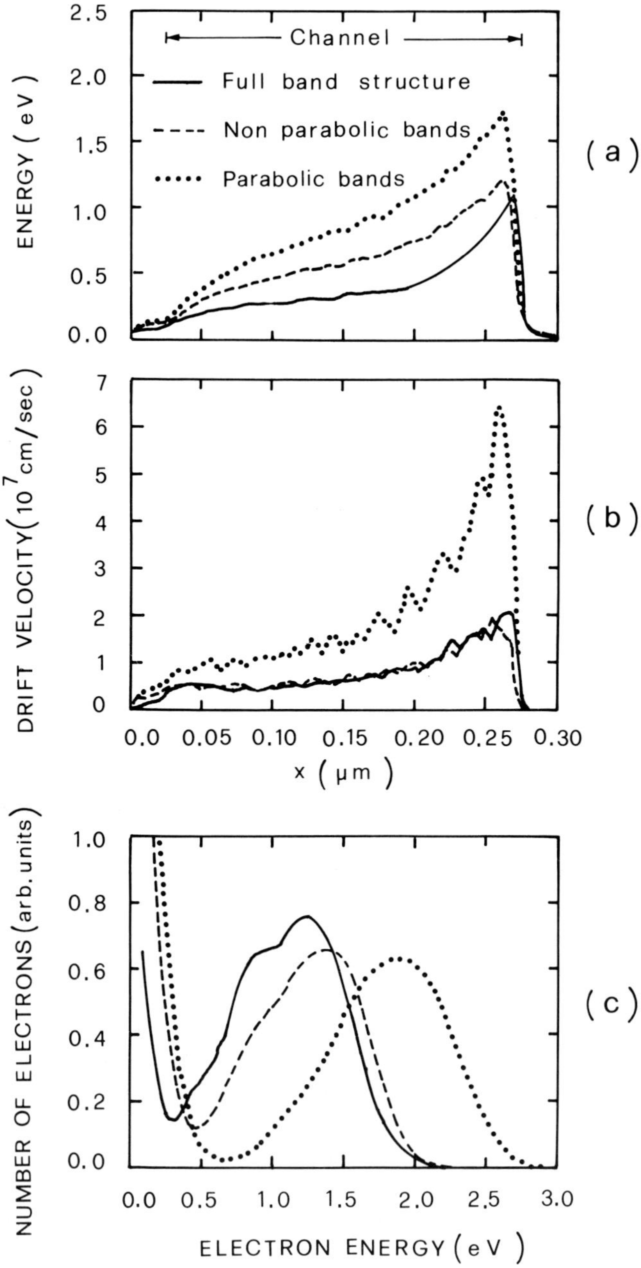

Fig. 6.54. Profile of the average electron (*a*) energy and (*b*) velocity for the 60 nm-channel simulated MOSFET at room temperature for different band structure models. The electron distribution at the drain end of the channel is also shown (*c*). The bias condition was $V_{GS} = V_{DS} = 2.5$ V [70]

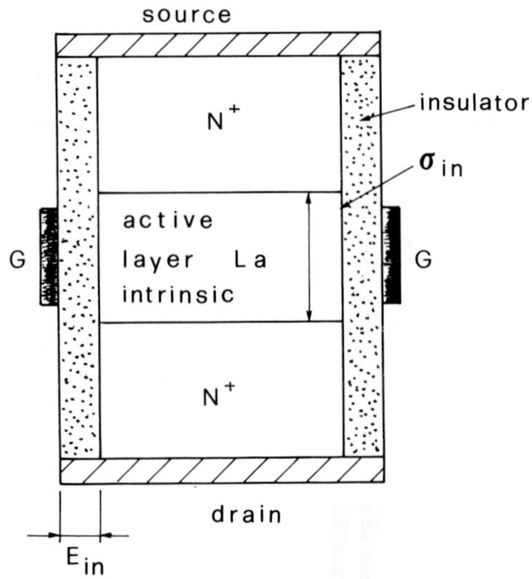

Fig. 6.55. One dimensional structure of the simulated injection modulated MISFET [71]

6.7.2 GaAs Injection-modulated MISFET

An extension of the simulation of injection FET's presented in section 6.3 has been performed by the same group [71] using the structure shown in Fig. 6.55. Here, an insulator is inserted between gate and channel, in order to prevent leakage current (and consequent breakdown) connected with the reduced gate-drain separation typical of injection FET's.
The Monte Carlo simulation is based on the same model outlined in section 6.3. It is used to model the response of the device as a function of technological parameters such as insulator layer thickness E_{in} and charge density σ_{in} at the interface between semiconductor and insulator. Figure 6.56 presents the variation of transconductance (a) and cutoff frequency f_T (b) as a function of E_{in} for two values of active-layer thickness. As the insulator thickness increases, g_m and f_T both decrease. This decrease, which is more pronounced for shorter active regions, is related to the attenuation of the gate control of the active layer in the presence of thick insulators. No strong dependence of the device performances on σ_{in} was found. The optimal structure was obtained with a channel thickness of 0.25 mm and an insulator layer of less than 500 Å.

6.7.3 Conclusions

The simulation of MOSFET's by Monte Carlo is progressing very fast. Due to the similarity of the two systems, HEMT and MOSFET simulation can proceed in parallel, positively influencing one another.

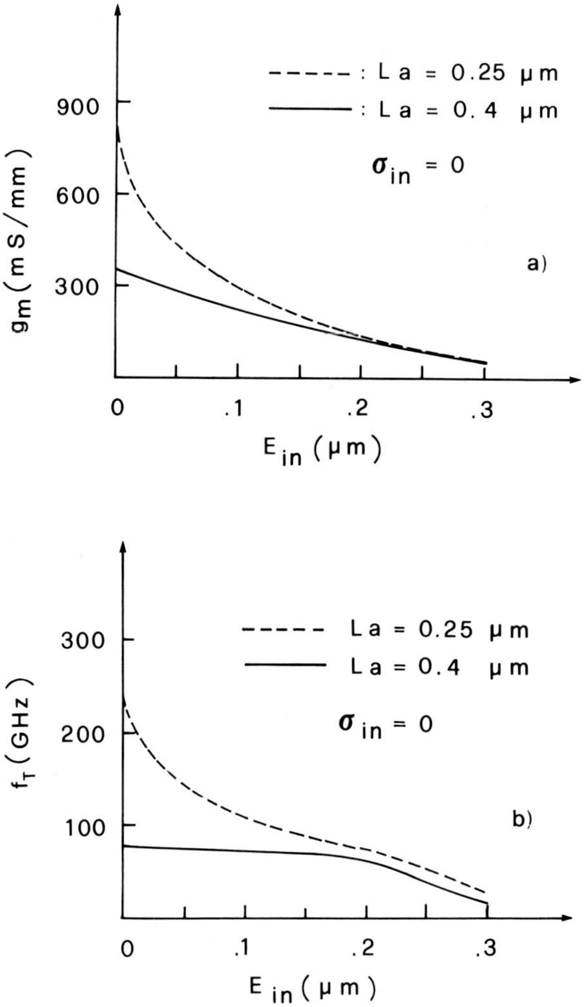

Fig. 6.56. (*a*) Transconductance and (*b*) cutoff frequency as a functon of insulator layer thickness for two different values of the active layer intrinsic thickness [71]

The major steps towards a full MC simulation of MOS structures (some of which are already under implementation) are:

(i) include the 2D transport model already developed by several authors into a self consistent simulation, in the same way as has already been started for the HEMT;

(ii) take into account phenomena connected with high density, in particular degeneracy and carrier-carrier interaction;

(iii) extend the multiplication technique developed by Ricco's et al. [64] to ensemble simulations;

(iv) develop realistic physical models for surface scattering, impact ionization and high-energy band states.

6.8 Hot Electron Transistors

The importance of a systematic study of the properties of ballistic semiconductor devices has long been recognized. We will outline here the Monte Carlo simulation of two different devices, the THETA (Tunneling Hot Electron Transfer Amplifier) [73] and the PDB (Planar Doped Barrier) transistor [74]. In both cases hot electrons are injected into a highly doped region. The interaction with the background electrons is one of the most important scattering sources that determines the performances of the device. The boundary condition for electron injection is also very important from the point of view of the simulation of the device.

6.8.1 *The THETA Device*

The device (see Fig. 6.57), is a heterostructure consisting of alternate layers of GaAs and AlGaAs; electrons are injected into the base via tunnel effect through the potential barrier between emitter and base. Because of the homogeneity of the electric field in the normal direction to the plane of Fig.

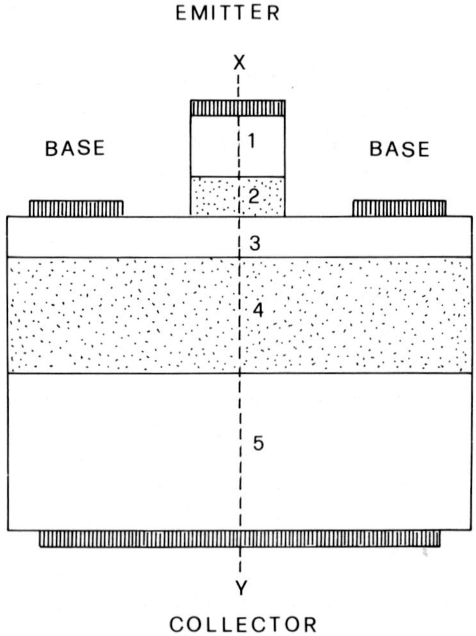

Fig. 6.57. Two dimensional structure of the simulated THETA device [73]

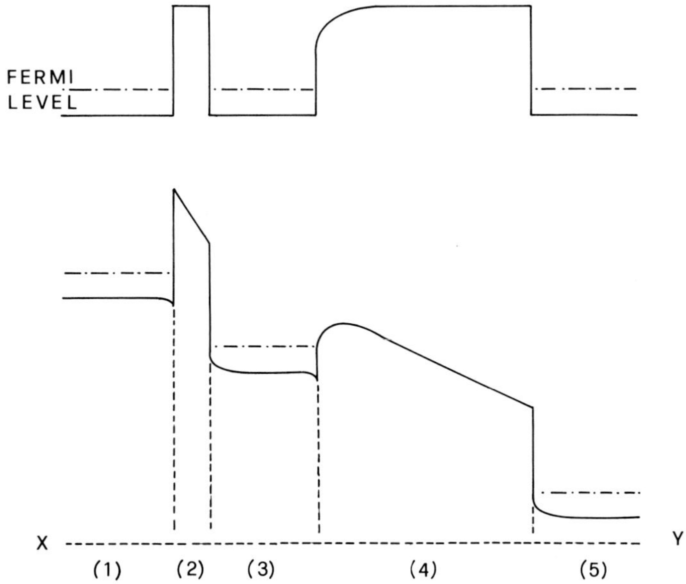

Fig. 6.58. Profile of the conduction band edge without and with applied bias [73]

6.57, it is possible to restrict ourselves to a two-dimensional real space simulation. In addition to electron degeneracy effects, the quantum structure of the potential barrier at the base collector interface has been taken into account; this is an important point in order to compare the model predictions with experimental data.

A two valley GaAs model is used for the simulation. The scattering mechanisms considered are with polar optical phonons, non-equivalent (Γ-L) and equivalent (L-L) intervalley phonons, acoustic phonons, and ionized impurities. The Monte Carlo algorithm is coupled to a Poisson solver in the way described earlier. In the THETA device the quantum transport properties are strongly affected by the shape of the conduction band edge which is shown, for the particular device that has been studied, in Fig. 6.58. This structure and the high doping density, which in the THETA device is of the order of 10^{18} cm^{-3}, introduced in order to obtain reasonable emitter currents with a bias smaller than the separation between Γ and L valleys, impose several quantum corrections necessary for an accurate description of electron dynamics in the device. The quantum effects are included in scattering rates (through the Pauli exclusion principle), in distribution functions and in transmission and reflection coefficients.

Some other important features induced by the peculiar structure of the THETA are listed below:

a) The doping in the base region produces a "sea" of cold degenerate electrons (in the simulated device about 10^5 electrons.) These electrons are confined in the base by the collector potential barrier; the presence of such electron "plasma" is an important energy or longitudinal momentum loss mechanism both on short and long range. Electron-electron and electron-plasmon scattering are considered as in Lugli and Ferry [99]. These two scatterings are implemented by splitting the coulomb potential in a long range part, responsible for the electron-plasmon scattering, and a short range part, responsible for the electron-electron scattering. The screening length, discriminating between the two different regimes, is the Fermi-Thomas length. To avoid unphysical long-range correlations in the electron-electron scattering the partners are chosen, if there are any, in the range of two Fermi-Thomas lengths.

To account for the Pauli exclusion principle which remarkably affects the dynamics of the highly degenerate particles considered, a technique is used in which the k-space is divided in elementary cells of arbitrary volume, the maximum occupation number (i.e. the maximum number of particles contained in agreement with Pauli exclusion principle) is obtained by counting the number of states contained in each cell. In order to consider particle inhomogeneities in the base (that is regions with different electron densities), four different portions of the base are considered, each characterized by a particular k-space matrix (see section 3.9.3).

b) The second relevant point is the presence of the emitter and collector barriers. The spectrum of injected electrons is obtained integrating numerically the Schroedinger equation for incident plane wave functions. In this way one obtains the transmission coefficient as a function of total electron energy $T_e(E)$ and of normal electron energy $T_e(E_n)$, where E_n is the energy associated with momentum component k_n normal to the collector barrier. Two approximations have been used: no alloy scatterings have been considered, such that the tunneling is perfectly elastic; the influence of the collector barrier has not been taken into account (resonant tunneling). Figure 6.59 shows the distribution of injected particles $N(E_n)$ at 4.2 K over a time T which, in terms of current density distribution, is given by:

$$N(E_n) = \int_0^T \int_S J(E_n)\, ds\, dt, \tag{6.8.1}$$

where S indicate the emitter surface area, and $J(E_n)$ is current distribution per unit surface

$$J(E_n) = \frac{m_e}{2\pi\hbar^3} T(E_n) \int_0^\infty f_E(E)[1 - f_B(E)]\, dE_t, \tag{6.8.2}$$

where f_E and f_B are, respectively, the electron distributions in the emitter and in the base region, and E_t is the energy associated with the momentum components parallel to the collector barrier.

c) The last element to analyse is the effect of a quantum collector barrier. As already mentioned, resonant tunneling has not been included in the model, thus it was possible to consider separately the collector barrier and compute numerically the transmission and reflection coefficients $T_c(E_n)$ and

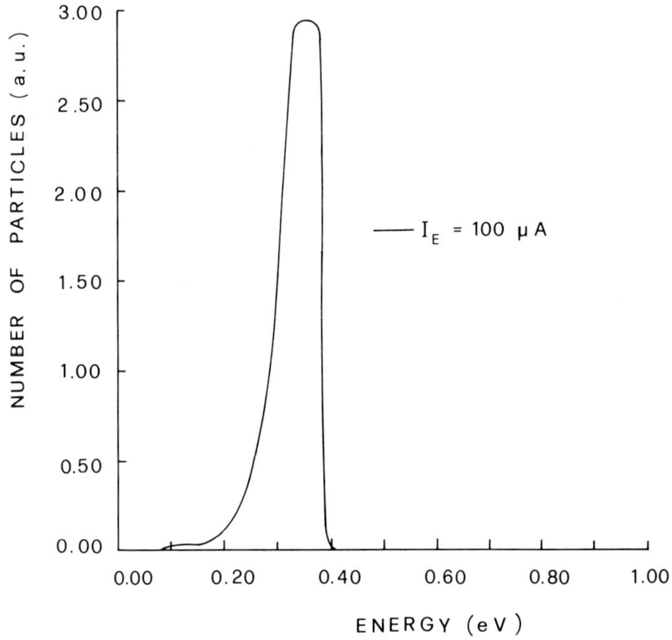

Fig. 6.59. Energy distribution of the electrons injected into the base [73]

$R_c(E_n)$. It is very likely that resonant tunneling has a considerable influence on THETA dynamics, but in any case the presence, in connection with the other effects previously described, of the quantum collector barrier, not considered in other simulations, permits a clearer interpretation of experimental results.

Figure 6.60 represents the collected electron spectrum for an emitter current of 100 μA and a collector barrier height of 170 meV. A comparison with the injected particle spectrum reported in Fig. 6.59 shows that the width of the distribution at half maximum, of about 65 meV, is of the same order as the injected electron distribution, and this is a clear signal of the ballistic behaviour of the injected electrons.

For the analysis of the characteristics curve few preliminary comments are needed. For a graded collector barrier in the THETA device the dependence between collector bias ΔV_c and barrier height ΔE_c is linear only for negative bias values and, in this case, is given by

$$\Delta E_c = \Phi_c - \Delta V_c, \tag{6.8.3}$$

where Φ_c is the collector barrier height for zero bias. For $\Delta V_c > 0$ this relation is no longer valid and is strictly connected with the grading profile which is, at present, not very well know. For Φ_c a value 215^{+35}_{-15} MeV, where the errors are related to the uncertainty on the Al mole fraction in AlGaAs. Figure 6.61 shows the theoretical energy distribution obtained using the linear approximation given by Eq. (6.8.3) (crosses), compared with experimental

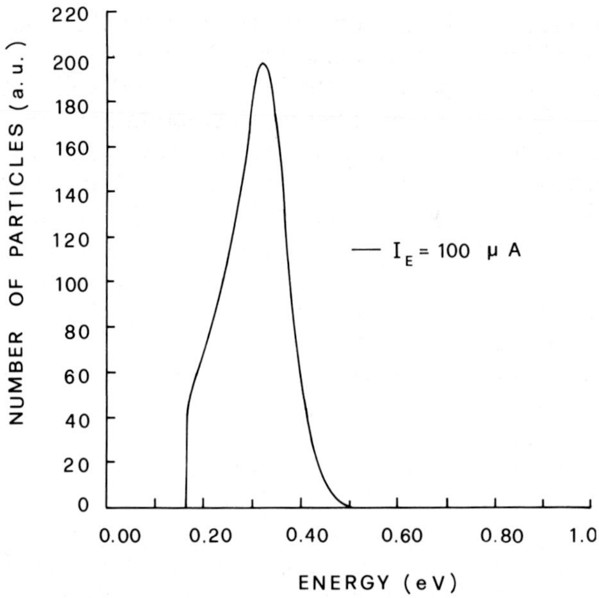

Fig. 6.60. Energy distribution of the collected electrons [73]

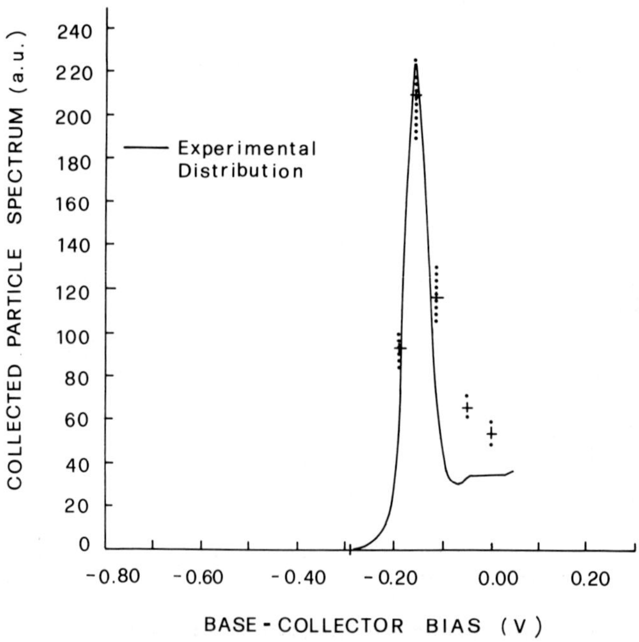

Fig. 6.61. Comparison of the experimental (solid line) and MC spectrum of collected electrons as a function of base-collector bias [73]

results (solid line). In this case one has to consider that the experimental curve is obtained indirectly, deriving the (I, V) diagram with respect to the collector bias. The two points in the interval (−1.0, 0.0) V are due to an enhancement originated by the phonon interaction and are not present in the experimental curve, probably because the real injected particle distribution is affected by interbarrier diffusion, and inelastic tunneling should then be considered. Analysing in more detail the (I, V) curve, one can see the small tail in the interval (−0.3, −0.2) V due to the corresponding tail in the electron distribution. The degeneracy of the electrons is now fundamental preventing about 85% of the scatterings via Pauli exclusion principle. As previously mentioned the comparison of the data is not straightforward for the positive collector bias because of the presence of the grading at the collector barrier. In this case it is not possible to use the linear relation (6.8.3); in fact the uncovered region between 0.0 and 1 V is actually still under investigation, in particular what concerns the grading profile. Using, at 1 V bias, the value for the collector barrier of 165 meV reported in the literature on the subject, values for the transfer ratio in good agreement with experimental results were obtained. In the same hypotheses Figs. 6.62 and 6.63, where we reported all Monte Carlo results, respectively, for $I_e = 60$ μA and $I_e = 100$ μA (crosses), shows that the point obtained for 1.0 V base-collector bias is in good agreement with experiment (solid lines). Then, assuming that the value of 165 meV is correct, this point provides a good check of the

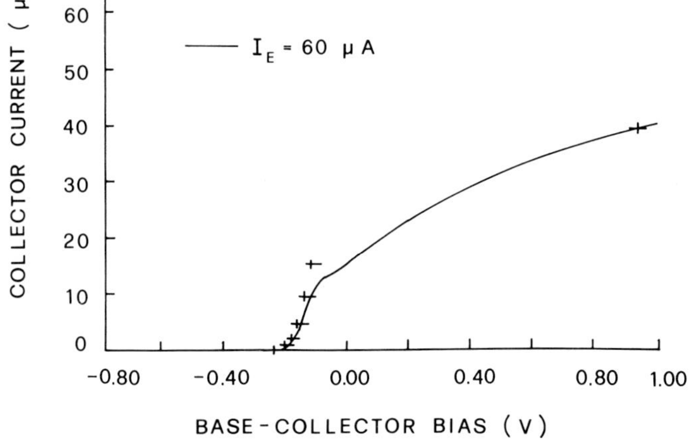

Fig. 6.62. Experimental (continuous curve) and MC (symbols) collector current versus base-collector voltage characteristics for a value of the emitter current of 60 μA [73]

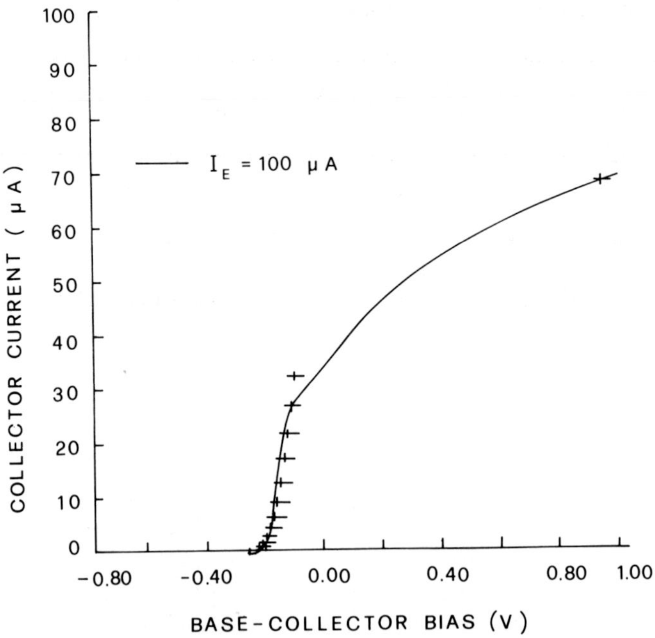

Fig. 6.63. Experimental (continuous) and MC (symbols) collector current versus base-collector voltage characteristics for a value of the emitter current of 100 μA [73]

model because in this case the effect of the resonant tunneling is quite irrelevant.

6.8.2 *GaAs FET with Hot-Electron Injection Structure*

A self-aligned GaAs MESFET with an n^+-AlGaAs source region has been simulated by Tomizawa et al. [44]. The self consistent MC method used is the same as previously described. The simulated device is shown in Fig. 6.64. The band-gap discontinuity at the interface between source and gate region is used as a launcher for electrons into the highly doped channel region. The probability of quantum reflection at the heterointerface is included in the simulation through the definition of a reflection coefficient. Thin n^{++} regions were used to represent Ohmic contacts to the source and drain regions. The Al content in the source region is taken as 0.2, which gives a band gap discontinuity of about 0.2 eV.

Figure 6.65 shows the distribution of electron energy (a), longitudinal component of electron velocity (b), and the mean electron velocity (c), along the device for a source-to-drain voltage of 0.1 V and unbiased gate. The electron energy is measured from the bottom of the conduction band. The straight line at about 0.3 eV indicates the bottom of the upper conduction band. In the vicinity of the heterojunction, the majority of electrons in the channel

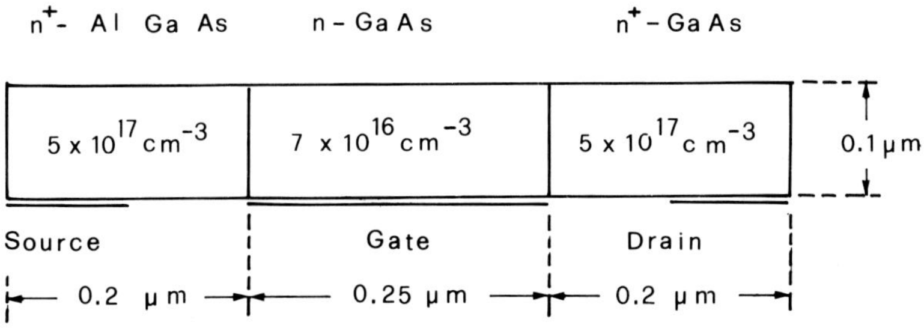

Fig. 6.64. One dimensional structure of the simulated FET with hot-electron injection [44]

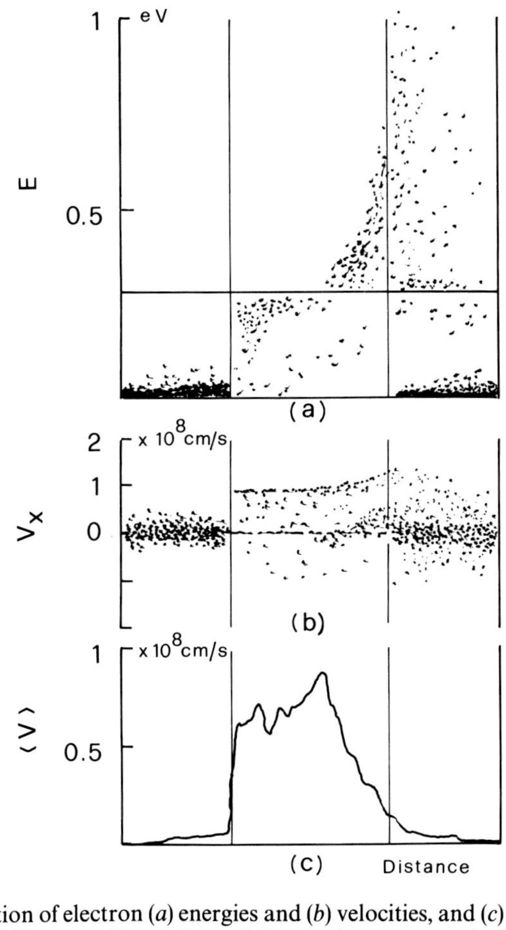

Fig. 6.65. Distribution of electron (*a*) energies and (*b*) velocities, and (*c*) profile of the average electron velocity for $V_{DS} = 1.0$ V and $V_{GS} = 0$ V. The dots reported here are only representative of the original figure in [44]

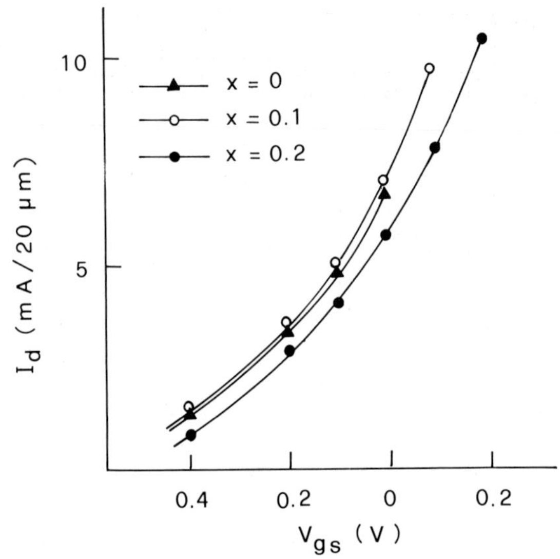

Fig. 6.66. Drain current versus gate voltage characteristics at $V_{DS} = 1.0$ V for three different values of the Al concentration in the injector structure [44]

have kinetic energies almost equal to the band discontinuity, and ballistic velocities ($\sim 9 \times 10^7$ cm/s) corresponding to the high energies. The high velocity of the injected electrons is sustained by the electric field through most of the channel region. Near the drain, many electrons are found with energies above the valley separation (see Fig. 6.65 a). Most of them are actually in the low-mobility L valley, as evidenced by the accumulation of low velocity electrons near the drain (see Fig. 6.65 b) and by the corresponding degradation of average velocity (see Fig. 6.65 c). Simulations with different Al contents x showed that the number of upper valley electrons increased with x for a fixed drain voltage.

The drain current I_D versus gate voltage V_D characteristics are plotted in Fig. 6.66 for $V_{DS} = 0.1$ V. Different Al compositions for the source regions were considered. The slope of the different curves (which gives the transconductance g_m of the device) is almost the same at the same drain current. The FET with layer Al content $x = 0.2$ exhibits a layer $g_m = 900$ mS/mm (vs. 830 mS/mm for $x = 0.1$) due to the higher drain current. A cutoff frequency f_T of around 220 GHz was found for both compositions. By comparing those values with the one obtained for conventional FET's with n^+ source ($g_m = 600$ mS/mm, $f_T = 160$ GHz), it is clear that the hot-electron injection leads to a drastic improvement in the device performance.

6.8.3 Planar-doped-Barrier Transistors

The transport mechanisms in an ideal planar-doped-barrier (PDB) structure have been analyzed by Littlejohn et al. [75]. A typical field profile of the

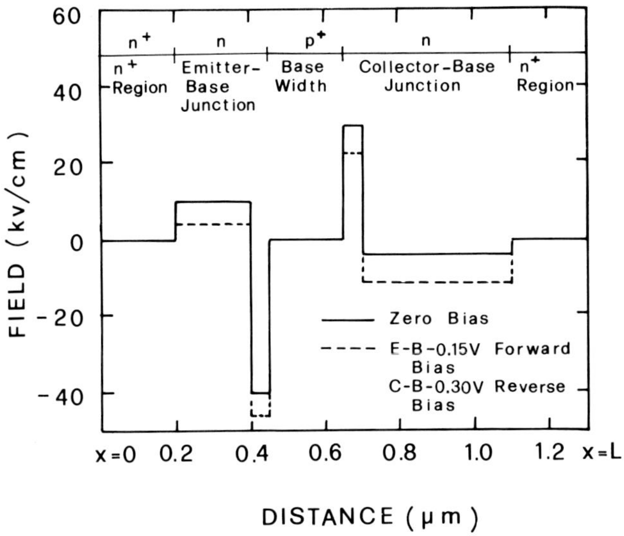

Fig. 6.67. One dimensional electric field distribution in the simulated planar doped barrier device with (dashed line) and without (solid line) bias voltages [75]

simulated structure is shown in Fig. 6.67. The dashed lines illustrate the field profile which results from an emitter-base forward bias of 0.15 V and a collector-base reverse bias of 0.3 eV. The equilibrium barrier heights are 0.2 eV for the emitter-base junction and 0.15 eV for the collector-base junction. The electron dynamics is followed with an ensemble Monte Carlo procedure performed on the given structure. The 3-valley GaAs model used for this simulation has been previously described in section 6.4 [32].

The ensemble average velocity for a transistor with 1000 Å base width and 300 Å width for the emitter and collector high field spikes regions is presented in Fig. 6.68 (curve *a*). The electric field distribution is also shown (curve *b*, with field polarity positive downward). The n^+ regions illustrated in Fig. 6.67 are assumed to have zero thickness for this simulation.

The electron average velocity is found to reach very high values in correspondence of the two field spikes at the emitter-base and collector-base junctions. Even inside the base, electrons exhibit a considerable overshoot, moving with velocity higher than 2×10^7 cm/sec. The transit time τ of the transistor was calculated as a function of base width by performing various simulations. Values ranging from 6.1 ps for a 2000 Å base width down to 4.8 ps for a 250 Å base were found. The corresponding frequency $f = 1/2\pi\tau$ (which constitute here a lower bound to the cutoff frequency) where, respectively, 26 and 33 GHz for a collector-base barrier height of 0.11 eV.

The PDB transistor has recently been used as a hot-electron spectrometer [100]. An ensemble Monte Carlo was performed by Wang et al. [76] to study the effect of scattering mechanisms at low temperatures on the electron energy distribution detected experimentally via the hot-electron spectroscopy. The simulated structure is shown in Fig. 6.69 As in the previous

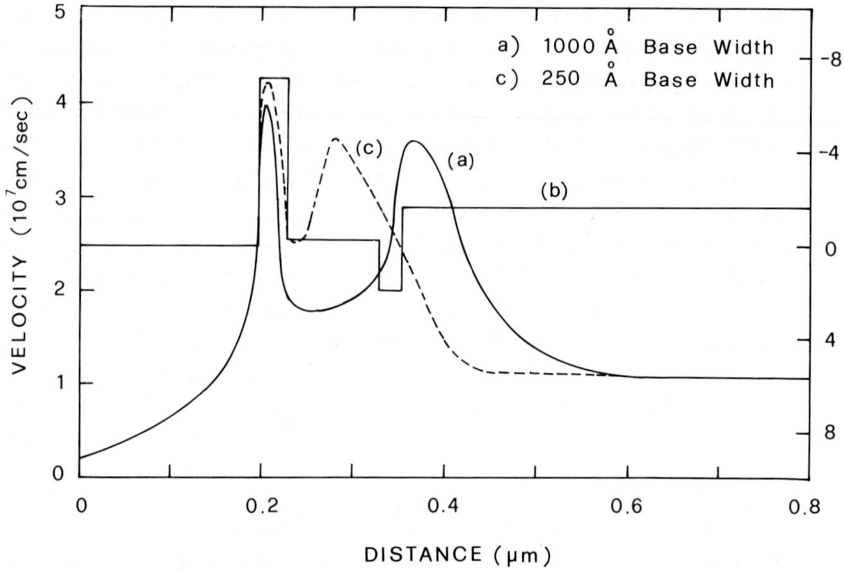

Fig. 6.68. Average electron energy for (*a*) 100 nm and (*c*) 25 nm based width, and (*b*) electric field as a function of distance [75]

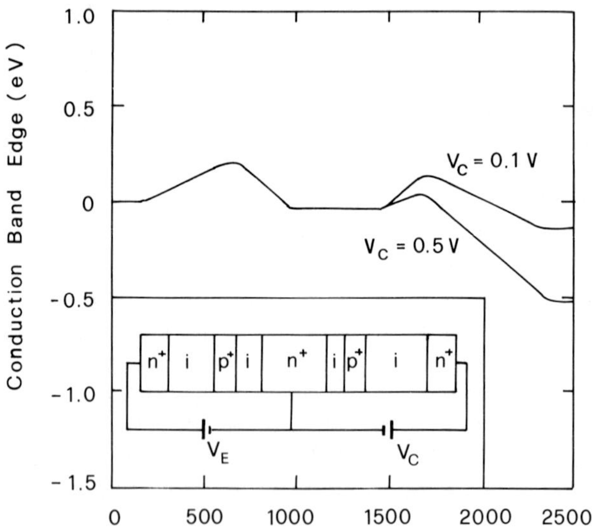

Fig. 6.69. Band diagram of the simulated PDB transistor for two different bias levels. The insert shows the device structure. The n^+ active layer has a width of 120 A and a doping density of 10^{18} cm^{-3} [76]

case, no self-consistency is used and the potential profile is kept fixed during the simulation. The entire band structure of GaAs is considered as described in section 6.4 [41]. In addition to the standard scattering mechanisms considered there, many particle effects (electron-electron, and electron-plasmon scatterings, degeneracy, plasmon-phonon coupling) are considered in the simulation. The device has a base width of 600 Å and a base doping concentration of 10^{18} cm^{-3}. Electrons are injected from the emitter mono-energetically, at about 0.26 eV. According to the simulation, the mean free path of the electrons in the base region at 4 K is around 620 Å. The forward-momentum distribution near the collectors plotted for various collector voltages are shown in Fig. 6.70. The distribution function depends significantly on the collector bias, due to the electron reflection by the collector barrier. The ballistic peak at the front of the distribution is formed by electrons that have suffered only small angle collisions going through the base. Since no quantum-mechanical reflection has been considered here, the high energy peak does not exhibit any dependency on the collector voltage. It was suggested by Hayes et al. [100] that the derivative of the collector current with respect to the collector voltage would directly give the electron distribution function. Such a suggestion is confirmed by the Monte Carlo results shown in Fig. 6.71. There the derivative $\partial I_c/\partial V_c$ is plotted as a function

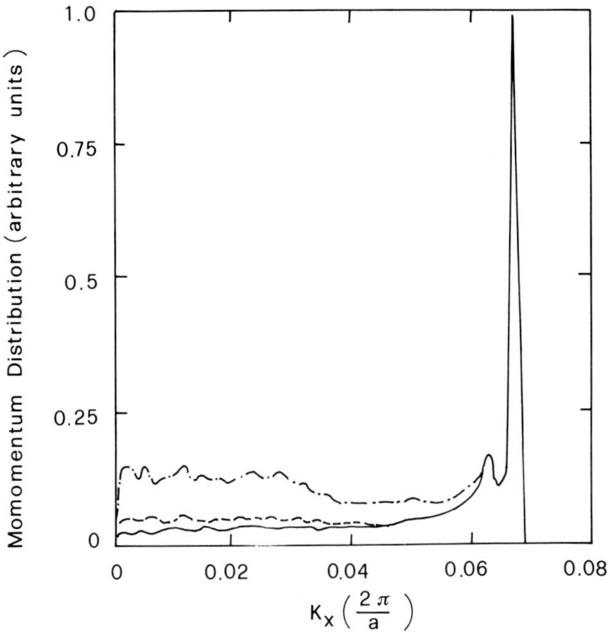

Fig. 6.70. Forward momentum distribution obtained from the MC simulation. The solid curve represent the distribution of the electrons before reaching the collector, the dashed curve and the dash-dotted curves include the contribution of both injected and reflected electrons at respectively $V_c = 0.7$ V and $V_c = 0.1$ V [76]

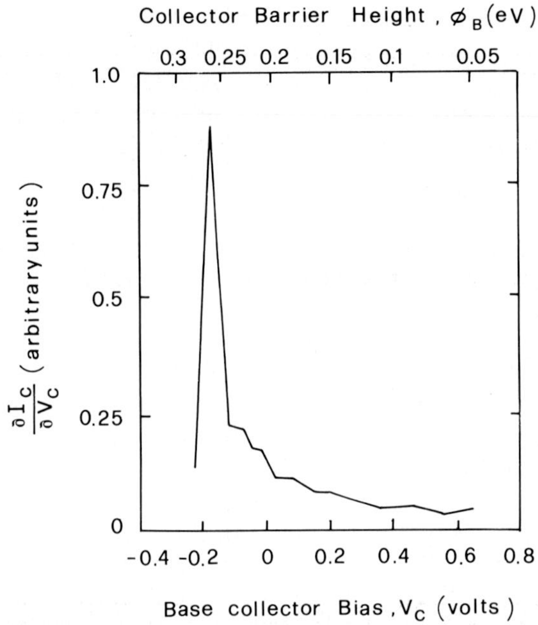

Fig. 6.71. Derivative of the collector current with respect to base collector voltage as a function of the base-collector voltage

of V_c. It was observed however that the derivative does not resolve the detailed features of the electron distribution function at the base-collector junction, since the distribution is disturbed by electron reflection at the collector barrier and therefore varies with the collector bias.

Similar results were obtained in a simulator performed by Long et al. [76]. The same authors showed [77] that improved performances can be obtained by using InGaAs for the base. This material has, in fact, a smaller mass than GaAs, which assures greater conductivities in the base, and a larger separation of the subsidiary minima, which allows a wider range of injection energies.

6.9 Permeable Base Transistor

An ensemble Monte Carlo study of a GaAs permeable base transistor (PBT) was performed using the hybrid method by Hwang et al. [78]. The cross section of the PBT showing the boundary conditions used in the simulation is given in Fig. 6.72. The portion of the device which is actually simulated is shown in Fig. 6.73. To avoid the time-consuming M.C. simulation of the low-field regions, the dashed line in Fig. 6.73 is taken as an injecting boundary. The MC carriers are initialized at that boundary with a carrier density n and a velocity v determined locally by the solution of the drift-diffusion

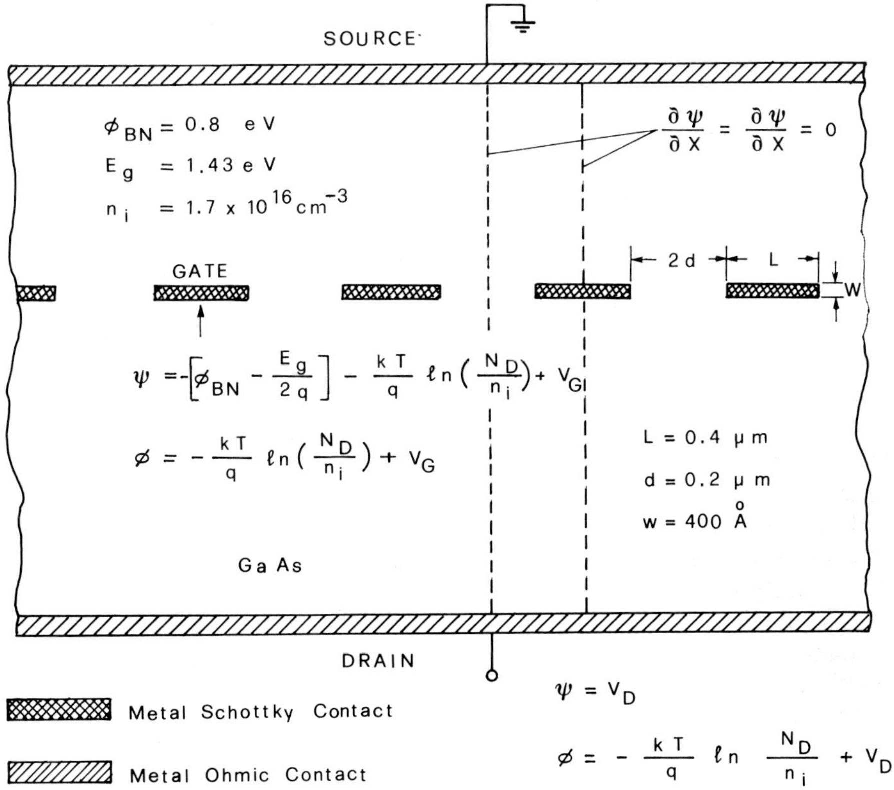

Fig. 6.72. Two dimensional structure of the simulated PBT showing the boundary conditions used in the numerical analysis [78]

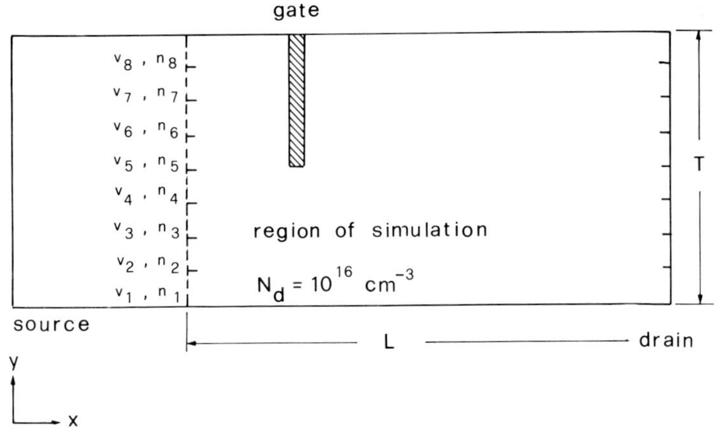

Fig. 6.73. Simulated portion of the PBT. The dotted lines show the location of the injecting boundaries, with the respective carrier density n and velocity v, determined from a drift diffusion simulator. Here $l = 0.7$ μm and $T = 0.2$ μm [78]

equation for the overall device. The injecting boundary, running from the centerline of the channel to the centerline of the gate, is subdivided into several sections, each characterized by a given value of n and v as shown in Fig. 6.73. A collecting boundary (also divided into small sections) is set at the drain contact. The gate contact is treated as an ideal absorbing boundary. Such an assumption is found to have a negligible influence on the MC results as long as the gate current is small compared to the other terminal currents. The MC model is the same as proposed by Hockney and coworkers [88]. By comparing the MC results for a uniformily doped PBT with the ones obtained by the conventional drift-diffusion approach, the authors found a considerable improvement of the device performance in the former case, which reveals the importance of non-static effects in this class of devices. It was then shown that an n^+-n-n^+ structure improves the PBT device performance by minimizing the series source and drain resistance. The comparison of the calculated cutoff frequency vs. gate-to-source voltage for uniformily-doped and selectively doped PBT devices is presented in Fig. 6.74. The n^+-n-n^+ profile is shown in the insert. The results of the drift-diffusion model for the two structures are also reported for comparison. Higher output currents and a better transconductance was found for the n^+-n-n^+ device. The microscopic analysis showed that the improved performance of this structure is partly related to the retarding field at the n^+-n junction, which prevents part of the transfer to the satellite valleys caused by the electric field at the drain end of the channel.

A self-consistent ensemble MC simulation of a submicron PBT was per-

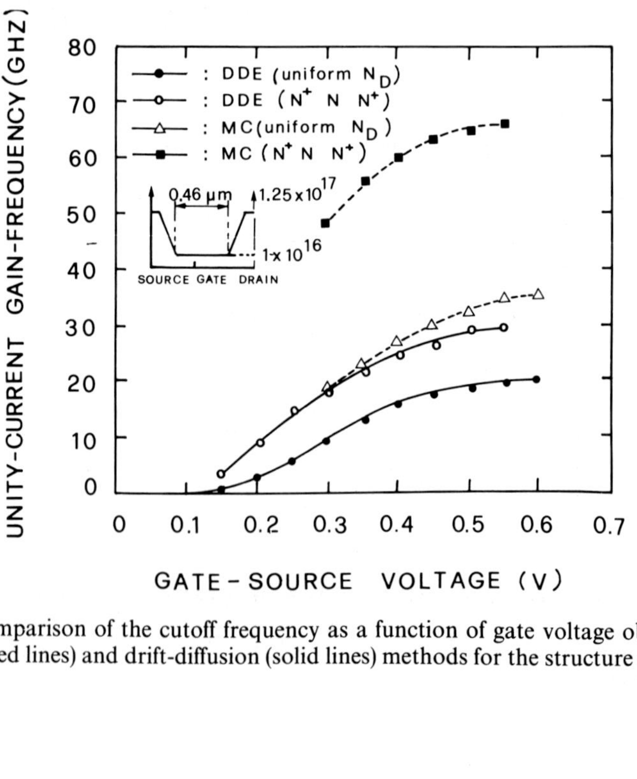

Fig. 6.74. Comparison of the cutoff frequency as a function of gate voltage obtained from the MC (dashed lines) and drift-diffusion (solid lines) methods for the structure shown in the insert [78]

formed by Awano et al. [79]. The same method had been previously applied to the study of GaAs MESFET's [24], and was presented in section 6.4. The two-dimensional model for the PBT is shown in Fig. 6.75. The base thickness L_B is 400 Å. Both the base width and the base spacing are $2d = 2000$ Å. The active n layer (with 10^{16} cm^{-3} doping density) is sandwiched between the two n^+ layer (with doping density equal to 5×10^{17} cm^{-3}). The emitter-

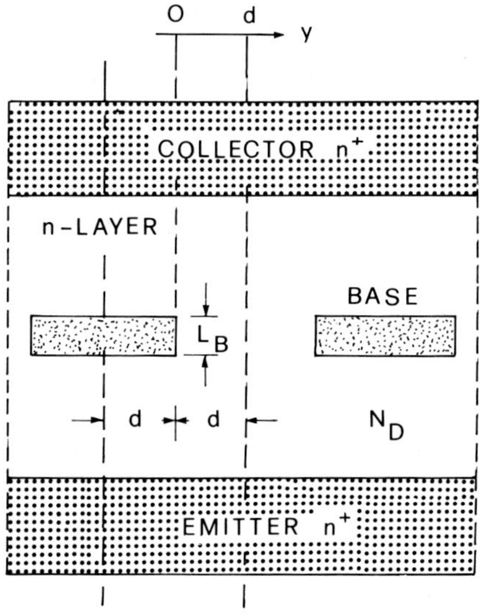

Fig. 6.75. Two dimensional structure of the simulated portion of the PBT [79]

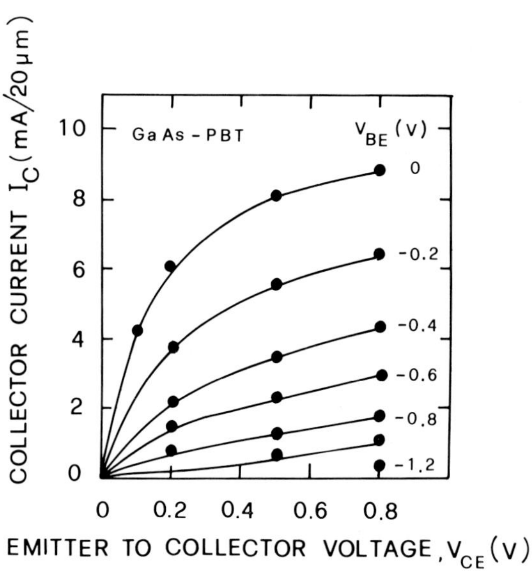

Fig. 6.76. Collector current versus emitter to collector voltage characteristics at various base to emitter voltages

collector distance is 0.25 μm. A unit structure delimited by the two dashed lines in Fig. 6.75 is used in the simulation. The lattice temperature is 77 K. The collector current I_C versus collector-to-emitter voltage V_{CE} characteristics are shown in Fig. 6.76. The base-to-emitter voltage V_{BE} includes a built-in potential of 0.8 V. The transconductance g_m and the cutoff frequency f_T of the simulated device at $V_{BE} = -0.2$ V and $V_{CE} = 0.8$ V are, respectively, 500 mS/mm and 150 GHz. The excellent performances are due to the fact that electrons exhibit quasi ballistic behavior through the whole active layer. The drawbacks are a relatively large collector conductance g_C of about 120 mS/mm at the same bias point and a large deviation of the threshold voltage V_{th} (equal to -1.4 V at $V_{CE} = 0.8$ V) from the pinch-off voltage V_p ($= qN_d d^2/2\varepsilon$, equal to -0.5 V), indicating the poor controllability of the base region. The physical basis of this phenomenon is the two-dimensional nature of the depletion region. Fig. 6.77 shows the equipotential lines (a) and the spatial distribution in the device at V_{CE} equal to 0.8 V for three values

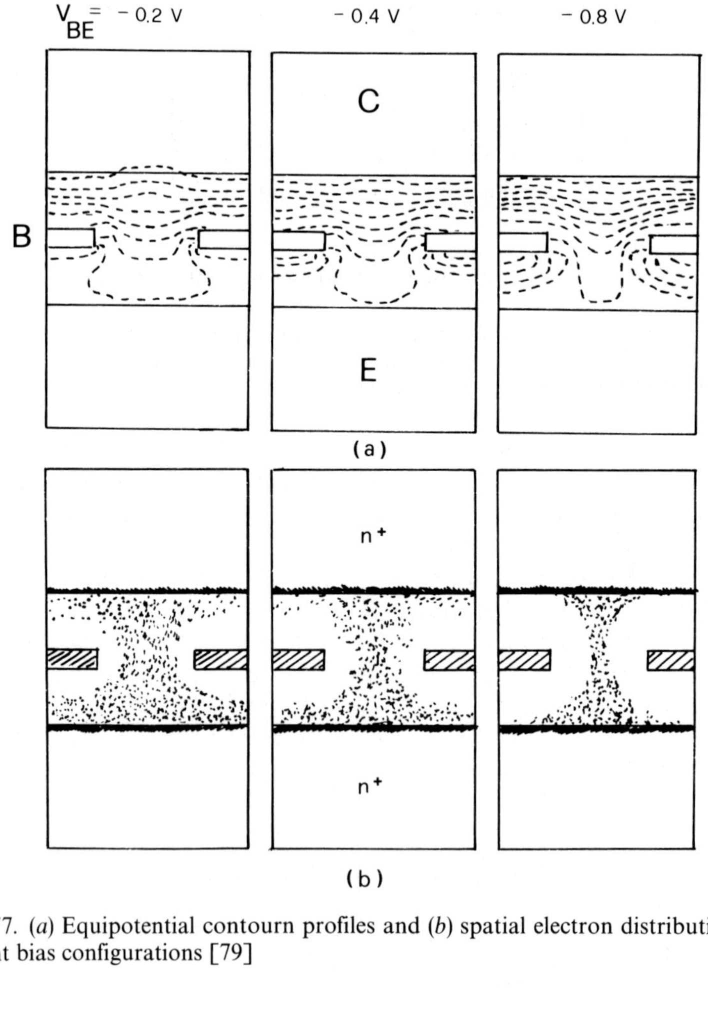

Fig. 6.77. (*a*) Equipotential contourn profiles and (*b*) spatial electron distribution for three different bias configurations [79]

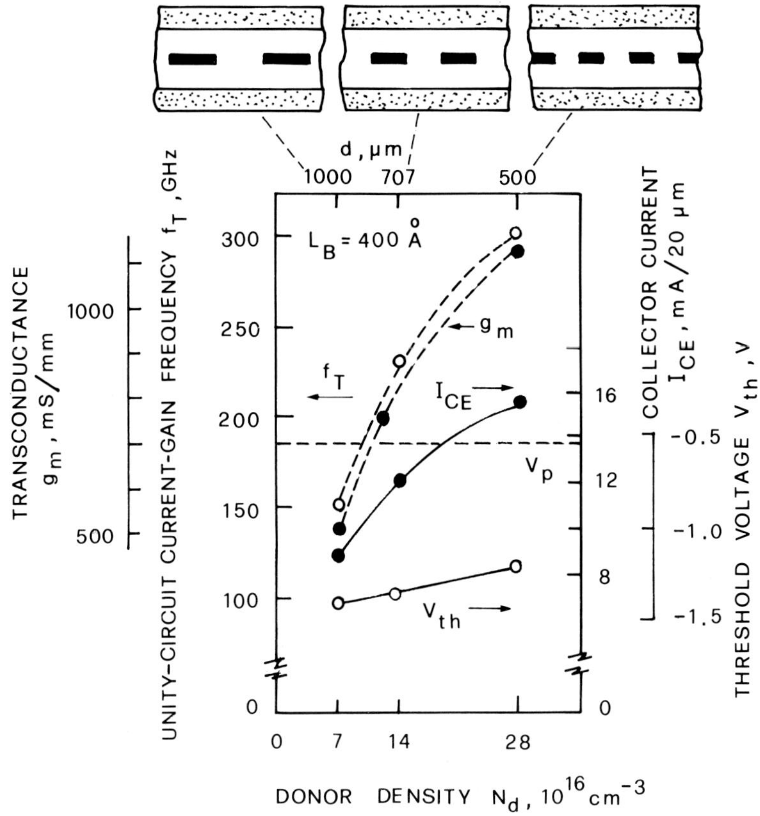

Fig. 6.78. Dependence of the cutoff frequency, collector current, transconductance, and threshold voltage of the simulated PBT on the donor density. Here $L_B = 40$ nm, $V_{BE} = -0.2$ V for the determination of g_m and f_T [79]

of the emitter-to-base voltages. It can be noticed that the depletion regions extend more easily towards the n^+ layers rather than towards the channel. Such an effect is even more pronounced as the aspect ratio of the conducting channel ($= L_B/d$) becomes smaller. A sharp improvement in the device performances was found at increasing doping density of the conducting channel. Fig. 6.78 presents the dependency on donor density of some of the most relevant quantities characterizing the device. The results are obtained by keeping the pinch-off voltage constant (that is changing the grating period d as the donor density is changed). The simulation indicates an increase in I_{CE}, g_m, and f_T as the density goes from 7×10^{16} cm^{-3} to 2.8×10^{17} cm^{-3}. However, the threshold voltage exhibits only a slight dependence on N_D.

6.10 Comparison with Traditional Simulators

As pointed out throughout the book, the Monte Carlo technique is a fairly new tool in the area of semiconductor device modeling. Traditionally, two alternative methods are used: the drift-diffusion (DD) model, and the hydrodynamic (HD) model. A detailed analysis of the two methods is beyond the scope of this book (a very good discussion can be found in [101] and [102]). It is nevertheless interesting to look at basic assumptions underlining the DD and HD models, and to compare their results with those obtained with a MC procedure.

All methods are based on Boltzmann transport equation (Eq. 2.5.2). By taking the first moments of the Boltzmann equation, we obtain (after few assumptions) three coupled equations:

$$\frac{\partial n}{\partial t} + \nabla \cdot (n\mathbf{v}) = 0, \tag{6.10.1}$$

$$\frac{\partial \mathbf{p}}{\partial t} + (\mathbf{v} \cdot \nabla)\mathbf{p} + e\mathbf{E} + \frac{1}{n}\nabla(nk_B T_e) = -\frac{\mathbf{p}}{\tau_p}, \tag{6.10.2}$$

$$\frac{\partial \omega}{\partial t} + \mathbf{v} \cdot \nabla\omega + e\mathbf{E} \cdot \mathbf{v} + \frac{1}{n}\nabla \cdot (k_B T_e n\mathbf{v}) - \frac{1}{n}\mathbf{V} \cdot (\mathcal{K}\nabla T_e) = -\frac{\omega - \omega_0}{\tau_e}, \tag{6.10.3}$$

where n, $\mathbf{p}$, and ω represent, respectively, the carrier density, average momentum and average energy. T_e denotes the carrier temperature, $\mathbf{E}$ the electric field, and $\mathcal{K}$ the thermal conductivity. Equation (6.10.1) is the usual current continuity equation, and expresses charge conservation (here no generation and recombination process is considered). Equations (6.10.2) and (6.10.3) express, respectively, momentum conservation (with $\mathbf{p} = m\mathbf{v}$) and energy conservation in the relaxation time approximation.

The DD approach assumes that carriers are always in equilibrium with the lattice ($T_e = T$), which leads to the usual equation, for steady-state conditions,

$$\mathbf{v} = \mu\mathbf{E} - \frac{1}{n}D\,\nabla n, \tag{6.10.4}$$

where the carrier mobility and diffusivity can be field-dependent quantities related through a generalized Einstein relation.

The HD approach uses all three equations (6.10.1)–(6.10.3), and relies on reasonable models for the relaxation times. Clearly, the HD scheme is far superior to the DD one, since it can account (when all terms of Eq. (6.10.3) are considered) for carrier heating and non-homogeneous distributions of the carrier temperature.

The MC procedure stands on an ever higher level, since it provides (even in non-homogeneous, non-stationary conditions) an exact solution of Boltzmann equation [57]. It correctly describes non local effects (in space $\mathbf{r}$ or time t) where an electron contributes to the current in $\mathbf{r}$ at t coming from a position $\mathbf{r}'$ at t' where the distribution function $f(\mathbf{r}', t')$ was different from $f(\mathbf{r}, t)$.

Unfortunately, the complexity and cost of each approach is inversely pro-

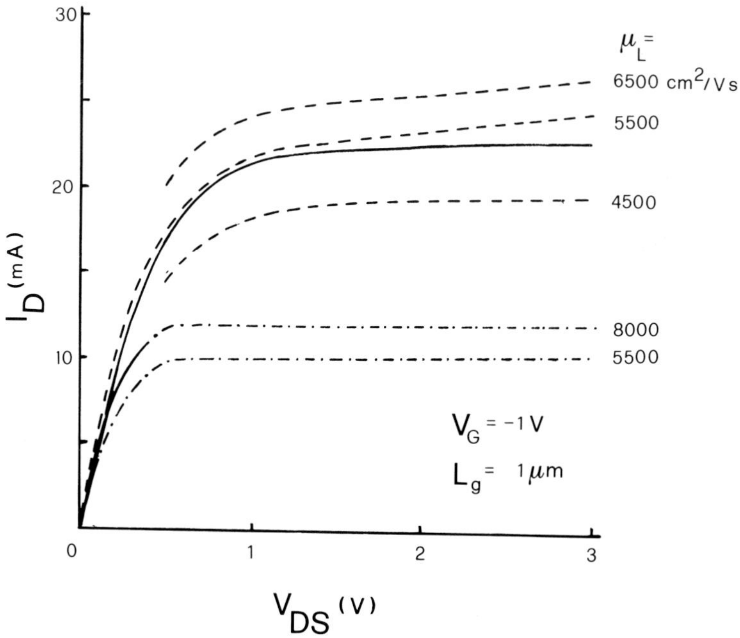

Fig. 6.79. Comparison of the drain current versus drain voltage characteristics for a GaAs MESFET obtained with three different methods [103]

portional to the refinement of the physical model it is based on. The use of one approach or another therefore depends on the specific device under investigation.

Figure 6.79 [103] shows the drain-current versus-drain-voltage for a GaAs MESFET with 1 μm gate (the device is the same studied by Warriner [104]). The results of the three different models are compared. The continuous line shows the MC curve, the dashed lines refer to a modified electron temperature (HD) model, and the dashed-dotted lines to a drift diffusion model.

The HD calculation have been performed using an empirical relation for the velocity-field curve, where the mobility μ_L enters as a parameter. The agreement between Monte Carlo and HD model is good for a value of μ_L equal to 5500 cm^2/Vs, although less drain-source conductance is seen. The DD model, on the contrary, does not predict large enough current even with a mobility of 8000 cm^2/Vs. This is a strong indication that the DD approach properly describes far-from-equilibrium situations, where velocity overshoot effects are present. Such inadequacy of the DD method has been pointed out by several studies (see for example [105, 106]).

The previous analysis has been confirmed by a comparative study done on a submicron Si-MOSFET [69]. Figure 6.80 shows the drain current versus effective channel length obtained with the three methods (continuous line MC; dashed line HD; dotted line DD). The disagreement is reduced with respect to the GaAs case because non-stationary effects are not as strong in

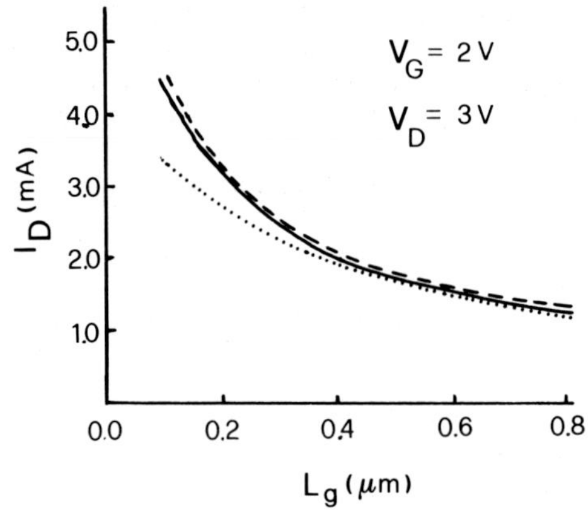

Fig. 6.80. Comparison of the drain current as a function of the effective gate length for a Si MOSFET as obtained with three different methods [69]

Si as they are in GaAs. As for the MESFET, the MC and HD results agree well even down to 0.1 μm channel length.

In conclusion, we have shown that the Monte Carlo simulation of semiconductor devices has made considerable progress over the last few years, and is today a very valuable tool in the area of devices modeling. Furthermore, the discussion of the last section indicates that MC is the best technique to study situations where non-stationary effects are important (as for example in submicron devices). It is safe to anticipate that, as the tendency toward the miniaturization of devices will continue in the future, MC simulators will progressively extend their applicability.
The success of MC in device modelling will ultimately depend on the compromise between two tendencies, one to use very sophisticated physical models (which lead to very costly, but extremely accurate simulations), the other to rely on simplified models, sacrifying a bit of accuracy for a reduced complexity and cost of the algorithms. Along this line, it will be extremely useful to be able to combine different methods, in order to fully exploit the potential of each approach.
As a final remark, we hope that this book will be of help to everybody entering the field of device simulation, and an incentive to those who are still skeptical about the capacity for the MC simulation to deal with real devices.

References

[1] Hockney, R. W., Warriner, R. A., Reiser, M.: Electron Lett. *10*, 484 (1974).
[2] Tomizawa, K., Awano, Y., Hashizume, N., Kawashima, M.: IEE Proc. *129*, 131 (1982).
[3] Blakey, P. A., Cherensky, S. C., Summers, P.: In: The Physics of Submicron Structures (Grubin, H. L., Hess, K., Iafrate, G. J., Ferry, D. K., eds.), p. 185. New York: Plenum Press. 1984.

[4] Hesto, P., Pone, J. F., Castagne, R.: J. Physique Lett. *C7*, 117 (1981).
[5] Zimmermann, J., Constant, E.: Solid-State Electron. *23*, 915 (1980).
[6] Lugli, P., Grondin, R. O., Ferry, D. K.: In: The Physics of Submicron Structures (Grubin, H. L., Hess, K., Iafrate, G. J., Ferry, D. K., eds.), p. 211. New York: Plenum Press. 1984.
[7] Hesto, P., Pone, J. F., Castagnè, R.: Appl. Phys. Lett. *40*, 405 (1982).
[8] Hesto, P., Pone, J. F., Castagnè, R., Pelouard, J. L.: In: The Physics of Submicron Structures (Grubin, H. L., Hess, K., Iafrate, G. J., Ferry, D. K., eds.), p. 101. New York: Plenum Press. 1984.
[9] Ghis, A., Boittiaux, B., Fauquembergue, R., Constant, E.: In: The Physics of Submicron Structures (Grubin, H. L., Hess, K., Iafrate, G. J., Ferry, D. K., eds.), p. 203. New York: Plenum Press. 1984.
[10] Castagnè, R.: In: High-Speed Electronics (Källbäck, B., Beneking, H., eds.), p. 2. Berlin-Heidelberg-New York: Springer. 1986.
[11] Hesto, P., Pone, J. F., Mouis, M., Castagnè, R.: J. Physique Lett. *43*, 229 (1982).
[12] Tomizawa, K., Awano, Y., Hashizume, N., Kawashima, M.: IEEE Proc. *132*, 37 (1985).
[13] Adachi, S., Kawashima, M., Kumabe, K., Yokoyama, K., Tomizawa, M.: IEEE Electron. Dev. Lett. *EDL-3*, 409 (1982).
[14] Bannov, N., Gruzinskis, V., Reklaitis, A., Ryzhij, V.: Solid-State Electron. *29*, 1207 (1986).
[15] Lugli, P., Ravaioli, U., Ferry, D. K.: Monte Carlo simulation of contacts in submicron devices. In: Proc., AIP Conf., No. 122, The Physics of VLSI, p. 162. Palo Alto: AIP. 1984.
[16] Ravaioli, U., Lugli, P., Osman, M., Ferry, D. K.: IEEE Trans. Electr. Dev. *ED32*, 2097 (1985).
[17] Baccarani, G., Mazzone, A. M.: Electron. Lett. *12*, 59 (1976).
[18] Maziar, C. M., Lundstron, M. S.: Electron. Lett. *23*, 61 (1987).
[19] Beard, S. J., Rees, M. D.: Electron. Lett. *17*, 811 (1981).
[20] Brennan, K.: IEEE Trans. Electron. Dev. *ED-32*, 2197 (1985).
[21] Brennan, K.: IEEE Trans. Electron. Dev. *ED-32*, 2467 (1985).
[22] Moglestue, C.: In: Proc., Int. Conf. Simulation of Semiconductor Devices and Processes (Board, K., Owen, D. R. J., eds.), p. 153. Swansea: Pineridge. 1984.
[23] Yoshii, A., Tomizawa, M., Yokoyama, K.: IEEE Trans. Electron. Dev. *ED-30*, 1376 (1983).
[24] Awano, Y., Tomizawa, K., Hashizume, N.: IEEE Trans. Electron. Dev. *ED-31*, 448 (1984).
[25] Yokoyama, K., Tomizawa, M., Yoshii A., Sudo, T.: IEEE Trans. El. Dev. *ED-32*, 2008 (1985).
[26] Yokoyama, K., Tomizawa, M., Yoshii, A.: IEEE El. Dev. Lett. *EDL-6*, 536 (1985).
[27] Tomizawa, M., Yoshii, A., Yokoyama, K.: Electr. and Comm. Japan *67-C*, 88 (1984).
[28] Awano, Y., Tomizawa, M., Hashizume N., Kawashima, M.: Electron. Lett. *19*, 20 (1983).
[29] Sanghera, G. S., Chryssafis, A., Moglestue, C.: IEE Proc. *127*, 203 (1980).
[30] Warriner, R. A.: Ph.D. Dissertation, Reading University, unpublished (1976).
[31] Moglestue, C.: COMPEL *1*, 7 (1982).
[32] Williams, C. K., Glisson, T. H., Hauser, J. R., Littlejohn, M. A., Abusaid, M. F.: Sol. St. Electron. *28*, 1105 (1985).
[33] Moglestue, C.: IEE Proc. *128*, 131 (1981).
[34] Moglestue, C.: IEE Proc. *130*, 275 (1983).
[35] Moglestue, C.: IEE Proc. *131*, 293 (1984).
[36] Moglestue, C.: IEE Proc. *132*, 217 (1985).
[37] Fauquembergue, R., Pernisek, M., Constant, E.: El. Lett. *18*, 670 (1982).
[38] Fauquembergue, R., Pernisek, M., Thobel, J. L., Boured, P.: In: Proc., Int. Conference on GaAs and Related Compounds, Crete, 1987.
[39] Moglestue, C.: IEEE Trans. Computer-Aided Design *CAD-4*, 536 (1985); Moglestue, C.: IEEE Trans. Computer-Aided Design *CAD-5*, 326 (1986).

[40] Yamada, Y., Ikeda, S., Shimojok, N.: In: Proc., XVII European Solid State Device Research Conference ESSDERC87 (Calzolari, P., Soncini, G., eds.), p. 111. Bologna: Tecnoprint. 1987.
[41] Wang, T., Hess, K.: J. Appl. Phys. *57*, 5336 (1985).
[42] Fauquembergue, R., Thobel, J. L., Descheerder, P., Pernisek, M., Wolf, P.: Solid-State Electron. *31*, 595 (1988).
[43] Tomizawa, M., Yoshii, A., Yokoyama, K.: IEEE El. Dev. Lett. *EDL5*, 362 (1984).
[44] Tomizawa, M., Hashizuma, N.: In: Proc., NASECODE IV Conference (Millet, J. J., ed.), p. 98. Dublin: Boole, 1985.
[45] Ravaioli, U., Ferry, D. K.: Superlattices and Microst. *2*, 75 (1986).
[46] Ravaioli, U., Ferry, D. K.: Superlattices and Microst. *2*, 377 (1986).
[47] Ravaioli, U., Ferry, D. K.: In High-Speed Electronics (Källbäck, B., Beneking, H., eds.), p. 136. Berlin-Heidelberg-New York: Springer. 1986.
[48] Ravaioli, U., Ferry, D. K.: IEEE Trans. Electr. Dev. *ED33*, 677 (1986).
[49] Mouis, M., Pone, J. F., Hesto, P., Castagnè R.: In: Proc., IEEE/Cornell Univ. Conf. on Advanced Concepts on High Speed Semiconductor Devices and Circuits. Ithaca: Cornell University. 1985.
[50] Castagnè, R.: Physica *134B*, 55 (1985).
[51] Mouis, M., Dolfus, P., Mougel, B., Pone, J. F., Castagnè, R.: In: High-Speed Electronics (Källbäck, B., Beneking, H., eds.), p. 35. Berlin-Heidelberg-New York: Springer. 1986.
[52] Pelouard, J. L., Castagnè, R., Hesto, P.: J. Opt. Soc. (to be published).
[54] Kizillyaly, I. C., Hess, K., Iafrate, G. J.: J. Appl. Phys. *61*, 2395 (1987).
[55] Kizillyaly, I. C., Hess, K.: Jap. J. Appl. Phys. *26*, 9 (1987).
[56] Kizillyaly, I. C., Hess, K., Iafrate, G. J., Smith, D.: In: Proc., NUMOS Conference, Los Angeles, 1986.
[57] Baccarani, G., Jacoboni, C., Mazzone, A. M.: Solid State Electr. *20*, 5 (1977).
[58] Park, Y. J., Navon, D. H., Tang, T. W.: IEEE Trans. Electron. Dev. *ED-31*, 1724 (1984).
[59] Tomizawa, K., Awano, Y., Hashizuma, N.: IEEE El. Dev. Lett. *EDL-5*, 362 (1984).
[60] Pelouard, J. L., Hesto, P., Castagnè, R.: Solid-State Electron. *31*, 333 (1988).
[61] Pelouard, J. L., Hesto, P., Castagnè, R.: In: Proc., IEEE/Cornell University Conference on Advanced Concept in High Speed Semiconductor Devices and Circuits, Ithaca, Cornell University. 1985.
[62] Pelouard, J. L., Hesto, P., Praseuth, J. P., Goldstein, L.: In: High-Speed Electronics (Källbäck, B., Beneking, H., eds.), p. 164. Berlin-Heidelberg-New York: Springer. 1986.
[63] Maziar, C. M., Klausmeier-Brown, M. E., Bandyopadhayay, S., Lundstrom, M. S., Datta, S.: IEEE Trans. Electron. Dev. *ED-33*, 881 (1986).
[64] Riccò, B., Sangiorgi, E., Venturi, F., Lugli, P.: In: Proc., IEDM, p. 559. Los Angeles, 1986.
[65] Sangiorgi, E., Riccò, B., Venturi, F.: IEEE Trans. Computer-Aided Des. *CAD-7*, 259 (1988).
[66] Venturi, F., Smith, R. K., Sangiorgi, E., Pinto, M. R., Riccò, B.: In: Simulation of Semiconductor Devices and Processes, Vol. 3 (Baccarani, G., Rudan, M., eds.). Bologna: Tecnoprint. 1988.
[67] Sangiorgi, E., Venturi, F., Riccò, B.: IEEE Trans. Computer-Aided Des. *7*, 259 (1988).
[68] Throngnumchai, K., Asada, K., Sugano, T.: IEEE Trans. El. Dev. *ED-33*, 1005 (1986).
[69] Tomizawa, M., Yokoyama, K., Yoshii, A.: IEEE Trans. Computer-Aided Des. *CAD-7*, 254 (1988).
[70] Fischetti, M. V., Laux, S. E.: Phys. Rev. *B38*, 9721 (1988).
[71] Fauquembergue, R., Pernisek, M., Constant, E.: Physica *129B*, 563 (1985).
[72] Mouis, M., Pone, J. F., Castagnè, R.: Solid-State Electron. *27*, 659 (1984).
[73] Antonelli, F., Lugli, P.: In: Proc., XVII European Solid State Device Research Conference ESSDERC87 (Calzolari, P., Soncini, G., eds.), p. 177. Bologna: Tecnoprint. 1987.
[74] Wang, T., Hess, K., Iafrate, G. J.: J. Appl. Phys. *59*, 2125 (1986).
[75] Littlejohn, M. A., True, R. J., Hauser, J. R., Golio, J. M.: J. Vac. Sci. Technolog. *B1*, 449 (1983).

[76] Wang, T., Hess, K., Iafrate, G. J.: J. Appl. Phys. *59*, 2125 (1986); Long, A. P., Beton, P. H., Kelly, M. J.: Semiconductor Sci. Technol. *1*, 63 (1986).
[77] Long, A. P., Beton, P. H., Kelly, M. J.: J. Appl. Phys. *62*, 1842 (1987).
[78] Hwang, C., Navon, D. H., Tang, T.: IEEE Trans. Electron Dev. *ED-34*, 154 (1987).
[79] Awano, Y., Tomizawa, K., Hashizume, N.: In: Proc., Int. Conference on GaAs and Related Compounds Biarritz. Inst. Phys. Conf. Ser. *74*, 623 (1984).
[80] Moll, J. L.: Physics of Semiconductors. New York: McGraw-Hill. 1966.
[81] Roderick, E. H.: Metal-Semiconductor Contacts. Oxford: Clarendon Press. 1980.
[82] Fawcett, W., Boardmann, A. D., Swain, S.: J. Phys. Chem. Solids *31*, 1963 (1970).
[83] Littlejohn, M. A., Hauser, J. R., Glisson, T. H.: J. Appl. Phys. *48*, 4587 (1977).
[84] Tang, J. Y., Hess, K.: J. Appl. Phys. *54*, 5139 (1983).
[85] Shah, J., Deveaud, B., Tsang, W. T., Gossard A. C., Lugli, P.: Phys. Rev. Lett. *59*, 2222 (1987).
[86] Yamaguchi, K., Asai, S., Kodera, H.: IEEE Trans. Electron. Dev. *ED-23*, 1283 (1976).
[87] Eastman, L. F., Stall, R., Woodward, D., Wood, C. E. C., Shur, M. S., Board, K.: Electron. Lett. *16*, 524 (1980).
[88] Hockney, R. W., Eastwood, J. W.: Computer Simulation Using Particles. New York: McGraw-Hill. 1981.
[89] Ando, T., Fowler, A. B., Stern, F.: Rev. Mod. Phys. *54*, 437 (1982).
[90] Brandt, A.: Math. Comp. *31*, 333 (1977).
[91] Delagebeaudef, D., Link, N.: IEEE Trans. Electron. Dev. *ED-29*, 955 (1982).
[92] Luryi, S.: Physica *134B*, 453 (1985).
[93] Sakaki, H.: Jpn. J. Appl. Phys. *21*, L381 (1982).
[94] Chelikowsky, J. R., Cohen, M. L.: Phys. Rev. *B14*, 556 (1976).
[95] Ridley, B. K.: J. Phys. *C10*, 1589 (1977).
[96] Keldysh, L. V.: Sov. Phys. JETP *21*, 1135 (1965).
[97] Sai-Halasz, G. A., Wordeman, M. R., Kern, D. P., Ganin, E., Rishton, S., Zicherman, D. S., Schmid, H., Polcari, M. R., Ng, H. Y., Restle, P. J., Chang, T. M. P., Dennard, R. H.: IEEE Electr. Dev. Lett. *EDL-8*, 463 (1987).
[98] Sai-Halasz, G. A., Wordeman, M. R., Rishton, S., Ganin, E., Kern, D. P.: to be published.
[99] Lugli, P. L., Ferry, D. K.: IEEE Electron. Dev. Lett. *EDL6*, 25 (1985).
[100] Hayes, J. R., Levi, A. F. J.: IEEE J. Quantum Electron. *QE-22*, 1744 (1986).
[101] Baccarani, G., Rudan, M., Guerrieri, R., Giampolini, P.: In: Process and Device Modeling (Engl, W. L., ed.). Elsevier/North-Holland. 1986.
[102] Selberherr, S.: Analysis and Simulation of Semiconductor Devices. Wien-New York: Springer. 1984.
[103] Curtice, W. R.: IEEE Trans. Electron. Dev. *ED-29*, 1942 (1982).
[104] Warriner, R. A.: Solid-State Electron. Dev. *1*, 105 (1977).
[105] Baccarani, G., Wordeman, M. R.: Solid-State Electron. *28*, 407 (1985).
[106] Snowden, C. M., Loret, D.: IEEE Trans. Electron. Dev. *ED-34*, 212 (1987).

Appendix A Numerical Evaluation of Some Integrals of Interest

In this Appendix we present an analytical approximation of integrals that involve the Bose-Einstein distribution function $N_q(x) = 1/[\exp(x) - 1]$ in terms of the dimensionless variable x.

With a good approximation N_q can be taken as

$$N_q(x) = \begin{cases} \dfrac{1}{x} - \dfrac{1}{2} + \sum\limits_{m=1}^{\infty} B_{2m} x^{2m-1}/(2m)!, & x \leqslant \bar{x}, \\ \exp(-x), & x > \bar{x}, \end{cases} \tag{A.1}$$

where B_{2m} are the Bernoulli numbers, and $\bar{x} < 2\pi$ in order to ensure the validity of the series expansion [1]. The best value of $\bar{x}$ depends slightly upon the number of terms considered in the power series, and can be taken as about 3.5.

For the case of electrons, by considering in Eq. (A.1) terms up to the ninth power, for the integrals in Eqs. (2.4.24)–(2.4.26) we obtain

$$F_1(x) = \begin{cases} \dfrac{x^2}{2} - \dfrac{x^3}{6} + \dfrac{x^4}{48} - \dfrac{x^6}{4320} + \dfrac{x^8}{241\,920} - \dfrac{x^{10}}{12\,096\,000} + \dfrac{x^{12}}{622\,702\,080}, & x \leqslant \bar{x}_a, \\ \dfrac{\bar{x}_a^2}{2} - \dfrac{\bar{x}_a^3}{6} + \dfrac{\bar{x}_a^4}{48} - \dfrac{\bar{x}_a^6}{4320} + \dfrac{\bar{x}_a^8}{241\,920} - \dfrac{\bar{x}_a^{10}}{12\,096\,000} + \dfrac{\bar{x}_a^{12}}{622\,702\,080} & \\ \quad + \exp(-\bar{x}_a)(\bar{x}_a^2 + 2\bar{x}_a + 2) - \exp(-x)(x^2 + 2x + 2), & x > \bar{x}_a, \end{cases} \tag{A.2}$$

$$G_1(x) = \begin{cases} \dfrac{x^2}{2} + \dfrac{x^3}{6} + \dfrac{x^6}{4320} + \dfrac{x^8}{241\,920} - \dfrac{x^{10}}{12\,096\,000} + \dfrac{x^{12}}{622\,702\,080}, & x \leqslant \bar{x}_e, \\ \dfrac{\bar{x}_e^2}{2} + \dfrac{\bar{x}_e^3}{6} + \dfrac{\bar{x}_e^4}{48} - \dfrac{\bar{x}_e^6}{4320} + \dfrac{\bar{x}_e^8}{241\,920} - \dfrac{\bar{x}_e^{10}}{12\,096\,000} + \dfrac{\bar{x}_e^{12}}{622\,702\,080} & \\ \quad + \exp(-\bar{x}_e)(\bar{x}_e^2 + 2\bar{x}_e + 2) - \dfrac{\bar{x}_e^3}{3} & \\ \quad - \exp(-x)(x^2 + 2x + 2) + \dfrac{x^3}{3}, & x > \bar{x}_e, \end{cases} \tag{A.3}$$

where $F_1(x)$ refers to absorption ($\bar{x}_a = 3.5$), and $G_1(x)$ refers to emission ($\bar{x}_e = 4$), respectively.

As regards the integrals entering Eqs. (2.4.30)–(2.4.32), we find by analogous calculations

$$F_2(x) = \begin{cases} \dfrac{x^3}{3} - \dfrac{x^4}{8} + \dfrac{x^5}{60} - \dfrac{x^7}{5040} + \dfrac{x^9}{272\,160} - \dfrac{x^{11}}{143\,305\,600} + \dfrac{x^{13}}{622\,720\,080}, & x \leqslant \bar{x}_a, \\ \dfrac{\bar{x}_a^3}{3} - \dfrac{\bar{x}_a^4}{8} + \dfrac{\bar{x}_a^5}{60} - \dfrac{\bar{x}_a^7}{5040} + \dfrac{\bar{x}_a^9}{272\,160} - \dfrac{\bar{x}_a^{11}}{143\,305\,600} + \dfrac{\bar{x}_a^{13}}{622\,702\,080} & \\ \quad + \exp(-\bar{x}_a)(\bar{x}_a^3 + 3\bar{x}_a + 6\bar{x}_a + 6) & \\ \quad - \exp(-x)(x^3 + 3x^2 + 6x + 6), & x > \bar{x}_a, \end{cases} \tag{A.4}$$

$$G_2(x) = \begin{cases} \dfrac{x^3}{3} + \dfrac{x^4}{8} + \dfrac{x^5}{60} - \dfrac{x^7}{5040} + \dfrac{x^9}{272\,160} - \dfrac{x^{11}}{143\,305\,600} + \dfrac{x^{13}}{622\,702\,080}, & x \leqslant x_e, \\ \dfrac{\bar{x}_e^3}{3} + \dfrac{\bar{x}_e^4}{8} + \dfrac{\bar{x}_e^5}{60} - \dfrac{\bar{x}_e^7}{5040} + \dfrac{\bar{x}_e^9}{272\,160} - \dfrac{\bar{x}_e^{11}}{143\,305\,600} + \dfrac{\bar{x}_e^{13}}{622\,702\,080} & \\ \quad + \exp(-\bar{x}_e)(\bar{x}_e^2 + 3\bar{x}_e^2 + 6\bar{x}_e + 6) - \dfrac{\bar{x}_e^4}{4} & \\ \quad - \exp(-x)(x^3 + 3x^2 + 6x + 6) + \dfrac{x^4}{4}, & x > \bar{x}_e, \end{cases} \tag{A.5}$$

For the case of holes, by considering only the first terms of the series in Eq. (A.1), we can give, for the integrals in Eqs. (2.4.40)–(2.4.44), the expressions:

$$F_3(x) = \begin{cases} 2x^2\left(1 - \dfrac{34}{105}\sqrt{2}x + \dfrac{x^2}{12}\right), & x \leqslant \dfrac{3}{\sqrt{2}}, \\ \dfrac{1017}{280} + \left(68 - \dfrac{2358}{x^2} + \dfrac{41\,931}{x^4}\right)\exp(-3) & \\ \quad -8\exp(-\sqrt{2}x)\left(x^2 + 4\sqrt{2}x + 28\right. & \\ \quad \left. + \dfrac{72\sqrt{2}}{x} + \dfrac{252}{x^2} + \dfrac{270\sqrt{2}}{x^3} + \dfrac{270}{x^4}\right), & x > \dfrac{3}{\sqrt{2}}, \end{cases} \tag{A.6}$$

$$F_4(x) = \begin{cases} x^3\left(\dfrac{136\sqrt{2}}{105} - x + \dfrac{44\sqrt{2}x^2}{315}\right), & x \leqslant \dfrac{3}{\sqrt{2}}, \\ \dfrac{801}{140} + \exp(-3)\left(312 - \dfrac{13\,248}{x^2} + \dfrac{300\,078}{x^4}\right) & \\ \quad -\exp(\sqrt{2}x)\left(8\sqrt{2}x^3 + 72x^2 + 288\sqrt{2}x\right. & \\ \quad \left. + 1824 + \dfrac{4320\sqrt{2}}{x} + \dfrac{14\,400}{x^2} + \dfrac{15\,120\sqrt{2}}{x^3} + \dfrac{15\,120}{x^4}\right), & x > \dfrac{3}{\sqrt{2}}, \end{cases} \tag{A.7}$$

$$G_3(x) = \begin{cases} 2x^2\left(1 + \dfrac{34\sqrt{2}}{105}x + \dfrac{x^2}{12}\right), & x \leqslant \dfrac{3}{\sqrt{2}}, \\ \dfrac{5913}{280} + \dfrac{136\sqrt{2}}{105}x^3 - 9\left(4 - \dfrac{162}{5x^2} + \dfrac{729}{7x^4}\right), & x > \dfrac{3}{\sqrt{2}}, \end{cases} \tag{A.8}$$

$$G_4(x) = \begin{cases} x^3\left(\dfrac{136\sqrt{2}}{105} + x + \dfrac{44\sqrt{2}}{315}x^2\right), & x \leqslant \dfrac{3}{\sqrt{2}}, \\[2ex] \dfrac{6371}{140} + 2x^4 - 81 + \dfrac{729}{x^2} - \dfrac{19\,683}{8x^4}, & x > \dfrac{3}{\sqrt{2}}. \end{cases} \tag{A.9}$$

References

[1] Gradshteyn, I. S., Ryzhik, I. M.: Table of Integrals, Series and Products. New York: Academic Press. 1966.

Appendix B
Generation of Random Numbers

We have seen that the Monte Carlo procedure is based on the generation of random numbers. Nowadays most computers contain routines that generate random numbers evenly distributed between 0 and 1.
In this Appendix we shall describe how random numbers are generated in general and we shall see how random numbers with any probability distribution can be obtained starting from a sequence of random numbers evenly distributed between 0 and 1. These numbers are designates in the present paper by the letter r.

Generation of Evenly Distributed Random Numbers

The method of generation most commonly used in the computer routines for the generation of evenly distributed random numbers is the multiplicative congruential method: the ith element r_i of the sequence is given by the previous element r_{i-1} by a relation such as

$$r_i = pr_{i-1} \pmod q), \tag{B.1}$$

where p and q are appropriate constants. The first element of the sequence (seed) must be given by the user. The numbers r_i of the sequence in Eq. (B.1) are obtained with a precise mathematical algorithm, and therefore they are not at all random; in fact, given the seed of the sequence, all its numbers are perfectly predictable. However, for "good" choices of the constants p and q, the sequences of r_i behave as random in the sense that they passed a large number of statistical tests of randomness. Numbers r of such a type are called pseudorandom numbers. They have the advantage over truly random numbers of being generated in a fast way and of being reproducible, when desired, especially for program debugging.

Generation of Random Numbers with Given Distributions

Random numbers x with a given probability distribution $f(x)$ in an interval (a, b) can be obtained starting from numbers r evenly distributed in the interval (0, 1) with different techniques [1, 2]. We shall describe here the three simplest of them, which are usually used in Monte Carlo transport calculations.

a) Direct Technique

If the function $f(x)$ is normalized to one in the interval of definition (a, b), let us call $F(x)$ the integral function of f. Then, given a number r, we correspondingly choose x_r such that

$$r = F(x_r) = \int_a^{x_r} f(x)\, dx. \tag{B.2}$$

The probability $P(x)\, dx$ that x_r obtained in this way lies within an interval dx around x is equal to the corresponding dF, since r has a flat distribution. Thus (see Fig. B.1)

$$P(x)\, dx = dF = f(x)\, dx, \tag{B.3}$$

as desired. If $f(x)$ is not normalized, then Eq. (B.2) must be replaced by

$$r = \int_a^{x_r} f(x)\, dx \bigg/ \int_a^b f(x)\, dx \tag{B.4}$$

with a constant $f(x)$, this technique provides the obvious formula for the generation of a random number evenly distributed between a and b:

$$x_r = a + (b - a) r. \tag{B.5}$$

Often, in real cases, the above simple direct technique cannot be used because it is not possible either to evaluate analytically the integral in

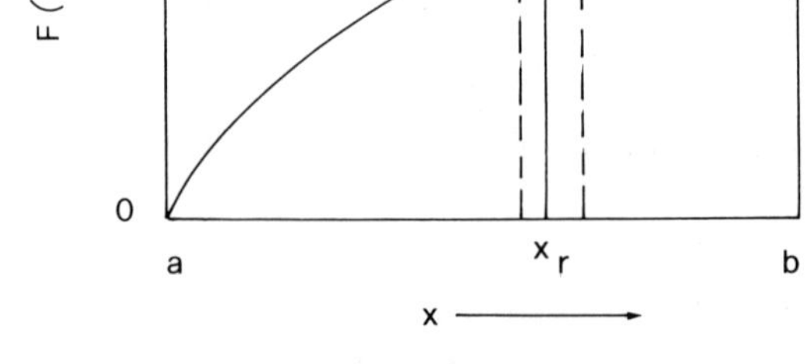

Fig. B.1. Direct technique for the generation of a random variable x_r with probability function $f(x) = dF(x)/dx$ starting from random numbers r evenly distributed between 0 and 1. (See text)

Eq. (B.2) or to solve with respect to x_r the equation which results from Eq. (B.2). In these cases one of the following techniques can be used.

b) Rejection Technique

Let C be a positive number such that

$$C \geqslant f(x) \tag{B.6}$$

in the whole interval (a, b), and let r_1 and r'_1 be two random numbers obtained with a flat distribution in (0, 1). Then

$$x_1 = a + (b - a)r_1 \qquad \text{and} \qquad f_1 = r'_1 C \tag{B.7}$$

are two random numbers obtained with a flat distribution in (a, b) and $(0, C)$, respectively. If

$$f_1 \leqslant f(x_1) \tag{B.8}$$

then x_1 is retained as choice of x, otherwise x_1 is rejected, and a new pair r_2, r'_2 is generated; the process is repeated until Eq. (B.8) is satisfied. Since for each pair r_i, r'_i a point with coordinates (x_i, f_i) is obtained from the uniform distribution in the rectangle $abCC$ of Fig. B.2, the probability that x_i within an interval dx around x will be accepted is proportional to the probability that x_i lies within this interval, proportional to dx, times the probability of accepting x_i, proportional to $f(x_i)$, as desired.
The above technique is always applicable, with any bounded $f(x)$ in a finite interval (a, b). However, when $f(x)$ is strongly peaked, many pairs of numbers might be generated before a successful trial, with a resulting large

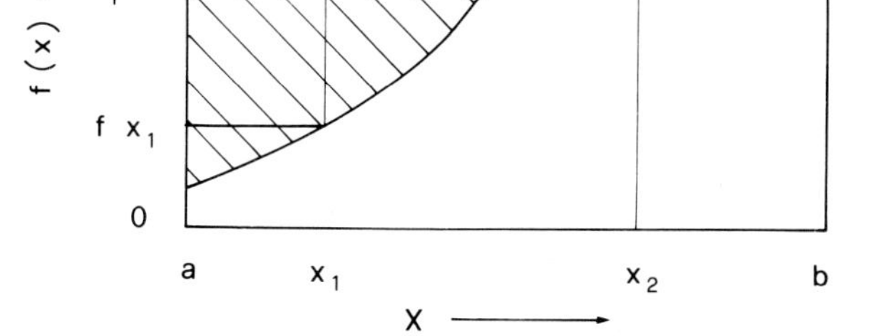

Fig. B.2. Rejection technique for the generation of a random variable x_r with probability function $f(x)$ starting from random numbers r evenly distributed between 0 and 1. (See text)

expense of computer time. The technique described below may overcome this difficulty by combining the two previous techniques.

c) Combined Technique

Let x_1 be a random number generated with the direct technique according to a distribution $g(x)$. If, furthermore, K is a constant such that

$$Kg(x) \geqslant f(x) \tag{B.9}$$

in the whole range (a, b) of interest, a new random number r_1 is generated in $(0, 1)$ and x_1 is accepted as a value of the random variable x if

$$r_1 Kg(x_1) < f(x_1). \tag{B.10}$$

In fact, in this way the probability of having an accepted x_1 within an interval dx around x is proportional to the probability that x_1 lies within this interval, proportional to $g(x)\,dx$, times the probability of accepting x_1, proportional to the ratio $f(x)/Kg(x)$. The final probability is therefore proportional to $f(x)\,dx$, as desired. A geometrical interpretation of the combined technique is shown in Fig. B.3: the selection of x_1 is equivalent to the generation of a point with flat distribution below the curve $Kg(x)$, while the condition of acceptance requires that the point lies in the area below the curve $f(x)$. If the curve $Kg(x)$ is not too far from $f(x)$, few attempts will be necessary per successful trial.
The rejection technique described in the previous section is a particular case of the combined technique, for $g = \text{const}$. The combined technique, however, can also be applied to unbounded functions or to functions defined in unbounded intervals.

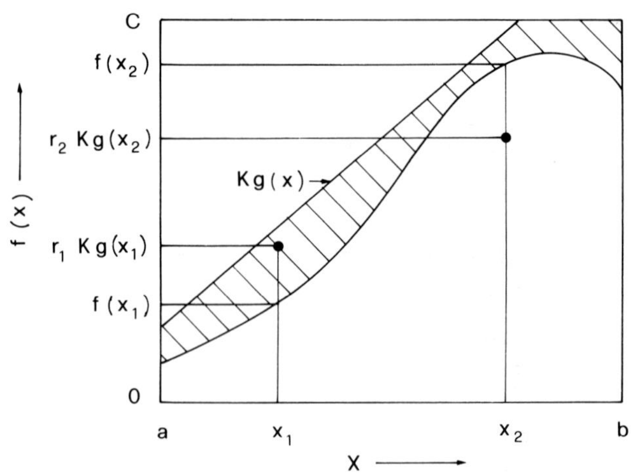

Fig. B.3. Combined technique for the generation of a random variable x, with probability function $f(x)$ starting from random numbers r evenly distributed between 0 and 1. (See text)

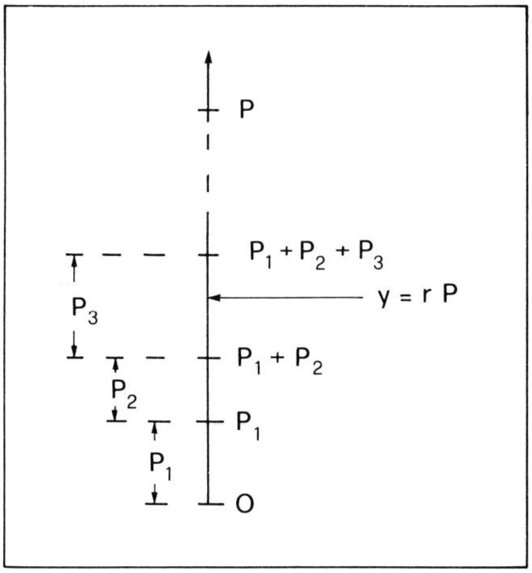

Fig. B.4. Generation of a random discrete variable i with probability function P_i starting from random numbers r evenly distributed between 0 and 1. (See text)

As we said above, other techniques can be used for special cases. As an example we may mention the use of the sum of a certain number of random numbers r to generate x according to a normal distribution [3]. For such a distribution the combined technique can also be used [4].

d) Discrete Case

When an event must be chosen among a given number of different possibilities or, in other words, when the variable x must be chosen from a discrete set, the direct technique can be used with $f(x)$ given by the sum of δ functions. Figure B.4 illustrates this case: if P_i is the probability of occurrence of the ith event x_i, then a random number

$$y = rP, \tag{B.11}$$

where

$$P = \sum_i P_i \tag{B.12}$$

is generated and compared successively with $P_1, P_1 + P_2, P_1 + P_2 + P_3, \ldots$. The jth event is chosen if j is such that the first of the above partial sums which is larger than y is $P_1 + P_2 + \cdots + P_j$. Figure B.4 shows immediately that the probability of choosing the jth event is proportional to P_j, as desired.

References

[1] Hammersley, J. M., Handscomb, D. C.: Monte Carlo Methods (Barlett, M. S., ed.). London: Methuen-Chapman and Hall. 1964.

[2] Buslenko, N. P., Golenko, D. I., Shreider, Yu. A., Sobol, I. M., Sragovich, V. G.: The Monte Carlo Method: The Method of Statistical Trials (Shreider, Yu. A., ed.). (Translated from Russian by G. J. Tee.) London: Pergamon Press. 1966.
[3] Eadie, W. T., James, F. E., Roos, M., Sadoulet, B.: Statistical Methods in Experimental Physics. Amsterdam: North-Holland. 1971.
[4] Price, P. J. Semiconductors and Semimetals, p. 249. New York: Academic Press. 1979.

Subject Index